AF294418

Encyclopaedia of Mathematical Sciences

Volume 65

Editor-in-Chief: R.V. Gamkrelidze

Springer

Berlin
Heidelberg
New York
Barcelona
Budapest
Hong Kong
London
Milan
Paris
Santa Clara
Singapore
Tokyo

M. A. Shubin (Ed.)

Partial Differential Equations VIII

Overdetermined Systems
Dissipative Singular Schrödinger Operator
Index Theory

Springer

Consulting Editors of the Series:
A.A. Agrachev, A.A. Gonchar, E.F. Mishchenko,
N.M. Ostianu, V.P. Sakharova, A.B. Zhishchenko

Title of the Russian edition:
Itogi nauki i tekhniki, Sovremennye problemy matematiki,
Differentsial'nye uravneniya s chastnymi proizvodnymi 8, Vol. 65,
Publisher VINITI, Moscow 1991

Cataloging-in-Publication Data applied for

Die Deutsche Bibliothek - CIP-Einheitsaufnahme

Partial differential equations / M. A. Shubin (ed.).‿- Berlin ;
Heidelberg ; New York ; Barcelona ; Budapest ; Hong Kong ;
London ; Milan ; Paris ; Santa Clara ; Singspore ; Tokyo :
Springer.
  Einheitssacht.: Differencial'nye uravnenija s častnymi proizvodnymi
  <engl.>
  Teilw. hrsg. von Yu. V. Egorov ; M. A. Shubin
  NE: Egorov, Jurij V. [Hrsg.]; Šubin, Michail A. [Hrsg.]; EST
  8. Overdetermined systems index of elliptic operators. - 1996
  (Encyclopaedia of mathematical sciences ; Vol. 65)
  ISBN-13: 978-3-642-48946-4
NE: GT

Mathematics Subject Classification (1991):
35N10, 35P99, 46N20, 47A53, 47B44, 58G05, 58G10, 58G12

ISBN-13: 978-3-642-48946-4   e-ISBN-13: 978-3-642-48944-0
DOI: 10.1007/978-3-642-48944-0

List of Editors, Authors and Translator

Editor-in-Chief

R. V. Gamkrelidze, Russian Academy of Sciences, Steklov Mathematical Institute,
ul. Vavilova 42, 117966 Moscow; Institute for Scientific Information (VINITI),
ul. Usievicha 20a, 125219 Moscow, Russia
e-mail: gam@ips.ac.msk.su

Consulting Editor

M. A. Shubin, Department of Mathematics, Northeastern University,
Boston, MA 02115, USA
e-mail: shubin@neu.edu

Authors

P. I. Dudnikov, Institute of Mathematics of the Ukrainian Academy of Sciences,
Kiev, Ukraine

B. V. Fedosov, Department of Mathematics, Moscow Institute of Physics and
Technology, Institutskij per. 9, 141700 Dolgoprudnyj, Moscow Region, Russia

B. S. Pavlov, Department of Mathematics, Auckland University,
Private bag 92019, Auckland, New Zealand
e-mail: pavlov@math. auckland.ac.nz

S. N. Samborski, Department of Mathematics, Caen University,
14032 Caen Cedex, France
e-mail: samborsk@univ-caen.fr

Translator

C. Constanda, Department of Mathematics, University of Strathclyde, Livingstone
Tower, 26 Richmond Street, Glasgow G1 1XH, Scotland, U.K.
e-mail: c.constanda@strath.ac.uk

Contents

I. Linear Overdetermined Systems of Partial Differential Equations. Initial and Initial-Boundary Value Problems

P. I. Dudnikov and S. N. Samborski

Translated from the Russian
by C. Constanda

Contents

Introduction

Consider a linear partial differential operator A that maps a vector-valued function $y = (y_1, \ldots, y_m)$ into a vector-valued function $f = (f_1, \ldots, f_l)$. We assume at first that all the functions, as well as the coefficients of the differential operator, are defined in an open domain Ω in the n-dimensional Euclidean space $\mathbb{R}^n$, and that they are smooth (infinitely differentiable). A is called an *overdetermined* operator if there is a non-zero differential operator A' such that the composition $A'A$ is the zero operator (and *underdetermined* if there is a non-zero operator A'' such that $AA'' = 0$). If A is overdetermined, then $A'f = 0$ is a necessary condition for the solvability of the system $Ay = f$ with an unknown vector-valued function y.

A simple example in $\mathbb{R}^3$ is the operator grad, which maps a scalar function y into the vector-valued function $(\partial y/\partial x_1, \partial y/\partial x_2, \partial y/\partial x_3)$. A necessary solvability condition for the system $\operatorname{grad} y = f$ has the form $\operatorname{curl} f = 0$. The operator curl, which maps $f = (f_1, f_2, f_3)$ into

$$\left(\frac{\partial f_3}{\partial x_2} - \frac{\partial f_2}{\partial x_3}, \frac{\partial f_1}{\partial x_3} - \frac{\partial f_3}{\partial x_1}, \frac{\partial f_2}{\partial x_1} - \frac{\partial f_1}{\partial x_2} \right),$$

is itself overdetermined, since $\operatorname{div} \operatorname{curl} = 0$, where

$$\operatorname{div}(h_1, h_2, h_3) = \frac{\partial h_1}{\partial x_1} + \frac{\partial h_2}{\partial x_2} + \frac{\partial h_3}{\partial x_3}.$$

Denoting by $C^\infty(\Omega, Y)$ the space of smooth functions on Ω with values in an Euclidean space Y, we arrive at the sequence of spaces and differential operators

$$C^\infty(\Omega, \mathbb{R}^1) \xrightarrow{\operatorname{grad}} C^\infty(\Omega, \mathbb{R}^3) \xrightarrow{\operatorname{curl}} C^\infty(\Omega, \mathbb{R}^3) \xrightarrow{\operatorname{div}} C^\infty(\Omega, \mathbb{R}^1) \to 0,$$

which is a *complex* (that is, the composition of any two consecutive operators is equal to the zero operator).

The generalisation to the case where Ω is a finite-dimensional manifold leads to the well-known *de Rham complex*

$$C^\infty(\Lambda^0(T^*(\Omega)) \xrightarrow{d} C^\infty(\Lambda^1(T^*(\Omega))$$
$$\xrightarrow{d} C^\infty(\Lambda^2(T^*(\Omega)) \xrightarrow{d} \cdots \xrightarrow{d} C^\infty(\Lambda^n(T^*(\Omega)), \qquad (0.1)$$

where $C^\infty(\Lambda^i(T^*(\Omega))$ is the space of smooth differential forms of degree i on Ω, that is, of the smooth cross-sections of the vector bundles of exterior forms $\Lambda^i(T^*\Omega)$ over Ω, which are constructed in terms of Ω. In this complex, all the operators d except the last one are overdetermined, and all except the first one are underdetermined.

4 P. I. Dudnikov and S. N. Samborski

In the general case, for a given differential operator $A_0 : C^\infty(\Omega, Y_0) \to C^\infty(\Omega, Y_1)$ there arises the problem of constructing a complex of differential operators

$$C^\infty(\Omega, Y_0) \xrightarrow{A_0} C^\infty(\Omega, Y_1) \xrightarrow{A_1} C^\infty(\Omega, Y_2) \to \cdots$$

and investigating its *cohomology* spaces, that is, the spaces

$$H_i = \operatorname{Ker} A_i / \operatorname{Im} A_{i-1}.$$

The most desirable would be a complex (called *exact*) whose cohomologies are zero; however, such a complex does not always exist. For example, the cohomologies of the complex (0.1) coincide with those of the manifold Ω, and the dimension of these cohomologies cannot be lowered for any complex of differential operators starting with the operator

$$d : C^\infty(\Lambda^0(T^*\Omega)) \to C^\infty(\Lambda^1(T^*\Omega)).$$

Thus, the problem is to associate every differential operator with a complex which is "best" in the sense that its cohomologies have the "smallest" dimension and the conditions $A_{i+1}f = 0$ yield a "full set" of necessary differential conditions on f for the system $A_i y = f$ to have solutions. The existence of such a complex for differential operators with constant coefficients has been well known for a long time.

In the case of operators with variable coefficients, such a complex was constructed in the 1960s by Spencer under some additional conditions of "non-degeneracy" of the coefficients. If all the functions and coefficients of A_0 are real-analytic, then the complex constructed in this way is *locally* (that is, in a sufficiently small neighbourhood of every point) *exact*. Consequently, in the real-analytic case the necessary compatibility conditions are also (locally) sufficient.

In the theory of partial differential equations, overdetermined and underdetermined operators play a similar role to that of non-invertible matrices in linear algebra. However, frequently they also serve as a natural instrument in the study of determined (that is, neither overdetermined, nor underdetermined) operators that do not belong to well-investigated classes such as elliptic, parabolic or hyperbolic. As an example, we consider the system of stationary Maxwell equations in $\mathbb{R}^3$

$$A_0(u, v) \equiv (\operatorname{curl} u + v, \operatorname{curl} u - v) = (f_1, f_2), \tag{0.2}$$

which is determined. Applying div to the equalities (0.2) and adjoining the equations thus obtained to the original system, we arrive at the system

$$A_0'(u, v) \equiv (\operatorname{curl} u + v, \operatorname{curl} v - u, \operatorname{div} v, -\operatorname{div} u = (f_1, f_2, h_1, h_2), \tag{0.3}$$

which is overdetermined: $A_1' A_0' = 0$, where

$$A_1'(f_1, f_2, h_1, h_2) = (\operatorname{div} f_1 - h_1, \operatorname{div} f_2 - h_2).$$

The system (0.2) and, under the condition $A_1'(f_1, f_2, h_1, h_2) = 0$, also the system (0.3), have the same set of smooth solutions. Thus, it is the same if we study the kernel and cokernel of the operator A_0, or the cohomologies of the complex

$$0 \to C^\infty(\Omega, \mathbb{R}^6) \xrightarrow{A_0'} C^\infty(\Omega, \mathbb{R}^8) \xrightarrow{A_1'} C^\infty(\Omega, \mathbb{R}^2) \to 0.$$

However, the latter is more convenient, since the operator A_0' is elliptic (in the sense of overdetermined systems, whose exact definition is given later).

The operators A_0 and A_0' correspond to two different forms of expression for the same mathematical object. Properties of systems of equations such as ellipticity, parabolicity, hyperbolicity and so on are only properties of the form of presentation and may appear or disappear when the form is changed. On the other hand, solvability properties remain unchanged; only the spaces where the systems are solvable change.

If the information concerning the solvability of a system is contained in the terms of "highest" order, then this immediately enables us to apply the well-developed techniques of the Fourier transform, a priori energy estimates, and perturbations. The above example of Maxwell's system shows that, written in a certain form, the system can be overdetermined, although in its original form it was not.

Now let Ω be a closed domain in $\mathbb{R}^n$ with a smooth boundary Γ. The usual object of study in mathematical physics is a *boundary value problem*, that is,

$$A_0 y = f, \quad By = g,$$

where A_0 is a differential operator and B is the composition of a differential operator and the operator of restriction to Γ of a mapping given on Ω. In general, the solvability of such a problem requires compatibility conditions of the form

$$\Phi(f, g) = 0.$$

The question is, in which class of operators should the "appropriate" operator Φ be sought? On the one hand, it is necessary to have an explicit, effective (that is, implementable in finitely many steps) procedure for obtaining Φ from the coefficients of the operators A and B. On the other hand, the conditions $\Phi(f, g) = 0$ must be sufficient, or close to sufficient (for example, in the sense of the finite-dimensional nature of the cohomology $\operatorname{Ker}\Phi/\operatorname{Im}(A, B)$), for the solvability in wide classes of the boundary value problems arising in mathematical physics. We turn to a simple example, namely, the Dirichlet problem for the operator grad in the domain $\Omega = \{x = (x_1, x_2, x_3) \in \mathbb{R}^3 : x_3 \geq 0\}$:

$$\operatorname{grad} y = f, \quad y|_\Gamma = g, \quad (f = (f_1, f_2, f_3)).$$

The obvious compatibility conditions are

$$\operatorname{curl} f = 0, \quad \frac{\partial g}{\partial x_1} - f_1|_\Gamma = 0, \quad \frac{\partial g}{\partial x_2} - f_2|_\Gamma = 0;$$

here and above the symbol $|_\Gamma$ denotes restriction to

$$\Gamma = \{x \in \Omega \subset \mathbb{R}^3 : x_3 = 0\}.$$

Generalising this example to the case of a manifold Ω with a boundary Γ and using the de Rham complex (0.1), we arrive at the complex

$$0 \to C^\infty(\Lambda^0(T^*(\Omega)) \overset{d_0,\gamma^0}{\to} C^\infty(\Lambda^1(T^*(\Omega)) \times C^\infty(\Lambda^0(T^*\Gamma))$$
$$\overset{\Phi_1}{\to} C^\infty(\Lambda^2(T^*\Omega)) \times C^\infty(\Lambda^1(T^*\Gamma)) \overset{\Phi_2}{\to} \cdots \to 0, \tag{0.4}$$

in which, denoting by γ_i the operator of restriction of an i-form $\varphi \in C^\infty(\Lambda^i(T^*\Omega))$ to the form $\gamma_i\varphi \in C^\infty(\Lambda^i(T^*\Omega))$ and by d'_i the operator in the de Rham complex for Γ, we have

$$\Phi_i(\varphi, \psi) = (d_i\varphi, d'_i\varphi - \gamma_i\psi).$$

(We remark that the cohomologies of this complex coincide with the corresponding cohomologies of the pair (Ω, Γ).)

The Φ_i in the complex (0.4) belong to the class of *differential boundary* (DB-) operators, that is, of mappings from $C^\infty(\Omega, Y_1) \times C^\infty(\Gamma, W_1)$ to $C^\infty(\Omega, Y_2) \times C^\infty(\Gamma, W_2)$ of the form

$$(f, g) \mapsto (A_{11}f, A_{21}f + A_{22}g),$$

where A_{11} and A_{22} are differential operators on Ω and Γ, respectively, and A_{21} is the composition of a differential operator on Ω and the operator of restriction to Γ of mappings defined on Ω.

We return to the question of the appropriate class where we should seek the compatibility operators. It turns out that for a large number of problems in mathematical physics this is the class of DB-operators. This conclusion is drawn from the following results, which form the subject matter of the present survey.

1. For each of the operators $(A, B) : C^\infty(\Omega, Y_0) \to C^\infty(\Omega, Y_1) \times C^\infty(\Gamma, W_0)$ of a boundary value problem which satisfies the condition of "non-degeneracy of the coefficients", there exists a complex of DB-operators

$$C^\infty(\Omega, Y_0) \overset{(A,B)}{\to} C^\infty(\Omega, Y_1) \times C^\infty(\Gamma, W_0)$$
$$\overset{\Phi_1}{\to} C^\infty(\Omega, Y_2) \times C^\infty(\Gamma, W_1) \overset{\Phi_2}{\to} \cdots \to 0, \tag{0.5}$$

which can be constructed in finitely many steps (within the framework of differentiation of the coefficients and linear algebra for $x \in \Omega$ fixed); we refer to this complex in the sequel.

2. Suppose that the coefficients of A and B and the boundary Γ are real-analytic, that U is a neighbourhood in $\mathbb{R}^n$ of the point $x \in \Gamma$, and that $\Omega' = \Omega \cap U$, $\Gamma' = \Gamma \cap U$ and $\mathfrak{A}(\Omega', H)$ ($\mathfrak{A}(\Gamma', H)$) are sets of real-analytic

vector-valued functions on Ω' (Γ') with values in a Euclidean space H. Then (0.5) generates the complex

$$\mathfrak{A}(\Omega',Y_0) \overset{(A,B)}{\to} \mathfrak{A}(\Omega',Y_1) \times \mathfrak{A}(\Gamma',W_0) \overset{\Phi_1}{\to} \mathfrak{A}(\Omega',Y_2) \times \mathfrak{A}(\Gamma',W_1) \overset{\Phi_2}{\to} \cdots \to 0,$$

which is exact for a sufficiently small neighbourhood U (this is a generalisation of the Cauchy-Kovalevska theorem).

3. Let $H^i(\Omega,Y)$ be the Hilbert Sobolev spaces of functions on Ω with values in Y (having square-integrable generalized derivatives up to order i), $H^{\mathrm{T}}(\Gamma,W) = \bigoplus_{i=1}^{m} H^{t_i}(\Gamma,W^{(i)})$ the Hilbert Sobolev spaces of functions with values in $W = \bigoplus_{i=1}^{m} W^{(i)}$, and $T = (t_1,\ldots,t_m)$ a multi-index. In this case, if the operator A is elliptic and (A,B) satisfies the coerciveness condition (the generalized Lopatinskij condition, defined rigorously in what follows), then the cohomologies of the complex

$$0 \to H^i(\Omega,Y_0) \overset{(A,B)}{\to} H^{i-k_1}(\Omega,Y_1) \times H^{i-T_1}(\Gamma,W_1)$$

$$\overset{\Phi_1}{\to} H^{i-k_2}(\Omega,Y_2) \times H^{i-T_2}(\Gamma,W_2) \overset{\Phi_2}{\to} \cdots \to 0,$$

generated by (0.5) (the numbers $k_1,k_2,\ldots$ and the multi-indices $T_1,T_2,\ldots$ are connected with the operator (A,B)) are finite-dimensional.

4. Let A be a parabolic operator (in the sense of overdetermined systems), and suppose that (A,B) is coercive (satisfies the *Lopatinskij condition*). Then the complex of anisotropic Hilbert Sobolev spaces generated by (0.5) is exact.

5. Let A be a hyperbolic operator (in the sense of overdetermined systems), and suppose that (A,B) satisfies the uniform Lopatinskij condition and does not contain overdetermination on the boundary (that is, Φ_1 is a differential operator). Then the complex of Hilbert spaces of the form $H_\gamma^{q,a}$ generated by (0.5) is exact (rigorous definitions are given later; these spaces are common in the theory of quadratic hyperbolic systems).

In the assertions 3, 4 and 5 we additionally assume that A is formally integrable, which is a property of the form of the operators. If this is not so, then the above assertions are true in different norms. These norms are found in the process of changing the form of the operator when we go over to an equivalent, formally integrable one (which is always possible). In general, the formal properties of an operator play an exceptionally important role even in the answer to the question of what kind of problems are characteristic to that operator. Outside the framework of typical systems, this is very seldom determined by the homogeneous dominant terms of a differential operator. Here is an example. We consider the operator $A : y \to \operatorname{curl} y + b \times y$, where b is a vector field in a domain $\Omega \subset \mathbb{R}^3$ with boundary Γ. If $b = 0$, then the kernel and cokernel of any boundary value problem for this operator are infinite-dimensional. If $b \neq 0$ but $\operatorname{curl} b = 0$, then there are *Noether* (that is, with finite-dimensional kernel and cokernel) boundary value problems. The explanation for this is that when we go over to a formally integrable operator, that is, when the form of the operator is changed, the operator becomes overdetermined elliptic. If $\operatorname{curl} b \neq 0$, then there may be cases where the corresponding equivalent, formally integrable operator is hyperbolic; consequently, for the given operator it is natural to prescribe not boundary conditions, but initial boundary conditions. Such a variety of possibilities (for the same dominant part) corresponds to the variety of formal properties of differential operators. Hence, an important part of this survey is devoted to a detailed discussion of the formal theory of linear differential operators considered from various points of view. Their logically independent presentation enables the reader who meets the subject for the first time to choose the approach that suits him best. In particular, to try to please those who are not very keen on probing the non-trivial depths of multilinear algebra, in Chap. 1, Sect. 3.3 and Sect. 3.6 we propose a new approach to the study of the formal theory.

Chapter 1
Complexes Associated with Differential Operators

§1. Jets and Differential Operators

Let Ω be a manifold with boundary, Γ its boundary, and E a vector bundle over Ω (if the class of smoothness of Ω and E is not specified, then we assume that both are infinitely differentiable). We denote by $C^r(E)$ and $C^\infty(E)$ the vector spaces of the cross-sections of E of smoothness class r and ∞, respectively, and by $T\Omega$ and $T^*\Omega$ the tangent and cotangent bundles of Ω. The fibre of the bundles over a point $x \in \Omega$ is written as $E|_x$, $T_x\Omega$ and $T_x^*\Omega$. We denote by $S^k(T^*\Omega)$ the bundle of symmetric k-forms over Ω, and by $\Lambda^k(T^*\Omega)$ the bundle of skew-symmetric k-forms.

We fix a point $x \in \Omega$ and say that two cross-sections s_1, $s_2 \in C^\infty(E)$ are *k-equivalent at x* if for any smooth curve $\varphi : \mathbb{R}^1 \to \Omega$ with $\varphi(0) = x$ and any smooth linear function $\psi : E \to \mathbb{R}^1$ on the fibres of E, the functions

$$\frac{d^\nu}{dt^\nu}[\psi \circ (s_2 - s_1) \circ \varphi](t)$$

are equal to zero for $t = 0$ and all $\nu = 0, 1, \ldots, k$.

Let $J^k(E)|_x$ be the collection of equivalence classes and $J^k(E)$ the disjunctive union for all $x \in \Omega$ of the $J^k(E)|_x$. The set $J^k(E)$ is endowed with a natural structure of vector bundle over Ω, called the *bundle of k-jets* of E (Vinogradov, Krasil'shchik and Lychagin (1986), Bourbaki (1971), Kuranishi (1967), and Pommaret (1978)). The equivalence class of a cross-section $s \in C^\infty(E)$ in $J^k(E)|_x$ is called the k-jet of s at x and is denoted by $j^k s(x)$.

The mapping $x \mapsto j^k s(x)$ is a cross-section of the bundle $J^k(E)$ and is called the *k-jet of the cross-section s*.

We consider the localisation of these objects. Let Ω be a domain in $\mathbb{R}^n$, and Y_1 and Y_2 Euclidean spaces. Then the bundles E and F can be identified with the direct products $\Omega \times Y_1$ and $\Omega \times Y_2$, respectively. A *bundle mapping* $E \to F$ is a mapping $f : \Omega \times Y_1 \to \Omega \times Y_2$ such that $f(x,y) = (x, F(x)y)$, where $F(x)$ is a linear operator from Y_1 to Y_2 for every $x \in \Omega$, that is, a family $\{F(\cdot)\}$ of linear operators depending on a parameter in Ω. The spaces of the cross-sections $C^r(E)$ and $C^\infty(E)$ of the bundle $E = \Omega \times Y$ are identified with the vector spaces of functions on Ω with values in Y and of the corresponding smoothness. In this case, that is, when $E = \Omega \times Y$, we use the notation $C^r(\Omega, Y)$ (instead of $C^r(\Omega \times Y)$), emphasising that under this identification we are dealing with mappings from Ω to Y. The Euclidean structure of $\mathbb{R}^n$ enables us to identify the tangent and cotangent vectors (that is, the elements of $T\Omega$ and $T_x^*\Omega$, respectively) with vectors in $\mathbb{R}^n$. Denoting by $L(\mathbb{R}^n, Y)$ $(L_{\mathrm{sym}}^k(\mathbb{R}^n, Y))$ the space of linear (k-linear symmetric) mappings from $\mathbb{R}^n$

to Y (from $\mathbb{R}^n \times \cdots \times \mathbb{R}^n$ to Y), for $E = \Omega \times Y$ we obtain the obvious identifications

$$S^k(T^*\Omega) = \Omega \times L^k_{\mathrm{sym}}(\mathbb{R}^n, Y),$$

$$J^k(E) = \Omega \times Y \times L(\mathbb{R}^n, Y) \times \cdots \times L^k_{\mathrm{sym}}(\mathbb{R}^n, Y).$$

A cross-section $f \in C^r(E)$ in the trivialisation $E = \Omega \times Y$ has the form $(x, f(x))$ and can always be further identified with the mapping $x \mapsto f(x)$ from Ω to Y, which we also denote by f.

The operator j^k that associates a function f with its k-jet is a differential operator of the form

$$j^k f(x) = (f(x), Df(x), \ldots, D^k f(x)), \tag{1.1}$$

where Df $(D^i f)$ is the derivative (ith derivative) of the mapping $f : \Omega \to Y$, which is a mapping from Ω to $L(\mathbb{R}^n, Y)$ $(L^i_{\mathrm{sym}}(\mathbb{R}^n, Y))$.

Using localisations, we now introduce two important mappings of jet bundles. (Their global transposition in the language of jets is a simple exercise.) Let $\varepsilon_k : S^k(T^*\Omega) \otimes E \to I^k(E)$ be an embedding and $\pi_{k,m} : I^k(E) \to I^m(E)$ $(k > m)$ a projection in localisations defined by

$$S^k(T^*\Omega) \otimes E = \Omega \times L^k_{\mathrm{sym}}(\mathbb{R}^n, Y) \ni (x, u)$$

$$\overset{\varepsilon_k}{\to} (x, 0, 0, \ldots, 0, u) \in \Omega \times Y \times L(\mathbb{R}^n, Y) \times \cdots \times L^k_{\mathrm{sym}}(\mathbb{R}^n, Y) = J^k(E),$$

$$J^k(E) = \Omega \times Y \times \cdots \times L^k_{\mathrm{sym}}(\mathbb{R}^n, Y) \ni (x, y, u_1, \ldots, u_k)$$

$$\overset{\pi_{k,m}}{\to} (x, y, u_1, \ldots, u_m) \in \Omega \times Y \times \cdots \times L^m_{\mathrm{sym}}(\mathbb{R}^n, Y) = J^m(E).$$

We obtain the bundle sequence

$$0 \to S^k(T^*\Omega) \otimes E \overset{\varepsilon_k}{\to} J^k(E) \overset{\pi_{k,k-1}}{\to} J^{k-1}(E) \to 0,$$

which is *exact*, that is, $\mathrm{Ker}\,\varepsilon_k = 0$, $\mathrm{Im}\,\varepsilon_k = \mathrm{Ker}\,\pi_{k,k-1}$, and $\mathrm{Coker}\,\pi_{k,k-1} = 0$.

Let Ω be a domain in $\mathbb{R}^n$ with coordinates $(x_1, x_2, \ldots, x_n)$, let $Y_1 = \mathbb{R}^{m_1}$ and $Y_2 = \mathbb{R}^{m_2}$, and let $a^\alpha_{ij} : \Omega \to \mathbb{R}^1$ be functions of smoothness class no less than k; here $j = 1, \ldots, m_1$, $i = 1, \ldots, m_2$, and $\alpha = (\alpha_1, \ldots, \alpha_n)$ is a multi-index. Then the linear differential operator A with coefficients a^α_{ij} is defined by

$$(Ay)_i(x) = \sum_{\substack{|\alpha| \le k \\ 1 \le j \le m_1}} a^\alpha_{ij}(x) \frac{\partial^{|\alpha|} y_j}{\partial x_1^{\alpha_1} \partial x_2^{\alpha_2} \ldots \partial x_n^{\alpha_n}}. \tag{1.2}$$

If there are localisations of the bundles $E = \Omega \times Y_1$ and $F = \Omega \times Y_2$, then, using the mapping $j^k : C^\infty(E) \to C^\infty(I^k(E))$ in its local form (1.2), we can rewrite this definition as

$$(Ay)(x) = p(x, A)(j^k y)(x),$$

where, for every fixed x, $p(x, A)$ is a linear mapping from $J^k(E)|_x = Y_1 \times L(\mathbb{R}^n, Y_1) \times \cdots \times L^k_{\text{sym}}(\mathbb{R}^n, Y_1)$ to $F|_x = Y_2$ whose matrix of coefficients $a^\alpha_{ij}(x)$ corresponds to the choice of bases in Y_1 and Y_2 (that is, to the identification of Y_i with $\mathbb{R}^{m_i}$, $i = 1, 2$).

Finally, if E and F are arbitrary vector bundles over Ω and $p : J^k(E) \to F$ is a vector bundle mapping, then we can define a linear differential operator A of order k by setting $y \mapsto Ay = pj^k y$. Locally (that is, for every local trivialisation of the bundles E and F), the differential operator is written in the form (1.2). To connect the notation, in this case we write $p(A)$ instead of p. In what follows it is also convenient to denote by $p(x, A)$ the mapping of fibres over x for a fixed $x \in \Omega$. Thus, $p(x, A) : J^k(E)|_x \to F|_x$.

By the *symbol* σA (or, as is also called, the *principal symbol*) of the operator $A = p(A)j^k$ we understand the bundle mapping $p(A)\varepsilon_k : S^k(T^*\Omega) \otimes E \to F$; if $\xi \in T^*_x\Omega$ is a fixed covector, then by the *symbol on the covector* ξ we understand the bundle mapping $E \to F$ defined for every $x \in \Omega$ by $y \mapsto (\sigma A)(x, \xi)y = \sigma A(x)(\xi \otimes \xi \otimes \cdots \otimes \xi)y$. The mapping $\sigma A(x, \xi)$ is also called the *principal homogeneous symbol on the covector* ξ *at the point* x.

Setting $x = (\xi_1, \xi_2, \ldots, \xi_n)$ in the localisations $E = \Omega \times Y_1$ and $F = \Omega \times Y_2$ for the differential operator A acting as in (1.2), we obtain

$$((\sigma A)(x, \xi)y)_i = \sum_{\substack{|\alpha|=k \\ 1 \le j \le m_1}} a^\alpha_{ij} \xi^\alpha y_j,$$

where $\xi^\alpha = \xi_1^{\alpha_1} \xi_2^{\alpha_2} \cdots \xi_n^{\alpha_n}$.

Thus, a *differential operator of order* k is a mapping from $C^{k+r}(E)$ to $C^r(F)$ ($0 \le r \le \infty$), and the mappings $p(A) : J^k(E) \to F$ and $\sigma^k A$, $\sigma^k A(\cdot, \xi)$ connected with it are mappings of finite-dimensional bundles (matrices depending on $x \in \Omega$ in the localisations of these bundles).

Along with a differential operator $A : C^\infty(E) \to C^\infty(F)$ (of course, instead of ∞ we may have finite numbers corresponding to the order of A) we also consider the differential operators $A^{(l)}$, $l \ge 1$, called its *lth prolongations*, acting from $C^\infty(E)$ to $C^\infty(J^l(F))$ and defined by the compositions $A^{(l)} = j^l \circ A$. In the localisations $E = \Omega \times Y_1$ and $F = \Omega \times Y_2$, the operator $A^{(l)}$ is obtained from A by means of the formula

$$(A^{(l)})(x) = ((Ay)(x), D(Ay)(x), \ldots, D^l(Ay)(x)).$$

§2. Complexes, Equivalence of Morphisms and Compatibility Morphisms

Since in what follows we are dealing with complexes consisting of various types of spaces and mappings, it is convenient to discuss separately the purely algebraic part, which makes use of the language of categories (MacLane

(1963)). We confine ourselves to subcategories of the category of vector spaces and linear mappings (over the field of real or complex numbers).

Let $\mathfrak{A}$ be such a category, so that its objects are vector spaces (denoted below by H, H', H_1, $H_2, \ldots$) and for every pair of objects (H_1, H_2) the set $\mathrm{Mor}(H_1, H_2)$ of morphisms consists of some set of linear operators from H_1 to H_2.

Example 1.1. The category $\mathrm{D}(\Omega)$. Let Ω be a manifold. The objects of the category $\mathrm{D}(\Omega)$ are the vector spaces $C^\infty(E)$ of infinitely differentiable cross-sections of the bundles E over Ω, and the morphisms are linear differential operators with infinitely differentiable coefficients. If $E = \Omega \times Y_1$ and $F = \Omega \times Y_2$, then $A \in \mathrm{Mor}(C^\infty(E), C^\infty(F))$ has the local form (1.2).

Example 1.2. The category $\mathrm{D}_a(\Omega)$. Let Ω be a real-analytic manifold. The objects of the category $\mathrm{D}_a(\Omega)$ are vector spaces of real-analytic cross-sections of real-analytic bundles over Ω, and the morphisms are linear differential operators with real-analytic coefficients.

Example 1.3. The category $\mathrm{DB}(\Omega)$. We now assume that the manifold Ω is closed and that its boundary Γ is an infinitely differentiable manifold. The objects of the category $\mathrm{DB}(\Omega)$ are the linear spaces $C^\infty(E) \times C^\infty(G)$, where E and G are bundles over Ω and Γ, respectively, and the morphisms are linear *differential boundary operators* with infinitely differentiable coefficients, that is, operators Φ of the form

$$\Phi(f, g) = (\Phi^{11} f, \gamma(\Phi^{21} f) + \Phi^{22} g),$$

where Φ^{11}, Φ^{21} and Φ^{22} are linear differential operators in the corresponding spaces and γ is the operator of restriction to Γ of a cross-section defined on Ω.

Example 1.4. The category $\mathrm{DB}_a(\Omega)$. Let Ω be a closed real-analytic manifold, and suppose that its boundary Γ is a real-analytic manifold. The objects of the category $\mathrm{DB}_a(\Omega)$ are pairs of vector spaces of real-analytic cross-sections of bundles over Ω and Γ, as in Example 1.3, and the morphisms are differential boundary operators with real-analytic coefficients.

Example 1.5. The category $\mathrm{D}_c(\Omega)$. Suppose that $\Omega \subset \mathbb{R}^n$. The category $\mathrm{D}_c(\Omega)$ is a subcategory of $\mathrm{D}(\Omega)$ (see Example 1.1), whose morphisms are differential operators with constant coefficients.

We fix a category $\mathfrak{A}$.

Definition 1.1. A sequence of objects $\{H_i\}$ and morphisms $\{A_i \in \mathrm{Mor}(H_i, H_{i+1})\}$

$$0 \to H_0 \xrightarrow{A_0} H_1 \xrightarrow{A_1} H_2 \xrightarrow{A_2} \cdots \tag{1.3}$$

is called a *complex* if $A_{i+1} A_i = 0$ for all $i = 0, 1, 2, \ldots$

For every $A \in \operatorname{Mor}(H_1, H_2)$, the sets $\operatorname{Ker} A = \{x \in H_1 : Ax = 0\}$ and $\operatorname{Im} A = \{y \in H_2 : \exists x \in H_1 \text{ such that } y = Ax\}$ are vector spaces; consequently, for the complex (1.3) we can define the quotient spaces $\mathcal{H}_j = \operatorname{Ker} A_i / \operatorname{Im} A_{i-1}$, called the *spaces of the cohomologies of the complex* (1.3). The complex (1.3) is called *exact* if its cohomologies are zero, that is, if $\operatorname{Ker} A_i = \operatorname{Im} A_{i-1}$ $(i \geq 0,\ A_{-1} = 0)$.

Definition 1.2. Two complexes

$$0 \to H_0 \xrightarrow{A_0} H_1 \xrightarrow{A_1} H_2 \xrightarrow{A_2} \cdots \tag{1.4}$$

and

$$0 \to H_0' \xrightarrow{A_0'} H_1' \xrightarrow{A_1'} H_2' \xrightarrow{A_2'} \cdots \tag{1.5}$$

are called *cochain equivalent* in the category $\mathfrak{A}$ if there are morphisms $p_i : H_i \to H_i'$, $q_i : H_i' \to H_i$, $s_i : H_{i+1} \to H_i$, and $s_i' : H_{i+1}' \to H_i'$ $(i = 0, 1, \ldots)$ such that

a) the diagram

$$
\begin{array}{ccc}
H_i & \xrightarrow{\quad A_i \quad} & H_{i+1} \\[2mm]
p_i \downarrow \uparrow q_i & & p_{i+1} \downarrow \uparrow q_{i+1} \\[2mm]
H_i' & \xrightarrow{\quad A_i' \quad} & H_{i+1}'
\end{array}
$$

is commutative, that is, $p_{i+1} A_i = A_i' p_i$ and $q_{i+1} A_i' = A_i q_i$;

b) there hold the equalities

$$q_i p_i - 1 - s_i A_i = 0, \quad p_i q_i - 1 - s_i' A_i' = 0.$$

For equivalent complexes, the corresponding cohomology spaces are isomorphic (in the category $\mathfrak{A}$).

In problems involving overdetermined systems, complexes are very seldom the original object of investigation. Usually only a morphism A_0 is given, and the problem consists in constructing a suitable complex starting with A_0. The next definition clarifies the meaning of the word "suitable" in the preceding sentence. Let the category $\mathfrak{A}$ be fixed.

Definition 1.3. A morphism $A_1 \in \operatorname{Mor}(H_1, H_2)$ is called a *compatibility morphism (operator)* for a morphism (operator) $A_0 \in \operatorname{Mor}(H_0, H_1)$ in $\mathfrak{A}$ if $A_1 A_0 = 0$ and for any $\alpha \in \operatorname{Mor}(H_1, H_2')$ satisfying the condition $\alpha A_0 = 0$ there is a $c \in \operatorname{Mor}(H_2, H_2')$ such that $\alpha = c A_1$.

Example 1.6. We set $\mathfrak{A} = D(\Omega)$, where $\Omega \subset \mathbb{R}^2$. Let the operator $A_0 : C^\infty(\Omega, \mathbb{R}^1) \to C^\infty(\Omega, \mathbb{R}^2)$ be defined by $y \mapsto A_0 y = (\partial y / \partial x_1, \partial y / \partial x_2) = (f_1, f_2)$. Then the operator $(f_1, f_2) \mapsto A_1(f_1, f_2) = \partial f_1 / \partial x_2 - \partial f_2 / \partial x_1$ is a compatibility operator for A_0.

Example 1.7. Let $\mathfrak{A} = \mathrm{D}(\Omega)$, where $\Omega \subset \mathbb{R}^3$. We define an operator A_0 : $C^\infty(\Omega, \mathbb{R}^3) \to C^\infty(\Omega, \mathbb{R}^3)$ by

$$A_0 : y \mapsto \operatorname{curl} y = f.$$

Then the operator $A_1 : f \mapsto \operatorname{div} f$ is a compatibility operator for A_0.

Definition 1.4. A complex

$$0 \to H_0 \xrightarrow{A_0} H_1 \xrightarrow{A_1} H_2 \xrightarrow{A_2} \cdots$$

is called a *compatibility complex* for a morphism A_0 if every morphism A_i for $i \geq 1$ is a compatibility morphism for A_{i-1}.

Example 1.8 (the *de Rham complex* (Wells (1973))). Let Ω be a smooth manifold and $\Lambda^r(T^*\Omega)$ ($r \geq 0$, $\Lambda^0(T^*\Omega) = \Omega \times \mathbb{R}^1$) the bundle of r-forms over Ω. For every point $x \in \Omega$ we choose a neighbourhood $U \subset \Omega$ of x in which $\Lambda^r(T^*\Omega)|_U$ can be trivialised. Then there is a finite set of r-forms $\{dx^\alpha : dx^\alpha \in C^\infty(\Lambda^r(T^*\Omega))\}$ such that in U every r-form $\omega \in C^\infty(\Lambda^r(T^*\Omega))$ can be uniquely represented in the form $\omega(x) = \sum\limits_\alpha \omega_\alpha(x)dx^\alpha$ ($x \in U$, $\omega_\alpha \in C^\infty(U, \mathbb{R}^1)$). We define a differential operator $d_U : C^\infty(\Lambda^r(T^*\Omega)|_U) \to C^\infty(\Lambda^{r+1}(T^*\Omega)|_U)$ by setting

$$d_U\omega(x) = \sum_{i=1}^{n} \frac{\partial \omega_\alpha(x)}{\partial x_i} \, dx_i \wedge dx^\alpha$$

in U, where $(x_1, x_2, \ldots, x_n)$ is some coordinate system in U. It is well known that the family of operators $\{d_U : U \subset \Omega\}$ defines globally a differential operator $d : C^\infty(\Lambda(T^*\Omega)) \to C^\infty(\Lambda^{r+1}(T^*\Omega))$ which satisfies the identity $d \circ d = 0$. The resulting complex of differential operators

$$0 \to C^\infty(\Lambda^0(T^*\Omega)) \xrightarrow{d} C^\infty(\Lambda^1(T^*\Omega)) \xrightarrow{d} \cdots \xrightarrow{d} C^\infty(\Lambda^r(T^*\Omega))$$
$$\xrightarrow{d} C^\infty(\Lambda^{r+1}(T^*\Omega)) \xrightarrow{d} \cdots \xrightarrow{d} C^\infty(\Lambda^n(T^*\Omega)) \to 0$$

is a compatibility complex for the differential operator $d : C^\infty(\Lambda^0(T^*\Omega)) = C^\infty(\Omega, \mathbb{R}^1) \to C^\infty(\Lambda^1(T^*\Omega))$.

Example 1.9. The *Dolbeault complex* is a compatibility complex for the operator $\bar\partial$. A detailed description of this important complex (which is not used in our survey) can be found in Wells (1973).

As has already been mentioned, the original object of study is usually the morphism A_0 itself. Consequently, on the set of morphisms we introduce an equivalence relation such that equivalent morphisms have simultaneous (cochain equivalent) compatibility complexes.

Definition 1.5. Two morphisms $A_0 \in \operatorname{Mor}(H_0, H_1)$ and $A_0 \in \operatorname{Mor}(H_0', H_1')$ are called *equivalent* if the complexes

$$0 \to H_0 \overset{A_0}{\to} H_1$$

and

$$0 \to H_0' \overset{A_0'}{\to} H_1'$$

are cochain equivalent.

Example 1.10. Let $\mathfrak{A} = \mathrm{D}(\Omega)$, $\Omega \subset \mathbb{R}^3$, and let the operators $A_0 : C^\infty(\Omega, \mathbb{R}^3) \to C^\infty(\Omega, \mathbb{R}^3)$ and $A_0' : C^\infty(\Omega, \mathbb{R}^3) \to C\infty(\Omega, \mathbb{R}^4)$ be defined by

$$A_0 : y \mapsto \operatorname{curl} y + y = f,$$
$$A_0' : y \mapsto (\operatorname{curl} y + y, \operatorname{div} y) = (f', g).$$

Then A_0 and A_0' are equivalent.

The differential operators that establish the equivalence of the corresponding complexes (Definition 1.2) are of the following form:

$p_0 = q_0 = \mathrm{Id}$, the identity mapping on the space $C^\infty(\Omega, \mathbb{R}^3)$;
$p_1 : C^\infty(\Omega, \mathbb{R}^3) \ni f \mapsto (f, \operatorname{div} f) \in C^\infty(\Omega, \mathbb{R}^4)$;
$q_1 : C^\infty(\Omega, \mathbb{R}^4) \ni (f', g) \mapsto f' \in C^\infty(\Omega, \mathbb{R}^3)$;
s_0 and s_0', the zero operators.

Example 1.11 (the system of *stationary Maxwell equations*). Let $\mathfrak{A} = \mathrm{D}(\Omega)$, where $\Omega \subset \mathbb{R}^3$. Then the operators A_0 and A_0' defined by

$$A_0(u, v) = (\operatorname{curl} u + v, \operatorname{curl} v - u)$$

and

$$A_0'(u, v) = (\operatorname{curl} u + v, \operatorname{curl} v - u, \operatorname{div} v, \operatorname{div} u)$$

are equivalent.

Example 1.12. Let $\mathfrak{A} = \mathrm{D}(\Omega)$, $\Omega \subset \mathbb{R}^n$, and let A and P be arbitrary operators. Then the operators A and (A, PA) are equivalent.

Examples 1.11 and 1.12 show that equivalent differential operators are different ways of expressing a mathematical object that describes one and the same physical or other process.

Proposition 1.1. *Suppose that two morphisms $A_0 \in \mathrm{Mor}(H_0, H_1)$ and $A_0' \in \mathrm{Mor}(H_0', H_1')$ are equivalent. If A_0 has a compatibility complex, then so does A_0'.*

Proposition 1.2. *If two morphisms A_0 and A_0' are equivalent and (1.4) and (1.5) are compatibility complexes for them, then these complexes are cochain equivalent.*

A differential operator A with constant coefficients always has a compatibility complex in the category $\mathrm{D}_c(\Omega)$ mentioned in Example 1.5 (Palamodov (1967)).

If a differential operator has variable coefficients, then it does not necessarily have a compatibility operator in the category $D(\Omega)$.

Example 1.13. Let $\mathfrak{A} = D(\mathbb{R}^1)$, and let A_0 be the operator of multiplication by a smooth function $\alpha(x)$ such that $\alpha(0) = 0$ for $x \le 0$ and $\alpha(x) > 0$ for $x > 0$; that is,

$$(A_0 y)(x) = \alpha(x) y(x).$$

A_0 does not have a compatibility operator in the category $D(\mathbb{R}^1)$. Indeed, if there were a compatibility operator A_1 for A_0, then from the equality $A_1 A_0 = 0$ and the form of the function $\alpha(x)$ it would follow that the coefficients of A_1 were equal to zero for $x > 0$. Then the operator $A_1' = A_1/x$, which has smooth coefficients, would satisfy the condition $A_1' A_0 = 0$. However, it can be shown that A_1' cannot be represented in the form $A_1' = C A_1$ for any differential operator C with smooth coefficients.

The non-existence of a compatibility operator in the category $D(\Omega)$ in this example is a consequence of the strong "degeneracy" of the coefficients of A_0. However, if we impose some non-degeneracy (regularity) conditions on these coefficients, then A_0 always has a compatibility operator. These regularity conditions are described in Sect. 3.

§3. Differential Operators with Variable Coefficients

3.1. Regularity Conditions. We introduce the regularity conditions mentioned at the end of Sect. 2.

Let E_0 and E_1 be bundles over Ω, and $A : C^\infty(E_0) \to C^\infty(E_1)$ a differential operator of order k.

We consider the families of spaces

$$\mathfrak{R}_{k+l}(x) = \operatorname{Ker} p(x, A^{(l)}) \subseteq J^{k+l}(E_0)|_x \quad (x \in \Omega).$$

Example 1.14. Let $\Omega \subset \mathbb{R}^2$, $E_0 = E_1 = \Omega \times \mathbb{R}^1$, and $Ay = \partial^2 y / \partial x_1^2 + \partial^2 y / \partial x_2^2$. We represent an element $l \in J^2(E_0) = \Omega \times \mathbb{R}^1 \times L(\mathbb{R}^2, \mathbb{R}^1) \times L^2_{\mathrm{sym}}(\mathbb{R}^2, \mathbb{R}^1)$ in the form $l = (x, y, p, A)$, where $x \in \Omega$, $y \in \mathbb{R}^1$, $p = (p_1, p_2) \in \mathbb{R}^2$, and $A = (a_{ij})$, $i, j = 1, 2$, $a_{12} = a_{21}$.

Then the family of spaces $\mathfrak{R}_2(x)$ corresponding to A is defined by

$$\mathfrak{R}_2(x) = \{ e = (x, y, p, A) \in J^2(E_0) : a_{11} + a_{22} = 0 \}.$$

Example 1.15. Let $\Omega \subset \mathbb{R}^3$, let $b(x) = (b^1(x), b^2(x), b^3(x))$ be a smooth vector field over Ω, and suppose that the differential operator $A : C^\infty(\Omega, \mathbb{R}^3) \to C^\infty(\Omega, \mathbb{R}^3)$ acts on vector fields according to the formula $A : y \mapsto \operatorname{curl} y + b \times y$, where $\times$ is the vector product in $\mathbb{R}^3$.

We describe the corresponding family of spaces $\mathfrak{R}_1(x)$. We represent the elements of the bundle $J^1(\Omega, \mathbb{R}^3)$ in the form

$$e = (x, \{y^i, \ i = 1, 2, 3\}, \{u^i_k, \ i, \ k = 1, 2, 3\}).$$

Then the family of spaces $\mathfrak{R}_1(x)$ is the set of collections $e = (x, \{y^i, \ i = 1, 2, 3\}, \{u^i_k, \ i, \ k = 1, 2, 3\})$ satisfying the conditions

$$u^i_k - u^k_i + b^i(x)y^k - b^k(x)y^i = 0 \quad (i, \ k = 1, 2, 3; \ i \neq k).$$

The points of the bundle $J^2(\Omega, \mathbb{R}^3) = \Omega \times \mathbb{R}^3 \times L(\mathbb{R}^3, \mathbb{R}^3) \times L^2_{\mathrm{sym}}(\mathbb{R}^3, \mathbb{R}^3)$ can be represented in the form

$$\begin{aligned}
e' = (x, &\{y^i, \ i = 1, 2, 3\}, \{u^i_k, \ i, \ k = 1, 2, 3\}, \\
&\{u^i_{ks} : u^i_{ks} = u^i_{sk}, \ i, \ k, \ s = 1, 2, 3\}).
\end{aligned}$$

The corresponding equations that describe $\mathfrak{R}_2(x) = \operatorname{Ker} p(x, A^{(1)})$ have the form

$$\begin{gathered}
u^i_{ks} - u^k_{is} + b^i(x)u^k_s - b^k(x)u^i_s + \frac{\partial b^i}{\partial x_s}(x)u^k - \frac{\partial b^k}{\partial x_s}(x)y^i = 0, \\
u^i_k - u^k_i + b^i(x)y^k - b^k(x)y^i = 0, \\
i, \ k, \ s = 1, 2, 3, \quad i \neq k.
\end{gathered} \tag{1.6}$$

Definition 1.7. A differential operator A is called *sufficiently regular* if for every x the dimensions of the subspaces $\mathfrak{R}_{k+l}(x)$ do not depend on x and the mappings $\pi_{i+r,i} : \mathfrak{R}_{i+r}(x) \to \mathfrak{R}_i(x)$ have a constant rank for every $r \geq 0$ and $i \geq k$.

Example 1.16. If $\Omega \subset \mathbb{R}^n$ and the operator A has constant coefficients, then, clearly, A is sufficiently regular.

Example 1.17. Let Ω, E_0, E_1, and A be the same as in Example 1.15. The dimensions of the subspaces $\mathfrak{R}_1(x)$ and $\mathfrak{R}_2(x)$ do not depend on x. From (1.6) it follows that

$$(\operatorname{curl} b, y) \equiv \left(\frac{\partial b^3}{\partial x_2} - \frac{\partial b^2}{\partial x_3} \right) y^1 + \left(\frac{\partial b^1}{\partial x_3} - \frac{\partial b^3}{\partial x_1} \right) y^2 + \left(\frac{\partial b^2}{\partial x_1} - \frac{\partial b^1}{\partial x_2} \right) y^3 = 0.$$

We consider the projection $\pi_{2,1} : \mathfrak{R}_2(x) \to \mathfrak{R}_1(x)$. It is easy to verify that the image of $\pi_{2,1}$ is the vector space of points $(x, y, u) \in \mathfrak{R}_1(x)$ for which $\operatorname{curl}(b, y) = 0$. From this it follows that the mapping $\pi_{2,1} : \mathfrak{R}_2(x) \to \mathfrak{R}_1(x)$ may have a constant rank only if
1) $\operatorname{curl} b(x) = 0$ for all $x \in \Omega$, or
2) $\operatorname{curl} b(x) \neq 0$ for all $x \in \Omega$.
It can be shown that in either case the operator A is sufficiently regular.

3.2. Formally Exact Complexes. Consider a complex of sufficiently regular differential operators

$$C^\infty(E_0) \xrightarrow{A_0} C^\infty(E_1) \xrightarrow{A_1} C^\infty(E_2), \tag{1.7}$$

where A_0 and A_1 are of order k_0 and k_1, respectively.

18   P. I. Dudnikov and S. N. Samborski

Definition 1.8. The complex (1.7) is called *formally exact* if for any integer $l \geq 0$ the complexes

$$J^{k_0+k_1+l}(E_0) \xrightarrow{p(A_0^{(l+k_1)})} J^{k_1+l}(E_1) \xrightarrow{p(A_1^{(l)})} J^l(E_2)$$

are exact.

We turn our attention to the fact that the terms in the complex (1.7) are infinite-dimensional spaces, while Definition 1.8 refers to a complex of finite-dimensional bundles for every x.

Proposition 1.3. *Every formally exact complex is a compatibility complex (in the sense of Definition 1.4, which means that we deal with the category $\mathrm{D}(\Omega)$ mentioned in Example 1.1).*

We use the proof of this simple assertion to illustrate how the concepts introduced so far can be applied. Let A' be an operator such that $A'A_0 = 0$. Without loss of generality, we may assume that the order k' of A' is not less than the order k_1 of A_1 (otherwise we would replace A' with the operator $j^l A'$ for some $l > 0$). We need to show that there is a differential operator C such that $A' = CA_1$. From the condition $A'A_0 = 0$ it follows that the sequence

$$J^{k_0+k'}(E_0) \xrightarrow{p(A_0^{(k')})} J^{k'}(E_1) \xrightarrow{p(A')} E_2'$$

is a complex, and its formal exactness implies that

$$\operatorname{Ker} p(A_1^{(k'-k_1)}) = \operatorname{Im} p(A_0^{(k')}) \subset \operatorname{Ker} p(A').$$

Hence, there is a bundle mapping (that is, a differential operator of order zero) $\varphi : J^{k_0-k_1}(E_2) \to E_2$ such that $p(A') = \varphi \circ p(A_1^{(k'-k_1)})$. From the commutative diagram

$$
\begin{array}{ccc}
C^\infty(J^{k'}(E_1)) & \xrightarrow{\;\;p(A')\;\;} & C^\infty(E_2') \\[2mm]
\big\uparrow{\scriptstyle j^{k'}} & \searrow{\scriptstyle p(A_1^{k'-k_1})} & \big\uparrow{\scriptstyle \varphi} \\[2mm]
 & & C^\infty(J^{k'-k_1}(E_2)) \\[2mm]
 & & \big\uparrow{\scriptstyle j^{k'-k_1}} \\[2mm]
C^\infty(E_1) & \xrightarrow{\;\;A_1\;\;} & C^\infty(E_2)
\end{array}
$$

it follows that $A' = p(A')j^{k'} = \varphi j^{k'-k_1} A_1 = CA_1$. Proposition 1.3 is proved.

Corollary 1.1. *Let*

$$C^\infty(E_0) \xrightarrow{A_0} C^\infty(E_1) \xrightarrow{A_1} C^\infty(E_2)$$

and

$$C^\infty(E_0) \xrightarrow{A_0} C^\infty(E_1) \xrightarrow{A'_1} (E'_2)$$

be two formally exact complexes in which the operators A_1 and A'_1 have the same order $p(A_1)$ and $p(A'_1)$ and are epimorphisms. Then there is a bundle isomorphism $\Psi : E_2 \to E'_2$ such that

$$(A'_1 y)(x) = \Psi((A_1 y)(x))$$

for all $x \in \Omega$ and $y \in C^\infty(E_1)$.

Indeed, from the proof of Proposition 1.3 it follows that there are operators Φ and Ψ of order zero such that $A'_1 = \Psi A_1$ and $A_1 = \Phi A'_1$. Since $p(\Phi A'_1) = \Phi \circ p(A'_1)$, $p(\Psi A_1) = \Psi \circ p(A'_1)$, and $p(A'_1)$ and $p(A_1)$ are epimorphisms, we deduce that

$$\Psi \circ \Phi = \mathrm{Id}_{E'_2}, \quad \Phi \circ \Psi = \mathrm{Id}_{E_2}.$$

This corollary enables us to "paste together" a globally defined compatibility operator on Ω from locally defined ones. Let $\Omega = \cup U_i$. Also, let $\Psi_i : U_i \times Y \to E_0|_{U_i}$ be local trivialisations of the bundle E_0 over Ω, $A : C^\infty(E_0) \to C^\infty(E_1)$ a differential operator, and $A_{(i)}$ the restriction of A to the cross-sections over U_i, and suppose that for every i

$$C^\infty(E_0|_{U_i}) \xrightarrow{A_{(i)}} C^\infty(E_1|_{U_i}) \xrightarrow{A'_{(i)}} C^\infty(E'_i),$$

where $E'_i = U_i \times Y'_{(i)}$, is a formally exact complex. Without loss of generality, we may assume that all the operators $A'_{(i)}$ are of the same order and that the mappings $p(x, A'_{(i)})$ are epimorphisms for $x \in \Omega$. For $i \neq j$, the restriction of the operator A to $U_i \cap U_j$ has two compatibility operators, namely, $A'_{(s)} : C^\infty(E_1|_{U_i \cap U_j}) \to C^\infty(U_i \cap U_j, Y'_{(s)})$, where $s = i$ or $s = j$. Then, by Corollary 1.1, there are isomorphisms $\alpha_{ij}(x) \cdot Y'_{(i)} \to Y'_{(j)}$, $x \in U_i \cap U_j$, which define the structure of the vector bundle E' over Ω. In addition, the operators $A'_{(i)}$ define uniquely (up to an isomorphism of E') a differential operator $A' : C^\infty(E_1) \to C^\infty(E')$ which, according to Proposition 1.3, is a compatibility operator for A.

We conclude this discussion with an important example of a formally exact complex (consequently, also a compatibility complex). In this complex, formal exactness means the formal exactness of any fragment consisting of three objects and their two connecting morphisms.

Example 1.18 (Spencer (1969)). Let E_0 be a bundle over Ω and $E_1 = J^k(E_0)$. For the differential operator $j^k : C^\infty(E_0) \to C^\infty(J^k(E_0))$ there are bundles $C_i^k(E_0)$ and first order differential operators D_i^k such that the sequence

$$C^\infty(E_0) \xrightarrow{j^k} C^\infty(J^k(E_0)) \xrightarrow{D_1^k} C^\infty(C_1^k(E_0)) \xrightarrow{D_2^k} \cdots \tag{1.8}$$

is a formally exact complex and the mappings

$$p(D_1^k) : J^1(J^k(E_0)) \to C_1^k(E_0), \quad p(D_i^k) : J^1(C_{i-1}^k(E_0)) \to C_i^k(E_0)$$

are isomorphisms. With these properties, the complex (1.8) is determined uniquely up to isomorphisms of the bundles $C_1^k(E_0)$.

3.3. Formal Integrability. Our immediate purpose is to describe a procedure for constructing a compatibility operator for a sufficiently regular differential operator.

We present two methods which, in essence, are intrinsically connected with each other. In each of them, the desired operator is replaced by an equivalent (in the sense of Definition 1.5) first-order operator with additional formal properties.

Definition 1.9. A sufficiently regular differential operator $A : C^\infty(E_0) \to C^\infty(E_1)$ of order k is called *formally integrable* if the mappings

$$\pi_{i+k+1,i+k} : \mathfrak{R}_{i+k+1}(x) \to \mathfrak{R}_{i+k}(x)$$

are surjections for all $i \geq 0$ (we recall that for $A = p(A)j^k$, $\mathfrak{R}_k$ is (in the sufficiently regular case) the bundle $\operatorname{Ker} p(A)$, and $\pi_{m,r} : J^m(E_0) \to J^r(E_0)$ is the projection defined in Sect. 1.1 of this chapter).

Definition 1.10. A sufficiently regular differential operator $A_0 : C^\infty(E_0) \to C^\infty(E_1)$ of order k is called *formally integrable* if for every differential operator $A' : C^\infty(E_1) \to C^\infty(E_2)$ of order k' such that the order of $A'A$ is less than $k + k'$, there is a differential operator $A'' : C^\infty(E_1) \to C^\infty(E_2)$ of order less than k' such that $A'A_0 = A''A_0$.

The formal integrability of an operator A_0 of order k means that for any $l \geq 1$, all the differential consequences of order $k + l$ of the relations $A_0 y = 0$ (that is, the equalities derived by differentiating up to any order, equating the mixed derivatives and using linear algebra for any x) can be obtained only by differentiating up to an order not higher than l and using linear algebra.

Example 1.19. Let $\Omega \subset \mathbb{R}^3$, let y be a vector field on Ω, and let $A_0 y = \operatorname{curl} y + \lambda y$, where $\lambda \in \mathbb{R}^1$. The operator A_0 is formally integrable only for $\lambda = 0$. If $\lambda \neq 0$, then A_0 is equivalent to the operator $A_0' y = (\operatorname{curl} y + ly, \operatorname{div} y)$, which is already formally integrable.

Example 1.20. Again, let $\Omega \subset \mathbb{R}^3$, let $b(x)$ be a vector field on Ω, and let $A_0 y = \operatorname{curl} y + b \times y$. The formal integrability of the operator A_0 is equivalent to the equality $\operatorname{curl} b(x) = 0$ for all $x \in \Omega$. If $\operatorname{curl} b(x) \neq 0$ for all $x \in \Omega$ (this condition ensures the sufficient regularity of A_0), then A_0 is equivalent to the operator

$$A_0' y = (\operatorname{curl} y + b \times y, \operatorname{grad}(\operatorname{curl} b, y), (\operatorname{curl} b, y)),$$

which is formally integrable ($(\cdot, \cdot)$ is the inner product in $\mathbb{R}^3$).

Example 1.21. Let A_0 be a differential operator with constant coefficients in $\Omega \subset \mathbb{R}^n$. If A_0 contains only terms of the highest order, then it is always formally integrable.

Example 1.22. Let $\Omega \subset \mathbb{R}^n$, and let $A_0 : C^\infty(\Omega, \mathbb{R}^m) \to C^\infty(\Omega, \mathbb{R}^m)$ be an operator with a *non-characteristic covector*, that is, a covector $\xi \in \mathbb{R}^n \setminus \{0\}$ such that the mapping $\sigma_\xi(x, A) : \mathbb{R}^m \to \mathbb{R}^m$ is an isomorphism for every $x \in \Omega$ (for example, A_0 is elliptic or hyperbolic (Egorov and Shubin (1987))). Then A_0 is formally integrable. Indeed, suppose that A_0 is of order k, and let A_1 be an operator of order $k' \geq 0$ such that the order of $A_1 A_0$ is less than $k + k'$. In a sufficiently small neighbourhood $U \subset \Omega$ of an arbitrary point $x_0 \in \Omega$ we choose a system of coordinates $(x_1, x_2, \ldots, x_n)$ such that the covector dx_n is non-characteristic for A_0 at every point $x \in U$. Then in U the operator A_0 can be written locally in the form $A_0 y = l(x) \cdot \partial^k y / \partial x_n^k + My$, where the differential operator M is of order k and does not contain the derivative $\partial^k y / \partial x_n^k$, and the matrix $l(x)$ is an isomorphism. Without loss of generality, we may assume that $l(x)$ is the identity matrix for all $x \in U$. Then $A_1 A_0 y = A_1(\partial^k y / \partial x_n^k) + A_1 My$. Since the order of $A_1 A_0$ is strictly lower than $k + k'$, we must have $A_1(\partial^k y / \partial x_n^k) = 0$ for any function $y \in C^\infty(\Omega, \mathbb{R}^m)$. Consequently, $A_1 = 0$, from which the formal integrability of A_0 follows immediately.

Proposition 1.4. *If $A_0 : C^\infty(E_0) \to C^\infty(E_1)$ is a sufficiently regular operator, then there exist a bundle F and a differential operator $p : C^\infty(E_1) \to C^\infty(F)$ that can be determined in finitely many steps and are such that*

(i) *the operator $A_0' = pA_0$ is formally integrable;*
(ii) $\operatorname{Ker} A_0' = \operatorname{Ker} A_0$;
(iii) *if E_2 is a vector bundle and*

$$C^\infty(E_0) \xrightarrow{A_0} C^\infty(E_1) \xrightarrow{A_1} C^\infty(E_2)$$

is a complex of differential operators, then there are a bundle F' and differential operators A_1' and Q such that the diagram

$$
\begin{array}{ccccc}
C^\infty(E_0) & \xrightarrow{\;A_0\;} & C^\infty(E_1) & \xrightarrow{\;A_1\;} & C^\infty(E_2) \\
\Big\| & & \Big\downarrow{p} & & \Big\downarrow{Q} \\
C^\infty(E_0) & \xrightarrow{\;A_0'\;} & C^\infty(F) & \xrightarrow{\;A_1'\;} & C^\infty(F') \,,
\end{array}
$$

is commutative and p induces an isomorphism of the cohomologies of its top and bottom rows.

Essentially, as in Examples 1.21 and 1.22, the action of the operator p can be represented as a writing up of the differential consequences of the equality $A_0 y = 0$. Hence, without loss of generality, we may assume that $p = (\mathrm{Id}, p')$. In this case, the assertions (ii) and (iii) in Proposition 1.5 follow immediately from the monomorphic nature of p. It is also easy to see that the operators A_0 and pA_0 are equivalent in the sense of Definition 1.5 in the category $\mathrm{D}(\Omega)$.

3.4. Involutiveness in the Sense of Spencer. In what follows we examine the important concept of involutiveness. In Sect. 3.5 and Sect. 3.6 this concept is introduced by means of other methods, which can be read independently of each other and of the method below.

Let V be a finite-dimensional vector space, $\Lambda^p V$ and $S^k V$ the exterior and symmetric degrees of V, respectively, and $\otimes$ the operation of tensor product.

We define the linear mapping

$$\delta : \Lambda^p V \otimes S^m V \to \Lambda^{p+1} V \otimes S^{m-1} V.$$

For $p = 0$, this mapping is the composition of the natural embedding of $S^m V$ in $\otimes^m V = V \otimes (\otimes^{m-1} V)$ and the mapping $V \otimes (\otimes^{m-1} V) \to V \otimes S^{m-1} V$ generated by the projection $\otimes^{m-1} V \to S^{m-1} V$. For $p > 0$, δ maps $w \otimes u$ into $\delta(w \otimes u) = (-1)^p w \wedge \delta u$, where $w \in \Lambda^p V$ and $u \in S^m V$. Writing $\delta(w \otimes u \otimes y) = \delta(w \otimes u) \otimes y$, we obtain the mapping

$$\delta : \Lambda^p V \otimes S^m V \otimes Y \to \Lambda^{p+1} V \otimes S^{m-1} V \otimes Y.$$

It is easy to see that $\delta^2 = 0$. Setting $V = T_x^* \Omega$ and $Y = E|_x$ for every $x \in \Omega$, we arrive at the mapping

$$\delta : \Lambda^p(T_x^* \Omega) \otimes S^m(T_x^* \Omega) \otimes E|_x \to \Lambda^{p+1}(T_x^* \Omega) \otimes S^{m-1}(T_x^* \Omega) \otimes E|_x,$$

which, by means of the above action in fibres, generates (for every i and k) the complex of morphisms of bundles over Ω

$$0 \to \Lambda^i(T^* \Omega) \otimes S^k(T^* \Omega) \otimes E \xrightarrow{\delta} \Lambda^{i+1}(T^* \Omega) \otimes S^{k-1}(T^* \Omega) \otimes E$$

$$\xrightarrow{\delta} \Lambda^{i+2}(T^* \Omega) \otimes S^{k-2}(T^* \Omega) \otimes E \to \cdots \to 0. \tag{1.9}$$

This complex is exact.

Now let $A : C^\infty(E_0) \to C^\infty(E_1)$ be a sufficiently regular differential operator of order k. Then the kernel of the morphism $\pi_{k+l,k+l-1} : \mathfrak{R}_{k+l} \to \mathfrak{R}_{k+l-1}$ is a bundle g_{k+l}, which is a sub-bundle of $S^{k+l}(T^* \Omega) \otimes E_0$ and is called the *symbolic bundle* or, simply, the *symbol*, of the operator A. By definition, the sequence of bundles

$$0 \to g_{k+l} \to \mathfrak{R}_{k+l} \xrightarrow{\pi_{k+l,k+l-1}} \mathfrak{R}_{k+l-1}$$

is exact.

If we denote by σ_l the symbol (in the sense of Sect. 1.2) of the differential operator $A^{(l)}$ (the lth extension of A), then σ_l is a bundle mapping from $S^{k+l}(T^* \Omega) \otimes E_0$ to $S^l(T^* \Omega) \otimes E_1$. Obviously, g_{k+l} is also the kernel of σ_l, that is, the sequence

$$0 \to g_{k+l} \to S^{k+l}(T^* \Omega) \otimes E_0 \xrightarrow{\sigma_l} S^l(T^* \Omega) \otimes E_1$$

is exact.

Since the g_{k+l} are sub-bundles of the $S^{k+l}(T^*\Omega) \otimes E_0$, they can be substituted in the tensor cofactors in (1.9) in place of the $S^{k+l}(T^*\Omega) \otimes E_0$. Then it turns out that the restrictions to $\Lambda^r(T^*\Omega) \otimes g_s$ of δ generate the correctly defined complex

$$0 \to g_{k+l} \xrightarrow{\delta} T^*\Omega \otimes g_{k+l+1} \xrightarrow{\delta} \Lambda^2(T^*\Omega) \otimes g_{k+l+2} \xrightarrow{\delta} \cdots$$
$$\xrightarrow{\delta} \lambda^l(T^*\Omega) \otimes g_k \to \delta(\Lambda^l(T^*\Omega) \otimes g_k) \to 0. \qquad (1.10)$$

Unlike (1.9), this complex does not have to be exact in all the terms, but it is in the first two.

Definition 1.11. An operator A of order k, its symbol σ, and its symbolic bundle g_k are called *involutive* if the complex (1.10) is exact for all $l \geq 0$.

Example 1.23. Let $\Omega \in \mathbb{R}^2$, $Y = \mathbb{R}^1$, and $Ay = (\partial^2 y/\partial x_1^2, \partial^2 y/\partial x_2^2)$. The fibre of the symbolic bundle g_2 at an arbitrary point $x \in \Omega$ is given in $S^2(T^*\Omega)|_x = L^2_{\mathrm{sym}}(\mathbb{R}^2, \mathbb{R}^1)$ by

$$g_2|_x = \{A = (a_{ij}) \in L^2_{\mathrm{sym}}(\mathbb{R}^2, \mathbb{R}^1), \; i, j = 1, 2 : a_{11} = a_{22} = 0\}.$$

Thus, g_2 is one-dimensional. It is easy to verify that the bundles g_{2+l} are non-zero for all $l > 0$. Rewriting (1.10) for $l = 2$ and $k = 2$, we obtain the complex

$$0 \to g_4 \xrightarrow{\delta} T^*\Omega \otimes g_3 \xrightarrow{\delta} \Lambda^2(T^*\Omega) \otimes g_2 \to 0,$$

which is not exact, since $g_2|_x = 0$, $g_3|_x = 0$, and $(\Lambda^2(T^*\Omega) \otimes g_2)|_x \simeq g_2|_x \neq 0$ for any point $x \in \Omega$. From this it follows that the operator A is not involutive. However, since g_{2+l} is the zero bundle for all $l > 0$, it is not difficult to convince ourselves that the first extension $A^{(1)}$ of A, defined by

$$A^{(1)}y = \left(\frac{\partial^3 y}{\partial x_1^3}, \frac{\partial^3 y}{\partial x_1^2 \partial x_2}, \frac{\partial^3 y}{\partial x_1 \partial x_2^2}, \frac{\partial^3 y}{\partial x_2^3}, \frac{\partial^2 y}{\partial x_1^2}, \frac{\partial^2 y}{\partial x_2^2} \right),$$

is an involutive operator.

This example is also an illustration of the following important result, called the *Poincaré δ-lemma* (by analogy with the Poincaré lemma on the local exactness of the de Rham complex (Example 1.8) for the operator of exterior differentiation).

Theorem 1.1. *There is an integer $\mu \geq 0$ depending only on the dimension n of the manifold Ω, the order k of the differential operator, and the dimension d of the fibre in the bundle E_0, such that the complex (1.10) is exact for all $l \geq \mu$ and $i \geq 0$.*

Thus, every sufficiently regular operator A has an involutive extension $A^{(\mu)} = j^\mu A$. We remark that if A is formally integrable, then so is $A^{(\mu)}$. Since it is obvious that $A^{(\mu)}$ is equivalent to A (in the sense of Definition 1.5),

Theorem 1.1 and Proposition 1.4 show that every sufficiently regular operator can be replaced by an equivalent one, formally integrable and involutive.

It is useful to find upper bounds for the number μ in Theorem 1.1 in terms of the numbers n, k and d occurring in the formulation of this theorem. Such estimates can be obtained from the relations

$$\mu(0, d, 1) = 0,$$

$$\mu(n, d, 1) = d\binom{a+n}{n-1} + a + 1, \quad \text{where} \quad a = \mu(n-1, d, 1),$$

$$\mu(n, d, k) = \mu(n, b, 1), \quad \text{where} \quad b = \sum_{\nu=0}^{k} \binom{\nu+n-1}{n-1} d.$$

We return to Example 1.18. The differential operator D_p^k in (1.8) acts according to the formula

$$D_p^k : C^\infty(\Lambda^p(T^*\Omega) \otimes J^k(E)) \to (\Lambda^{p+1}(T^*\Omega) \otimes J^{k-1}(E))$$

and is determined uniquely from the following conditions:

(i) every cross-section s of the bundle $\Lambda^i(T^*\Omega) \otimes I^k(E)$ with differential form Φ of order j (that is, a cross-section of the bundle $\Lambda^j(T^*\Omega)$) satisfies the equality

$$D_{i+j}^k(\Phi \wedge s) = d\Phi \wedge \pi_{k,k-1}s + (-)^j \Phi \wedge D_i^k;$$

(ii) the complex

$$0 \to C^\infty(E) \xrightarrow{j^k} C^\infty(J^k(E)) \xrightarrow{D_0^k} C^\infty(T^*\Omega \otimes J^{k-1}(E))$$

is exact.

Now let $A = p(A)j^k : C^\infty(E_0) \to C^\infty(E_1)$ be a formally integrable operator (if A is only sufficiently regular, then we need to replace it with an equivalent, formally integrable one). As before, suppose that the bundles $\mathfrak{R}_k \subset J^k(E_0)$ and $\mathfrak{R}_{k+l} \subset K^{k+l}(E_0)$ correspond to the operators A and $A^{(l)}$, respectively.

Considering the commutative (in view of the properties (i) and (ii) of D) diagram

$$
\begin{array}{ccc}
C^\infty(\Lambda^i(T^*\Omega) \otimes J^{k+l+1}(E_0)) & \xrightarrow{p_{l+1}(A)} & C^\infty(\Lambda^I(T^*\Omega) \otimes J^{l+1}(E_1)) \\
\downarrow{\scriptstyle D_i^{k+l+1}} & & \downarrow{\scriptstyle D_i^{l+1}} \\
C^\infty(\Lambda^{i+1}(T^*\Omega) \otimes J^{k+1}(E_0)) & \xrightarrow{p_l(A)} & C^\infty(\Lambda^{i+1}(T^*\Omega) \otimes J^l(E_1)),
\end{array}
$$

whose rows consist of cross-section mappings generated by the bundle mappings, we convince ourselves that D_p^k generates a differential operator

$$D : C^\infty(\Lambda^i(T^*\Omega) \otimes \mathfrak{R}_{k+l+1}) \to C^\infty(\Lambda^{i+1}(T^*\Omega) \otimes \mathfrak{R}_{k+l}).$$

The complex

$$0 \to \theta \xrightarrow{j^m} C^\infty(\mathfrak{R}_m) \xrightarrow{D} C^\infty(\Lambda(T^*\Omega) \otimes \mathfrak{M}_{m-1}) \xrightarrow{D} C^\infty(\Lambda^2(T^*\Omega) \otimes \mathfrak{R}_{m-2})$$

$$\xrightarrow{D} \cdots \xrightarrow{D} C^\infty(\Lambda^n(T^*\Omega) \otimes \mathfrak{R}_{m-n}) \to 0, \tag{1.11}$$

whose cohomologies do not depend on m for m sufficiently large, is called the *first Spencer complex* generated by the operator A (or by its corresponding "equation" $\mathfrak{R}_m$). It is also called the *first Spencer resolvent* of the solutions of a homogeneous equation; however, in the commonly accepted sense, this complex is not the resolvent for an arbitrary operator A (consequently, by definition, it is not an exact complex).

We define the *second Spencer complex* ("*resolvent*") as follows.

For $m \geq \mu$, where μ is the number in Theorem 1.1, we introduce the sets

$$C^0 = C^0_m = \mathfrak{R}_m, \quad C^i = C^i_m = (\Lambda^i(T^*\Omega) \otimes \mathfrak{R}_m)/\delta(\Lambda^{i-1}(T^*\Omega) \otimes g_{m+1}).$$

Then C^i are bundles. We consider the diagram

$$
\begin{array}{ccccccccc}
0 & \longrightarrow & \Lambda^i(T^*\Omega) \otimes g_{m+1} & \longrightarrow & \Lambda^i(T^*\Omega) \otimes \mathfrak{R}_{m+1} & \longrightarrow & \Lambda^i(T^*\Omega) \otimes \mathfrak{R}_m & \longrightarrow & 0 \\
& & \downarrow{\scriptstyle -\delta} & & \downarrow{\scriptstyle D} & & \downarrow{\scriptstyle \tilde{D}} & & \\
0 & \longrightarrow & \delta(\Lambda^i(T^*\Omega) \otimes g_{m+1}) & \longrightarrow & \Lambda^{i+1}(T^*\Omega) \otimes \mathfrak{R}_m & \longrightarrow & C^{i+1} & \longrightarrow & 0
\end{array}
$$

in which the rows are exact and the mapping $\tilde{D}$ in the right-hand column is generated by D so that the diagram is commutative. From this diagram it follows that $\tilde{D}$ acts "through" C^i, that is, it generates a differential operator

$$D^i : C^\infty(C^i) \to C^\infty(C^{i+1}).$$

The complex

$$0 \to \theta \xrightarrow{j^m} C^\infty(C^0) \xrightarrow{D^0} C^\infty(C^1) \xrightarrow{D^1} \cdots \xrightarrow{D^{n-1}} C^\infty(C^n) \to 0, \tag{1.12}$$

generated by the formally integrable differential operator $A : C^\infty(E_0) \to C^\infty(E_1)$ and called the *second Spencer complex*, is determined uniquely (up to a bundle isomorphism) by the following properties:

(i) $C^0 = C^0_m = \mathfrak{R}_m$;

(ii) for any i ($0 \leq i \leq n-1$), the mapping $D^i : C^\infty(C^i) \to C^\infty(C^{i+1})$ is a first-order differential operator corresponding to the mapping $\rho^i : J^1(C^i) \to C^{i+1}$ of vector bundles whose symbol is the epimorphism $\sigma(D^i) : T^*\Omega \otimes C^i \to C^{i+1}$ generated by the mapping $T^*\Omega \otimes \Lambda^i(T^*\Omega) \to \Lambda^{i+1}(T^*\Omega)$;

(iii) the sequence

$$0 \to \theta \xrightarrow{j^m} C^\infty(C^0) \xrightarrow{D^0} C^\infty(C^1)$$

is exact;

(iv) the sequence (1.12) is formally exact.

3.5. Involutiveness in the Sense of Kuranishi. Below we give another definition of involutiveness for first-order operators, which is equivalent to the preceding one.

Let $A : C^\infty(E_0) \to C^\infty(E_1)$ be a first-order differential operator, $g_1 \subset T^*\Omega \otimes E_0$ its symbolic bundle, and $g_2 \subset S^2(T^*\Omega) \otimes E_0$ the first extension of g_1. For every point $x \in \Omega$ we define numbers $\tau^i(x)$ $(i = 0, 1, 2, \ldots, \dim \Omega - 1)$ by

$$\tau^i(x) = \min_{H_i} \dim(g_1|_x \cap (H_i \otimes E_0|_x)),$$

where H_i are $(n - i)$-dimensional subspaces of $T^*_x \Omega$.

It is not difficult to show that

$$\dim g_2|_x \leq \sum_{i=0}^{n-1} \tau^i(x).$$

Definition 1.12. A basis $(t_1, t_2, \ldots, t_n)$ in $T^*_x \Omega$ is called *quasi-regular* for g_1 at $x \in \Omega$ if

$$\dim g_2|_x = \sum_{i=0}^{n-1} \tau(x, (t_{i+1}, t_{i+2}, \ldots, t_n)),$$

where $\tau(x, (t_{i+1}, t_{i+2}, \ldots, t_n)) = \dim(g_1|_x \cap (\{t_{i+1}, \ldots, t_n\} \otimes E_0|_x))$ (here $\{t_{i+1}, \ldots, t_n\}$ is the subspace of $T^*_x \Omega$ generated by the vectors $t_{i+1}, t_{i+2}, \ldots, t_n$).

Definition 1.13. A system of coordinates $(x_1, x_2, \ldots, x_n)$ in Ω is called *regular* for a differential operator A if at every point $x \in \Omega$ the covectors $dx_1, dx_2, \ldots, dx_n$ form a quasi-regular basis for g_1. The operator A is called *involutive* if there is a regular system of coordinates in the neighbourhood of every point $x \in \Omega$.

3.6. Commutation Relations and Compatibility Operators. In the Spencer complexes, the information concerning the original operator $A : C^\infty(E_0) \to C^\infty(E_1)$ is "encoded" in the structure of the bundles (that is, in the relations that define them as sub-bundles and quotient bundles), while the differential operators are uniformised in the limit. When localising, we obtain "simple" differential operators, but on a set of functions connected with non-differential linear relations. The elimination of these relations lowers the number of functions, while "complicating" the differential operators. Consequently, at least from the practical point of view, it is useful to know how to construct the coefficients of the operators in a compatibility complex of A in a local trivialisation, directly from the coefficients of A. We describe such a technique, based on the use of commutation relations characteristic to formally integrable differential operators.

The concept of involutiveness introduced in Sect. 3.4 and Sect. 3.5 is defined below in a different way, independent of those followed earlier. We also remark that the discussion in Sect. 3.2 shows that we can concentrate on local arguments when constructing compatibility operators. Consequently, we describe the operator A locally in a suitable form. It is convenient to reduce the problem to first-order operators; to do so, we make use of a modification of the standard method, which enables us to lower the differentiation order by introducing new functions representing the derivatives of the original ones. More precisely, we proceed as follows. Let $A : C^\infty(E_0) \to C^\infty(E_1)$ be a differential operator of order k. We define a first-order differential operator $\bar A : C^\infty(J^{k-1}(E_0)) \to C^\infty(E_1)$ so that $\bar A j^{k-1} = A$. We now introduce a first-order operator $A' : C^\infty(J^{k-1}(E_0)) \to C^\infty(E_1 \times C^1(E_0))$, where $C^1(E_0)$ is the bundle in Example 1.18, by writing

$$A'y = (\bar A y, D_1^k y),$$

where D_1^k is the operator in the complex (1.8).

Example 1.24. Let Ω be a domain in $\mathbb{R}^2$ and A the Laplace operator $Ay = \partial^2 y/\partial x_1^2 + \partial^2 y/\partial x_2^2$.

The cross-sections of the bundle $J^1(\Omega \times \mathbb{R}^1)$ can be represented in the form $s = (y, y_1, y_2) \in \mathbb{R}^3$. The operators $j^1 : C^\infty(\Omega, \mathbb{R}^1) \to C^\infty(J^1(\Omega \times \mathbb{R}^1))$ and $D_1^1 : C^\infty(J^1(\Omega \times \mathbb{R}^1)) \to C^\infty(C^1(\Omega \times \mathbb{R}^1))$ are defined by

$$j^1 : y \mapsto (y, \partial y/\partial x_1, \partial y/\partial x_2),$$
$$D_1^1 : (y, y_1, y_2) \mapsto (\partial y/\partial x_1 - y_1, \partial y/\partial x_2 - y_2, \partial y_1/\partial x_2 - \partial y_2/\partial x_1).$$

(We mention that the first two equations in the equality $D_1^1 s = 0$ correspond to the introduction of new unknown functions in the standard method of passing to a first-order operator, while the third one describes the symmetry of the mixed derivatives.)

Taking $\bar A : C^\infty(J^1(E_0)) \to C^\infty(E_1)$ to be $\bar A . (y, y_1, y_2) \mapsto \partial y_1/\partial x_1 + \partial y_2/\partial x_2$, we find that the first-order operator A' is given by

$$A'(y, y_1, y_2) = \left(\frac{\partial y_1}{\partial x_1} + \frac{\partial y_2}{\partial x_2}, \frac{\partial y}{\partial x_1} - y_1, \frac{\partial y}{\partial x_2} - y_2, \frac{\partial y_1}{\partial x_2} - \frac{\partial y_2}{\partial x_1} \right).$$

It is easy to convince ourselves that the operators A and A' are equivalent in the category $D(\Omega)$ and that if A is sufficiently regular, then so is A'.

In what follows we assume that in Ω we have selected a coordinate neighbourhood U with coordinates $(x_1, x_2, \ldots, x_n)$ and trivialisations over U of the bundles $E_0 = U \times Y$ and $E_1 = U \times W$, where Y and W are Euclidean spaces. Also, suppose that in U we have chosen one coordinate—x_n, say—so that the local form (1.2) of the differential operator A contains $\partial/\partial x_n$ explicitly. Then Y and W can be expressed as direct sums of subspaces, namely $Y = Y^+ \oplus Y^-$ and $W = W^+ \oplus W^-$ (where $\dim Y^+ = \dim W^+$), so that, after applying a suitable (for every x in a sufficiently small neighbourhood $U' \subset U$) isomorphism between Y^+ and W^+, we can rewrite the operator A in the form

$$Ay = A(y^+, y^-) = \begin{cases} \dfrac{\partial y^+}{\partial x_n} + L(y^+, y^-), \\ M(y^+, y^-) \equiv M^+ y^+ + M^- y^-, \end{cases} \tag{1.13}$$

where L and M are differential expressions that do not contain $\partial/\partial x_n$. The most important fact of the formal theory is that in every equivalence class (in the category $D(\Omega)$) we can choose a representative whose local form (1.13) in an appropriate coordinate system has the property that the kernel $\operatorname{Ker} M$ of M is mapped by the operator $y \mapsto \partial y^+/\partial x_n + Ly$ into $\operatorname{Ker} M^+$. Then such a "partition" of A enables us to disentangle a complex system containing implicit connections between functions and their mixed partial derivatives. Furthermore, "almost all" coordinates are "appropriate coordinates' (the meaning of this phrase is clarified below) if the operator is expressed "correctly". First, we illustrate this by means of simple examples, after which, assuming the above statement to be true (in the form of specific commutation relations), we draw conclusions from it regarding the construction of a compatibility operator. Finally, we show how to choose the representative in an equivalence class (that is, how to rewrite the operator) so that these commutation relations are always satisfied.

Example 1.24. We define the trivial bundles $E_0 = \mathbb{R}^2 \times \mathbb{R}^m$ and $E_1 = \mathbb{R}^2 \times \mathbb{R}^{2m}$ and consider the first-order differential operator defined by

$$A_0 y = \begin{cases} \dfrac{\partial y}{\partial x_1} + L_1(x)y = f_1, \\ \dfrac{\partial y}{\partial x_2} + L_2(x)y = f_2, \end{cases} \tag{1.14}$$

where (x_1, x_2) is a coordinate system in $\mathbb{R}^2$ and $L_1(x)$ and $L_2(x)$ are $m \times m$-matrices. We also assume that the *Frobenius conditions*

$$\frac{\partial L_1}{\partial x_2} - \frac{\partial L_2}{\partial x_1} - L_1 L_2 + L_2 L_1 = 0$$

are satisfied. These conditions are equivalent to the commutation relations

$$\left(\frac{\partial}{\partial x_1} + L_1\right)\left(\frac{\partial}{\partial x_2} + L_2\right) = \left(\frac{\partial}{\partial x_2} + L_2\right)\left(\frac{\partial}{\partial x_1} + L_1\right). \tag{1.15}$$

From (1.15) it follows that the system (1.14) is solvable only if f_1 and f_2 satisfy the compatibility conditions

$$\left(\frac{\partial}{\partial x_1} + L_1\right)f_2 - \left(\frac{\partial}{\partial x_2} + L_2\right)f_1 = 0.$$

It is easy to see that

$$A_1(f_1, f_2) = \left(\frac{\partial}{\partial x_1} + L_1\right)f_2 - \left(\frac{\partial}{\partial x_2}\right)f_1$$

is a compatibility operator for the operator A in (1.14).

Example 1.24 admits the following important generalisation, which plays a crucial role in the local construction of compatibility operators.

Let A_0 be a sufficiently regular first-order differential operator in a local coordinate system $(x_1, x_2, \ldots, x_n)$ in $U \subset \Omega$, written in the form

$$A_0 y = \begin{cases} \dfrac{\partial y^+}{\partial x_n} + L_0 y = f, \\ M_0^+ y^+ + M_0^- y^- = g, \end{cases} \tag{1.16}$$

where L_0, M_0^+ and M_0^- are differential expressions in terms of $(\partial/\partial x_1, \partial/\partial x_2, \ldots, \partial/\partial x_{n-1})$. We assume that there is a differential operator L_1 containing differentiation only with respect to $x_1, x_2, \ldots, x_{n-1}$ and such that

$$M_0^+ \left(\frac{\partial y^+}{\partial x_n} + L_0 y \right) = \left(\frac{\partial}{\partial x_n} + L_1 \right) (M_0^+ y^+ + M_0^- y^-). \tag{1.17}$$

Then, obviously, the system (1.16) is solvable only if the compatibility condition

$$\frac{\partial g}{\partial x_n} + L_1 g - M_0^+ f = 0$$

is satisfied.

Proposition 1.5. *Let A_0 be a differential operator expressed locally in the form (1.16), and suppose that the commutation relations (1.17) are satisfied with some operator L_1. If M_1 is a compatibility operator for the operator $y \mapsto M_0 y = M_0^+ y^+ + M_0^- y^-$, then A_1 defined by*

$$A_1(f, g) = \left(\frac{\partial g}{\partial x_n} + L_1 g - M_0^+ f, M_1 g \right)$$

is a compatibility operator for A_0.

We remark that the coordinate x_n occurs in M_0 only as a parameter; therefore, the use of Proposition 1.5 in the construction of a compatibility operator lowers the dimension of the argument (and, at the same time, the number of equations in the system). Consequently, it is possible (when the commutation relations are satisfied at every step) to reduce the dimension of x successively so that after at most $n - 1$ steps the problem reduces to the construction of a compatibility operator for a differential operator of the form

$$y \mapsto \frac{\partial y_s^+}{\partial x_s} + By,$$

where B does not contain differentiation with respect to x_s. Obviously, the compatibility operator in such a case is the zero operator.

We need to find conditions under which the commutation relations (1.17) hold for a differential operator in an appropriate coordinate system.

The next assertion shows that formal integrability is such a condition, which enables us to reduce the verification of (1.17) to that of the commutation relations only for the symbols of operators. This is an important step, since it moves the problem from the complicated (in view of the abundance of implicit relations) differential algebra to the somewhat simpler linear algebra.

Proposition 1.6. *Let $A : C^\infty(E_0) \to C^\infty(E_1)$ be a formally integrable first order differential operator written locally in a neighbourhood $U \subset \Omega$ in the form (1.16), and suppose that there is a symbol $\tau(x, \xi')$ depending smoothly on $x \in U$ and such that*

$$\sigma M^+(x, \xi')(\xi_n y^+ + \sigma L_0(x, \xi')y) = (\xi_n + \tau(x, \xi'))\sigma M_0(x, \xi')y, \qquad (1.18)$$

where $\xi' = (\xi_1, \xi_2, \ldots, \xi_{n-1})$ and $\sigma M_0^+(x, \xi')$, $\sigma M_0(x, \xi')$ and $\sigma L_0(x, \xi')$ are the symbols of the corresponding operators at a point $x \in U$ for a covector $\xi' \neq 0$.

Then there is a differential operator L_1 such that $\tau(x, \xi')$ is the symbol of L_1 and the commutation relations (1.17) are satisfied.

In view of Proposition 1.6, we restrict our attention to the symbols of differential operators.

Definition 1.14. Let $\sigma A_0(x, \xi)$ be the symbol of a differential operator A_0 of order k. The symbol σA_0 (and the differential operator A_0) are called *involutive* if for every symbol $\sigma'(x, \xi)$ of order $k_1 > 1$ and such that $\sigma'(x, \xi)\sigma A_0(x, \xi) = 0$, there is a first-order symbol $\bar\sigma(x, \xi)$ and a symbol $\tau(x, \xi)$ of order $k_1 - 1$ such that $\sigma(x, \xi)\sigma A_0(x, \xi) = 0$ and $\sigma'(x, \xi) = \tau(x, \xi)\bar\sigma(x, \xi)$.

It is useful to compare this definition with that of formal integrability (Definition 1.10).

We assert that the commutation relations (1.18) hold for every involutive differential operator in an appropriate coordinate system, and proceed to construct such a system.

Definition 1.15. A covector $\xi \in T_x^*\Omega$ is called *quasi-regular* for the symbol σA_0 (and the operator A_0) at a point $x \in \Omega$ if

$$\dim \operatorname{Ker} \sigma A_0(x, \xi) = \min_{\eta \neq 0} \dim \operatorname{Ker} \sigma A_0(x, \eta),$$

and *non-characteristic* if $\operatorname{Ker} \sigma A_0(x, \xi) = 0$.

Obviously, every non-characteristic covector is quasi-regular.

Definition 1.16. An operator A_0 is said to have a *constant defect* if the number $\operatorname{Ker} \sigma A_0(x, \xi)$ does not depend on (x, ξ) ($\xi \neq 0$). Every covector of an operator with a constant defect is quasi-regular.

Example 1.25. If $\Omega \subset \mathbb{R}^3$ and the operator A_0 is defined by $A_0(y) = \operatorname{curl} y$, then A_0 has a constant defect.

Proposition 1.7. *If $\sigma(x, \xi)$ is an involutive symbol and $(x_1, x_2, \ldots, x_n)$ a coordinate system in $U \subset \Omega$ such that the covector dx_n is quasi-regular for $\sigma(x, \xi)$ at every point $x \in U$, then $\sigma(x, \xi)$ satisfies the commutative relations (1.18) in U.*

Definition 1.17. A differential operator $A : C^\infty(E_0) \to C^\infty(E_1)$ is called *normalised* if

(i) A is a first-order operator;

(ii) A is formally integrable;

(iii) A is involutive;

(iv) the symbol $\sigma A : T^*\Omega \otimes E_0 \to E_1$ of A is surjective.

The first three conditions have already been discussed; the fourth one means that among the relations $Ay = 0$ there are no (explicit or implicit) algebraic (that is, non-differential) ones between the functions $(y_1, y_2, \ldots, y_m) = y$. If such relations exist, then, of course, we may exclude them, thus reducing the number of functions $y_1, y_2, \ldots, y_m$. In addition, the transformed operator is equivalent to the original one in the category $D(\Omega)$.

Theorem 1.2. *Every sufficiently regular differential operator A can be transformed after finitely many steps (without leaving the framework of differential and linear algebra in the bundle fibres) into a normalised operator equivalent to it in the category $D(\Omega)$.*

In practice, the transformation mentioned in Theorem 1.2 reduces to the completion of the following successive steps.

The first step is going over to a formally integrable operator by adjoining the missing relations; the second one is going over to an equivalent involutive operator by means of an extension of a sufficiently large order (that is, A is replaced by an operator $A^{(m)} = j^m A$, where m never exceeds the number μ estimated after Theorem 1.1). Obviously, the involutive operator obtained in this way is formally integrable. The third step consists in going over to an equivalent first-order operator, as described at the beginning of this section. Finally, the fourth step (the elimination of non-differential relations) was discussed before Theorem 1.2.

Proposition 1.8. *If $A_0 : C^\infty(E_0) \to C^\infty(E_1)$ is a normalised differential operator, then for every point $x \in \Omega$ there is a neighbourhood $U \subset \Omega$ such that $E_0|_U \cong U \times Y$ and $E_1|_U \cong U \times W$, and in which a coordinate system $(x_1, x_2, \ldots, x_n)$, called a regular system, can be chosen so that*

(i) *there is a decomposition $Y_1 \oplus Y_2 \oplus \cdots \oplus Y_{n+1}$ of the space Y such that A can be written in U in the form*

$$A_0 : y \mapsto \left\{ A_{(i)}y = \frac{\partial y_i^+}{\partial x_i} + L_i y_i^+ + M_i(y_{i+1}, \ldots, y_{n+1}) \right\} \qquad (1.19)$$

$$(i = 1, 2, \ldots, n),$$

where $Y_i^+ = \bigoplus_{j \leq i} Y_j$, $y_i \in Y_i$, $y_i^+ \in Y_i^+$, and the differential expressions L_i (M_i) do not contain partial derivatives with respect to $x_1, \ldots, x_n$ $(x_{i+1}, \ldots, x_n)$;

(ii) *for every $i \leq n$ the operator $y \mapsto (A_{(1)}y, A_{(2)}y, \ldots, A_{(i)}y)$ (where the variables $x_{i+1}, \ldots, x_n$ occur as parameters) is normalised and the covector dx_i is quasi-regular for it.*

In the set of coordinate systems $(x_1, x_2, \ldots, x_n)$ in U equipped with the natural topology, the regular coordinate systems for a normalised operator A_0 form a dense open set.

If we fix a basis $(e_1, e_2, \ldots, e_m)$ in Y, then we can always renumber it $(e_1', e_2', \ldots, e_m')$ so that the decomposition $Y = \oplus Y_i$ has the form

$$Y_1 = (e_1', e_2', \ldots, e_{k_1}'),$$
$$Y_2 = (e_{k_1+1}', e_{k_1+2}', \ldots, e_{k_2}'), \ldots, Y_{n+1} = (e_{k_n+1}', e_{k_n+2}', \ldots, e_m')$$
$$(1 \leq k_1 \leq k_2 \leq \ldots \leq k_n \leq m).$$

For a dense open set of bases this renumbering can be arbitrary. Consequently, in the "general" situation the reduction of the operator A_0, written in terms of local coordinates, to the form (1.19) does not require a change of coordinates, nor even a change of unknown functions. A special choice of coordinates enables us to simplify the form of the operator even further. For example, we can arrange things so that the operators M_i do not depend on $y_{i+2}, y_{i+3}, \ldots, y_{n+1}$.

The form (1.19) of a normalised operator A in a regular coordinate system is called the *normal Cartan form*.

From Propositions 1.6, 1.7 and 1.8 we immediately deduce the following assertion.

Proposition 1.9. *If* (1.19) *be the normal Cartan form of a normalised operator A_0, and the operators $A_{(i)}^+$ and $A_{(i)}^-$ are defined by*

$$(A_{(i)}, A_{(i-1)}, \ldots, A_{(1)})y = A_{(i)}^+ y_{i+1}^+ + A_{(i)}^-(y_{i+2}, y_{i+3}, \ldots, y_{n+1}),$$

where $y_i(x) \in Y_i$ and $y_i^+(x) \in Y_i^+$ for $x \in U$, then there are first-order differential operators $\hat{A}_{(i)}$ such that

$$\hat{A}_{(i)}(A_{(i-1)}, \ldots, A_{(1)}) = A_{(i-1)}^+ A_{(i)}.$$

The operators $A_{(i)}$ have the form $f \mapsto \dfrac{\partial f}{\partial x_i} + \cdots$, where the dots denote terms that do not contain $\partial f / \partial x_i$. These terms are found from the commutation relations, for example, by the method of undetermined coefficients.

The next assertion follows from Propositions 1.5 and 1.9.

Theorem 1.3. *Let* (1.19) *be the normal Cartan form for a normalised first-order operator $A_0 : C^\infty(U, Y) \to C^\infty(U, W)$, let $W_1 \oplus W_2 \oplus \cdots \oplus W_n$ be the corresponding decomposition of the space W so that $A_{(i)}$ acts from $C^\infty(U, Y)$*

to $C^\infty(U, W)$, and let $A^+(i)$ and $\hat{A}^+_{(i)}$ be the operators in Proposition 1.9. Then the operator A_1, which acts from $C^\infty(U, W)$ to $C^\infty\left(U, \bigoplus_{j=1}^{n} \bigoplus_{i=1}^{n-j} W_i\right)$ according to the formula

$$(f_1, f_2, \ldots, f_n) \mapsto \hat{A}_{(n)}(f_1, f_2, \ldots, f_{n-1}) - A^+_{(n-1)} f_n, \ldots,$$
$$\hat{A}_{(3)}(f_1, f_2) - A^+_{(2)} f_3, \hat{A}_{(2)} f_1 - A^+_{(1)} f_2),$$

where $f_i(x) \in W_i$ for $x \in U$, is a compatibility operator for A_0.

The compatibility operator A_1 for the normalised operator A_0, which is computed according to Theorem 1.3, is written in the normal Cartan form (the coordinate system $(x_1, x_2, \ldots, x_n)$ is regular for it). This enables us to write for it a compatibility operator A_2, and so on.

Thus, we obtain the compatibility complex

$$C^\infty(E_0) \xrightarrow{A_0} C^\infty(E_1) \xrightarrow{A_1} C^\infty(E_2) \xrightarrow{A_2} \ldots \to 0. \tag{1.20}$$

If A_0 is an operator with a constant defect and s is the number of non-zero operators among $A_{(n)}, A_{(n-1)}, \ldots, A_{(1)}$ in the normal Cartan form, then the compatibility complex (1.20) for A_0 contains exactly s non-zero terms. Of course, this complex is formally exact (Definition 1.8).

The construction of a compatibility complex for an arbitrary sufficiently regular operator A consists in first going over (according to Theorem 1.2) to an equivalent normalised operator A_0, then constructing its compatibility complex by repeatedly applying Theorem 1.3, and, finally, going over to the complex

$$C^\infty(E_0) \xrightarrow{A} C^\infty(E_1) \xrightarrow{A'} C^\infty(E_2) \to \ldots, \tag{1.21}$$

which, by Proposition 1.2, is cochain equivalent to the compatibility complex for A_0.

We remark that formal integrability and involutiveness are essential properties of the form in which the operator is written, and even the order of the compatibility operator depends essentially on the number of differentiations required (by Proposition 1.4 and Theorem 1.1) to pass from A to an equivalent, formally integrable and involutive operator.

At the same time, the requirements that the operator should be of first order and that the symbol should be an epimorphism in the definition of a normalised operator are of a purely technical nature and are made for the sake of brevity of the presentation. In concrete problems, it is not always convenient to replace the operator by an equivalent first-order one. This can be avoided if we make use of the commutation relations that hold for formally integrable operators of higher orders. Such relations can be found in Dudnikov and Samborski (1981).

3.7. The Real-Analytic Case. We consider the category $D_\alpha(\Omega)$, and let $\mathfrak{A}(\Omega, Y)$ be the set of real-analytic functions defined on an open set $\Omega \subset \mathbb{R}^n$ with values in an Euclidean space Y. Then for a sufficiently regular differential operator $A : \mathfrak{A}(\Omega, Y) \to \mathfrak{A}(\Omega, Y_1)$ and every $\Omega' \subset \Omega$ the complex (1.21) generates the complex

$$\mathfrak{A}(\Omega', Y_0) \xrightarrow{A_0} \mathfrak{A}(\Omega', Y_1) \xrightarrow{A'} \mathfrak{A}(\Omega', Y_2) \to \cdots \tag{1.22}$$

Theorem 1.4. *If $x \in \Omega$ and Ω' is a sufficiently small neighbourhood of x homeomorphic to an open ball in $\mathbb{R}^n$, then the complex (1.22) is exact.*

In other words, the system $A_0 y = f$ is locally solvable if and only if $A_1 f = 0$. We emphasize once more that the operator A' is constructed from A_0 in finitely many steps.

Theorem 1.4 is easily proved by induction (over $n = \dim \Omega$)) and the application of the commutation relations (1.7). Indeed, by Theorem 1.2 and Proposition 1.2, we may assume that A_0 is normalised and has the local form (1.16). According to the inductive assumption, the complex

$$\mathfrak{A}(\Omega'', Y) \xrightarrow{M_0} \mathfrak{A}(\Omega'', G) \xrightarrow{M_1} \mathfrak{A}(\Omega'', W), \tag{1.23}$$

where $M_0 y = M_0^+ y^+ + M_0^- y^-$ and Ω'' is a neighbourhood of the point $x^0 \in \Omega$, is exact for any x_n sufficiently close to x_n^0 (we recall that, by (1.16), x_n occurs in the operators M_0 and M_1 as a parameter).

If $A_1(f, g) = 0$, where A_1 is the compatibility operator defined in Proposition 1.5, then $M_1 g = 0$, which, in view of the exact complex (1.23), defines a function $(x_1, x_2, \ldots, x_n) \mapsto (z^+(x_1, x_2, \ldots, x_n), z^-(x_1, x_2, \ldots, x_n))$ that satisfies the equation $M_0 z = 0$.

By the Cauchy-Kovalevska theorem, there is a function

$$(x_1, x_2, \ldots, x_n) \to (y^+(x_1, \ldots, x_n), y^-(x_1, \ldots, x_n))$$

satisfying the top equation in (1.16) and such that y^- coincides with z^- everywhere and y^+ coincides with z^+ for $x_n = x_n^0$. It remains to show that this new function y satisfies the bottom equation in (1.16) for all x_n, and not just for $x_n = x_n^0$.

If we denote the function $M_0^+ y^+ + M_0^- y^- - g$ by w, then we see that $w = 0$ for $x_n = x_n^0$.

In view of the relations $\partial g / \partial x_n + L_1 g - M_0^+ f = 0$ (which follow from the equality $A_1(f, g) = 0$) and the commutation relations (1.17), we find that $(\partial / \partial x_n + L_1) w = 0$ for all x in a sufficiently small neighbourhood of the point x^0. Consequently, $w = 0$ in this neighbourhood.

The exactness in the remaining terms of the complex (1.22) is shown in the same way. The theorem is proved.

Following Cartan, we describe the set of solutions of the homogeneous equation $Ay = 0$. For a normalised operator A, such an explicit description is obtained by means of its normal Cartan form (1.19).

Let $A : \mathfrak{A}(E_0) \to \mathfrak{A}(E_1)$ be a normalised differential operator which in some regular coordinate system $(x_1, x_2, \ldots, x_n)$ in a neighbourhood U has the normal Cartan form

$$Ay = \left\{ A_{(i)}y = \frac{\partial y_i^+}{\partial x_i} + L_i y_i^+ + M(y_{i+1}, \ldots, y_{n+1}) \right\} = 0, \qquad (1.24)$$

where $i = 1, 2, \ldots, n$, $y = y_1 \oplus y_2 \oplus \cdots \oplus y_{n+1}$ and $y_i^+ = \bigoplus_{j \leq i} y_j$.

The following problem, called the *Cauchy-Cartan problem*, was studied by Cartan. We consider the functions

$$y_{n+1}^0 \text{ of variables } x_1, x_2, \ldots, x_n \text{ with values in } Y_{n+1},$$
$$y_n^0 \text{ of variables } x_1, x_2, \ldots, x_{n-1} \text{ with values in } Y_n,$$
$$\vdots$$
$$y_2^0 \text{ of variable } x_1 \text{ with values in } Y_2,$$

and let y_1^0 be an arbitrary vector in Y_1.

The problem consists in finding a solution of the system (1.24) such that in the neighbourhood of the point $(x_1^0, x_2^0, \ldots, x_n^0)$ we have

$$y_{n+1}(x_1, x_2, \ldots, x_n) = y_{n+1}^0(x_1, x_2, \ldots, x_n),$$
$$y_n(x_1, x_2, \ldots, x_{n-1}, x_n^0) = y_n^0(x_1, x_2, \ldots, x_{n-1}),$$
$$\vdots$$
$$y_2(x_1, x_2^0, x_3^0, \ldots, x_n^0) = y_2^0(x_1),$$
$$y_1(x_1^0, x_2^0, \ldots, x_n^0) = y_1^0.$$

Theorem 1.5. *If $A : \mathfrak{A}(F_0) \to \mathfrak{A}(F_1)$ is a normalised differential operator and the Cauchy-Cartan data $(y_{n+1}^0, y_n^0, \ldots, y_1^0)$ in a regular system of coordinates in U are real-analytic functions of their variables, then for every f satisfying the condition $A_1 f = 0$ (A_1 is a compatibility operator for A) the Cauchy-Cartan problem with data $(y_{n+1}^0, y_n^0, \ldots, y_2^0, y_1^0)$ has a unique locally real-analytic solution.*

This theorem is a special case, corresponding to linear equations, of the *Cartan-Köhler theorem*, which also holds for quasi-linear equations (in the above formulation, with $f = 0$).

Theorem 1.5 is easily proved (in the same way as Theorem 1.4), by applying the commutation relations (1.17) inductively.

3.8. Additional Comments. If we do not consider the real-analytic case, as in Sect. 3.7, but the C^∞-case, then the solvability questions become very complicated. The local exactness of the compatibility complex holds for an operator A with constant coefficients in various spaces (a detailed account of this may be found in Palamodov (1967)). For operators vith variable coefficients, including those with "good" (in the sense of the above arguments) formal properties, this is not always so. The first example of this kind was given by Lewy (1957), and it features an operator A which is determined (that is, it has a zero compatibility operator). The equation $Ay = f$ turns out to be locally solvable only for a real-analytic right-hand side f. Using the notation $z = x_1 + ix_2$, $u = u_1 + iu_2$, $f = f_1 + if_2$, and $\partial/\partial\bar{z} = (\partial/\partial x_1 + i\partial/\partial x_2)/2$, we can write this operator as

$$\frac{\partial u}{\partial \bar{z}} + iz\frac{\partial u}{\partial x_2} = \frac{1}{2}f.$$

More details about this example can be found in Lewy (1957) and Spencer (1969); other similar examples are given in Hörmander (1963).

Systems of the form

$$\frac{\partial y}{\partial x_i} = F_i(x, y) \quad (i = 1, 2, \ldots, n) \tag{1.25}$$

have been studied in detail.

The Cauchy problem $y(x^0) = y^0$ has a unique solution not only in the case of real-analytic functions, but also in spaces of functions with limited smoothness. This result, called the *Frobenius theorem*, holds because if we proceed with the construction of the solutions of the system (1.25) as in the proof of the Cartan-Köhler theorem (that is, inductively over i), then the differential equations obtained at every step are ordinary ones, for which existence theorems are available.

If in (1.25) the vector-valued function y depends not only on the variables $x_1, \ldots, x_n$, but also on the additional oness $x_{n+1}, x_{n+2}, \ldots, x_m$, and the expressions on the righ-hand side contain $\partial/\partial x_{n+1}, \ldots, \partial/\partial x_m$, then the Frobenius theorem cannot be applied. In this case it is frequently convenient to regard y as a mapping from $\{x_1, x_2, \ldots, x_n\}$ to a Banach space B consisting of functions of $x_{n+1}, \ldots, x_m$, and $F_i(x, \cdot)$ as an unbounded mapping to B. The necessary existence theorems in this case are corollaries to the assumptions made about the operators $F_i(x, \ldots)$, usually formulated in terms of spectral properties. The case when the mappings $y \mapsto F_i(x, y)$ are linear can be found in Krein and Shikhvatov (1970) and Sysoev (1974); the case when they are non-linear is discussed in Samborski (1977). The Cartan-Köhler theorem (Samborski (1980)) has also been considered in Banach spaces. In this case, the data y_i^0 of the Cauchy-Cartan problem posed for the original involutive system are elements of functional Banach spaces B_j consisting of functions of the variables $x_1, x_2, \ldots, x_{j-1}$. In Samborski (1980) constructions are indicated of operators in the spaces $\bigoplus_{l \leq j} B_j$, whose infinitesimal character is equivalent to the solvability of the Cauchy-Cartan problem in the set of smooth functions with values in B_i. Sometimes perturbation theory can be used to obtain the "principal part" of these operators explicitly (Samborski (1980)).

§4. Differential Boundary Operators

4.1. Compatibility Operators. Let E_1 and E_2 be smooth vector bundles over a smooth manifold Ω with boundary Γ, and G_0 and G_1 smooth vector bundles over Γ. In Example 1.3 we introduced the category $\mathrm{DB}(\Omega)$ whose morphisms are differential boundary (DB-) operators. We recall that a *DB-operator* acts from $C^\infty(E_0) \times C^\infty(G_0)$ to $C^\infty(E_1) \times C^\infty(G_1)$ according to the formula

$$\Phi : (f, g) \mapsto (\Phi^{11} f, \Phi^{21} f + \Phi^{22} g),$$

where Φ^{11} and Φ^{22} are differential operators and Φ^{21} is the composition of a differential operator and the operator γ of restriction to Γ of a cross-section over Ω. If $G_0 = 0$, then a DB-operator, which in this case has the form (A, B), is called a *boundary value problem operator*.

Applying Definition 1.5 to the category $\mathrm{DB}(\Omega)$, we speak of equivalent DB-operators. As in the case of differential operators, the question arises of choosing a representative from every equivalence class, whose form is most convenient for constructing compatibility operators and studying the solvability of boundary value problems.

Definition 1.17. A DB-operator $\Phi : C^\infty(E_0) \times C^\infty(G_0) \to C^\infty(E_1) \times C^\infty(G_1)$ is called *normalised* if the differential operator $\Phi^{11} : C^\infty(E_0) \to C^\infty(E_1)$ is normalised (in the sense of Definition 1.17) and the boundary operator $\Phi^{21} : C^\infty(E_0) \to C^\infty(G_1)$ contains only differentiation in directions tangent to the boundary.

Proposition 1.10. *Every DB-operator $\Phi : C^\infty(E_0) \times C^\infty(G_0) \to C^\infty(E_1) \times C^\infty(G_1)$ whose component $\Phi^{11} : C^\infty(E_0) \to C^\infty(E_1)$ is sufficiently regular is equivalent in the category $\mathrm{DB}(\Omega)$ to a normalised DB-operator.*

We do not indicate the explicit form of the mappings involved in the equivalence mentioned in Proposition 1.10 (this is a simple exercise), but make only the following clarifying remarks. First, we may assume that the differential operator Φ^{11} is formally integrable and involutive, and that the order of Φ^{11} is strictly higher than that of Φ^{21}; otherwise, we replace Φ^{11} with the operator $J^l \Phi^1$ with a suitable l, which is equivalent to it in the category $\mathrm{D}(\Omega)$. Going over to an equivalent first-order operator by treating the derivatives of the original functions as new ones, we now find that, in general, Φ^{21} does not contain differentiations, which means that it is normalised.

To construct compatibility operators, we identify the "tangent part" of a differential operator on a manifold Ω with boundary.

Let A be a first-order differential operator, and let $(x_1, x_2, \ldots, x_n)$ be a coordinate system such that in a neighbourhood $U \subset \Omega$ of the point $x \in \Gamma = \partial\Omega$ the boundary Γ is described locally by the equation $x_n = 0$.

As was discussed in Sect. 3.3, the differential operator A can be written in the form

$$(Ay)(x) = \left(\frac{\partial y^+}{\partial x_n}(x) + L(x)y(x), M(x)y(x) \right),$$

where the differential operator $M(x) = M(x_1, \ldots, x_n)$ contains differentiation only with respect to $x_1, x_2, \ldots, x_{n-1}$ and the variable x_n occurs in it as a parameter.

Definition 1.19 (local). The differential operator $y \mapsto M(x_1, x_2, \ldots, x_{n-1}, 0)y$ (containing differentiation with respect to $x_1, x_2, \ldots, x_{n-1}$) is called the *tangent part* of A, and is denoted by A^τ.

Covering the boundary $\Gamma = \partial\Omega$ with neighbourhoods, we obtain the tangent part of A in each of them; "pasting" these parts together defines uniquely a global differential operator A^τ on Γ. We can convince ourselves of this immediately, but prefer to deduce it as a corollary to the definition of the global operator A^τ using the language of jets.

Let γ be the operator of restriction of a cross-section of a bundle over Ω to a cross-section over Γ. The embedding ξ of the boundary Γ in Ω induces in the cross-sections of the jet bundles a mapping $\zeta_0 : J^1(E_0)|_\Gamma \to J^1(E_0|_\Gamma)$ such that

$$\zeta_0((j^1 S)|_x) = j^1(\gamma S)(x) \quad (x \in \Gamma, \ S \in C^\infty(E_0)).$$

Let $A = p(A)j^1$ be a first-order differential operator from $C^\infty(E_0)$ to $C^\infty(E_1)$, let the bundle $\bar{E}_1$ over Γ be defined by

$$\bar{E}_1 = E_1|_\Gamma / (p(A)\operatorname{Ker}\zeta_0),$$

and let ζ' be the projection $E_1 \to \bar{E}_1$. We can define uniquely a mapping $\bar{p} : J^1(E_0|_\Gamma) \to \bar{E}_1$ such that $\bar{p}\zeta_0 = \zeta' p(A)$.

Definition 1.20 (global). The differential operator $A^\tau : C^\infty(E_0|_\Gamma) \to C^\infty(\bar{E}_1)$ defined by $A^\tau = \bar{p}\gamma^1$ is called the *tangent part* of A.

Example 1.26. Let d_Ω be the operator in the de Rham complex for a manifold Ω with a smooth boundary Γ. Then the tangential part of d_Ω is the operator d_Γ in the de Rham complex for the manifold Γ.

If the differential operator A is normalised, then so is A^τ.

Denoting by γ the operator of restriction of the cross-sections of bundles over Ω to cross-sections over Γ, we can rewrite every normalised (in the sense of Definition 1.18) DB-operator $\Phi : C^\infty(E_0) \times C^\infty(G_0) \to C^\infty(E_1) \times C^\infty(G_1)$ in the form

$$\Phi(f, g) = (Af, B\gamma f + Cg),$$

where B is a differential operator in the cross-sections of the bundles over Γ (that is, which contains differentiation only in directions tangent to Γ). In the cross-sections of the bundles over Γ we define a differential operator $\bar{\theta} : C^\infty(E_0|_\Gamma) \times C^\infty(G_0) \to C^\infty(\bar{E}) \times C^\infty(G_1)$ by

$$\bar{\theta}(f', g) = (A^\tau f', Bf' + Cg). \tag{1.26}$$

Definition 1.21. A normalised operator Φ is called *regular* if the operator $\bar{\theta}$ defined by (1.26) is sufficiently regular.

Thus, the regularity of a DB-operator Φ is defined by two conditions: 1) the sufficient regularity of the differential operator Φ^{11}, and 2) the sufficient regularity on Γ of the operator $\bar{\theta}$ determined uniquely in terms of Φ^{11} and Φ^{21}. We obtain this operator $\bar{\theta}$ by going over from Φ^{11} to an equivalent normalised operator. Of course, the property of regularity of the operator can also be defined in terms of the original operators Φ^{11} and Φ^{21} without going over to an equivalent normalised operator, but for the sake of brevity we give the following definition.

Definition 1.22. A DB-operator Φ is called *regular* if the operators Φ^{11} and $\bar{\theta}$ are regular, where $\bar{\theta}$ is defined by (1.26) in which A, B and C are the components of a normalised DB-operator equivalent to Φ.

We begin the construction of a compatibility operator. It suffices to consider regular normalised DB-operators. Then, as usual, we can use Propositions 1.1 and 1.10, which enable us to determine (and write out explicitly) a compatibility operator for an arbitrary regular DB-operator.

Let $\Phi : (y, w) \mapsto (Ay, B\gamma y + Cw)$ be a regular normalised DB-operator, A^{τ} the tangent part of Φ (Definition 1.20), and $\bar{\theta}$ the operator defined by (1.26).

Let $\bar{\theta}'$ be a compatibility operator for θ and ζ' the bundle mapping introduced above, that is, the projection $E_1 \to \bar{E}_1$.

Then the DB-operator Φ_1' given by

$$\Phi_1'(f, g) = (A_1 f, \bar{\theta}'(\zeta', f, g))$$

is a natural candidate for the role of compatibility operator for Φ. However, we do not stop at Φ_1', but construct a slightly "adjusted" version of it. We give an example to illustrate what we mean.

Example 1.27. (The Dirichlet problem for the operator grad). Let $\Omega = \{(x_1, x_2, x_3) \in \mathbb{R}^3 : x_3 \geq 0\}$; the boundary Γ is given by the equation $x_3 = 0$. The tangent part A^{τ} of the operator $A = $ grad coincides with the two-dimensional gradient, that is,

$$A^{\tau} y = \left(\frac{\partial y}{\partial x_1}, \frac{\partial y}{\partial x_2} \right).$$

The mapping ζ' associates a vector-valued function $f = (f_1, f_2, f_3)$ on Ω with the vector-valued function $\zeta' f = (F_1|_\Gamma, f_2|_\Gamma)$ on Γ. Hence, the operator $\bar{\theta}$ is defined by

$$\theta y = \left(\left(\frac{\partial y}{\partial x_1}, \frac{\partial y}{\partial x_2} \right), y \right),$$

where y is a scalar function on Γ.

Correspondingly, the operator $\bar{\theta}'$ has the form

$$\bar{\theta}'(f_1', f_2', g) = \left(\frac{\partial f_1'}{\partial x_2} - \frac{\partial f_2'}{\partial x_1}, \frac{\partial g}{\partial x_1} - f_1', \frac{\partial g}{\partial x_2} - f_2' \right).$$

Finally, we write out the operator Φ_1':

$$\Phi_1'(f_1, f_2, f_3, g) = (\operatorname{curl}(f_1, f_2, f_3), \partial(f_1|_\Gamma)/\partial x_2 - \partial(f_2|_\Gamma)/\partial x_1,$$
$$\partial g/\partial x_1 - f_1|_\Gamma, \partial g/\partial x_2 - f_2|_\Gamma).$$

From this it is obvious that in the compatibility conditions $\Phi_1'(f, g) = 0$, the equality $\partial(f_1|_\Gamma)/\partial x_2 - \partial(f_2|_\Gamma)/\partial x_1 = 0$ is superfluous, since it follows from the condition $\operatorname{curl} f = 0$. The "adjustment" mentioned above involves the elimination of such conditions. The adjusted operator in this example has the form

$$\Phi_1(f, g) = \left(\operatorname{curl} f, \frac{\partial g}{\partial x_1} - f_1|_\Gamma, \frac{\partial g}{\partial x_2} - f_2|_\Gamma \right).$$

In the general case, the compatibility operator $\bar{\theta}'$ also contains superfluous components, which need to be eliminated. To this end, we remark that $\bar{\theta}'$ can always be rewritten in the form

$$\bar{\theta}'(f', g) = ((A^\tau)'f', \theta'(f', g)),$$

where $(A^\tau)'$ is a compatibility operator for A^τ and θ' does not contain relations only between the components of f', that is, θ' cannot be rewritten in the form

$$\theta'(f', g) = (\theta_1'f', \theta_2'(f', g))$$

with a non-zero θ_1'.

We now define an operator

$$\Phi_1 : C^\infty(\Omega, Y_1) \times C^\infty(\Gamma, W_1) \to C^\infty(\Omega, Y_2) \times C^\infty(\Gamma, W_2)$$

by

$$\Phi_1(f, g) = (A_1 f, \theta'(\zeta' f, g)). \tag{1.27}$$

(We recall that locally, in a regular coordinate system $(x_1, x_2, \ldots, x_n)$ in which the boundary Γ is given by $x_n = 0$ and A has the form $y \mapsto (\partial y^+/\partial x_n + Ly, My) = (f_1, f_2)$, the mapping ζ' associates with the vector field $f = (f_1, f_2)$ the restriction $f_2|_\Gamma$ to Γ of its component f_2.)

Proposition 1.11. *If Φ is a regular normalised DB-operator and the normal to the boundary Γ is quasi-regular (Definition 1.15) for Φ^{11}, then the DB-operator Φ_1 defined by (1.25) is a compatibility operator for Φ (in the categories* $\mathrm{DB}(\Omega)$ *and* $\mathrm{DB}_a(\Omega)$*).*

The operator Φ_1 for a regular normalised operator is itself regular and normalised, and the conormal to Γ, which is quasi-regular for Φ^{11}, has the same property for $(\Phi_1)^{11}$. This enables us to extend the construction of compatibility operators. Thus, we arrive at the compatibility complex

$$C^\infty(E_0) \times C^\infty(G_0) \xrightarrow{\Phi} C^\infty(E_1) \times C^\infty(G_1) \xrightarrow{\Phi_1} C^\infty(E_2) \times C^\infty(G_2) \to \cdots$$

Example 1.28. The compatibility complex of the Dirichlet problem for the operator d of differentiation of 0-forms on a manifold Ω with boundary has the form

$$0 \to C^\infty(\Lambda^0(T^*\Omega)) \xrightarrow{\Phi_1} C^\infty(\Lambda^1(T^*\Omega)) \times C^\infty(\Lambda^0(T^*\Gamma))$$
$$\xrightarrow{\Phi_2} C^\infty(\Lambda^2(T^*\Omega)) \times C^\infty(\Lambda^1(T^*\Gamma)) \xrightarrow{\Phi_3} \cdots,$$

where $C^\infty(\Lambda^i(T^*\Omega))$ and $C^\infty(\Lambda^i(T^*\Gamma))$ are the spaces of smooth differential forms of degree i on the manifolds Ω and Γ, respectively, and the operator

$$\Phi_i : C^\infty(\Lambda^i(T^*\Omega)) \times C^\infty(\Lambda^{i-1}(T^*\Gamma)) \to C^\infty(\Lambda^{i+1}(T^*\Omega)) \times C^\infty(\Lambda^i(T^*\Gamma))$$

is defined by $\Phi_i(f,g) = (d_\Omega f, \gamma f - d_\Gamma g)$; here d_Ω and d_Γ are the operators of differentiation of forms on Ω and Γ, respectively, and γ is the operator of restriction of forms to the boundary.

Example 1.29. Let Ω be a domain in $\mathbb{R}^3$ with boundary Γ, $f \in C^\infty(\Omega, \mathbb{R}^3)$, and $h \in C^\infty(\Omega, \mathbb{R}^1)$; also, let N_Γ be the normal bundle over Γ and $g \in C^\infty(N_\Gamma)$ a cross-section of the normal bundle. We consider the boundary value problem

$$A : y \to (\operatorname{curl} y, \operatorname{div} y) = (f, h), \quad By \equiv y_n = g,$$

where y_n is the normal component of the vector y on Γ. The compatibility complex of this problem has the form

$$0 \to C^\infty(\Omega, \mathbb{R}^3) \xrightarrow{(A,B)} C^\infty(\Omega, \mathbb{R}^4) \times C^\infty(N_\Gamma) \xrightarrow{\Phi_1} C^\infty(\Omega, \mathbb{R}^1) \to 0,$$

where $\Phi_1(f, g, h) = \operatorname{div} f$.

Example 1.30. We consider another boundary value problem for the operator A in the preceding example, namely, $By \equiv y_\tau = g$, where y_τ is the tangent component of the vector y on Γ and $g \in C^\infty(T\Gamma)$ a smooth vector field tangent to Γ. We denote by D_Γ the differential operator that maps g into the cross-section $D_\Gamma g$ of the normal bundle (in this case, into a scalar function on Γ) according to the formula

$$D_\Gamma g = (\beta \circ d_\Gamma \circ \alpha)g,$$

where $\alpha : T\Gamma \to \Lambda^1(T^*\Gamma)$ and $\beta : \Lambda^{n-1}(T^*\Gamma) \to N_\Gamma = \Gamma \times \mathbb{R}^1$ are identifications described by the relations $\partial/\partial x_i \to dx_i$ and $dx_1 \wedge dx_2 \wedge \ldots \wedge dx_n \to 1$, respectively.

The compatibility complex of the operator (A, B) has the form

$$0 \to C^\infty(\Omega, \mathbb{R}^3) \xrightarrow{(A,B)} C^\infty(\Omega, \mathbb{R}^4) \times C^\infty(T\Gamma)$$
$$\xrightarrow{\Phi} C^\infty(\Omega, \mathbb{R}^1) \times C^\infty(N_\Gamma) \to 0,$$

where $\Phi(f, h, g) = (\operatorname{div} f, D_\Gamma g - f_n)$ and f_n is the normal component of the field f on Γ.

Thus, a compatibility operator for a normalised DB-operator Φ can be constructed by means of (1.27). If Φ is regular but not normalised, then it needs to be replaced by an equivalent normalised operator Φ' (Proposition 1.10), for which the compatibility complex is constructed. According to Proposition 1.1, from this complex we then construct the compatibility complex for Φ. This enables us to associate with every regular DB-operator Φ a complex of DB-operators, which is its compatibility complex.

4.2. The Real-Analytic Case. We continue to use the notation in Sect. 3.7. Let Ω be a domain in $\mathbb{R}^n$ with a real-analytic boundary Γ, and let U be an open set in $\mathbb{R}^n$. We denote by Ω' and Γ' the intersections $\Omega \cap U$ and $\Gamma \cap U$, respectively.

If a DB-operator (A, B) has real-analytic coefficients and is regular, then, according to the scheme set forth in Sect. 4.1, it generates a complex

$$\mathfrak{A}(\Omega', Y_0) \overset{(A,B)}{\to} \mathfrak{A}(\Omega', Y_1) \times \mathfrak{A}(\Gamma', W_0)$$
$$\overset{\Phi_1}{\to} \mathfrak{A}(\Omega', Y_2) \times \mathfrak{A}(\Gamma', W_1) \overset{\Phi_2}{\to} \cdots \tag{1.28}$$

for every open set U such that $U \cap \Gamma \neq \emptyset$.

Theorem 1.6. *If the conormal to the boundary Γ at a point $x \in \Gamma$ is quasi-regular for a normalised operator $\tilde{A}$ equivalent to A, then the complex (1.28) is exact for a sufficiently small neighbourhood U that includes x and is homeomorphic to a ball in $\mathbb{R}^n$.*

In other words, the condition $\Phi_1(f, g) = 0$ (which can be constructed in finitely many steps) is necessary and sufficient for the solvability of the boundary value problem $Ay = f$, $By = g$ in the real-analytic case with a quasi-regular normal.

The next example shows that, in general, the assumption that the conormal is quasi-regular for $\tilde{A}$ cannot be replaced by the same assumption for A. However, if the conormal is non-characteristic for A, then this can be done.

Example 1.31. Suppose that $A : C^\infty(\mathbb{R}^3, \mathbb{R}^3) \to C^\infty(\mathbb{R}^3, \mathbb{R}^4)$ acts according to the formula

$$y = (y_1, y_2, y_3) \mapsto Ay = (\partial y_1/\partial x_1 + \partial y_2/\partial x_2,$$
$$\partial y_1/\partial x_2 - \partial y_2/\partial x_1 + y_3, \partial y_1/\partial x_3, \partial y_2/\partial x_3).$$

For this operator every covector is quasi-regular. Going over to an equivalent operator $\tilde{A}$, we find that

$$\tilde{A}y = \left(\frac{\partial y_1}{\partial x_1} + \frac{\partial y_2}{\partial x_2}, \frac{\partial y_1}{\partial x_2} - \frac{\partial y_2}{\partial x_1} + y_3, \frac{\partial y_1}{\partial x_2}, \frac{\partial y_2}{\partial x_2}, \frac{\partial y_3}{\partial x_3} \right),$$

that is, the covector dx_3 is non-characteristic and dx_1 and dx_2 cease being quasi-regular.

Outline of the proof of Theorem 1.6. We may assume that (A, B) is a normalised DB-operator which locally has the form

$$A : y \mapsto (\partial y^+ / \partial x_n + L_0 y, M_0 y) = (f_1, f_2) = f,$$
$$B : y \mapsto \beta(y|_\Gamma) = g,$$

where dx_n is the conormal.

Then $\bar{\theta} = (M_0|_\Gamma, \beta)$, and, by Theorem 1.4, the compatibility complex

$$\mathfrak{A}(\Gamma', Y') \xrightarrow{\bar{\theta}} \mathfrak{A}(\Gamma', Y'') \xrightarrow{\bar{\theta}'} \mathfrak{A}(\Gamma', Y'''),$$

where Y', Y'' and Y''' are Euclidean spaces arising in the construction of the operators $\bar{\theta}$ and $\bar{\theta}'$, is exact for some neighbourhood U. Consequently, if $\Phi(f, g) = 0$, then there is a function z on Γ' such that $M_0|_\Gamma z = f_2|_\Gamma$ and $\beta z = g$. Taking z as the Cauchy data for the equation

$$\frac{\partial y^+}{\partial x_n} + L_0 y = f_1,$$

we obtain a function y in Ω', which is the desired local solution of the problem $Ay = f$, $By = g$ if we show that $M_0 y = f_2$ in Ω' and not just on Γ'. To this end, we consider the difference $w = M_0 y - f_2$, which is equal to zero on Γ' and, in view of the explicit form (1.27) of the compatibility operator, Proposition 1.5 and the commutation relations (1.27), satisfies the equation

$$\frac{\partial w}{\partial x_n} + \cdots = 0,$$

where the dots denote terms that do not contain $\partial / \partial x_n$. Hence, $w = 0$. Exactness in the remaining terms of the complex (1.25) is proved absolutely analogously.

A special form of local boundary value problems that do not contain overdetermination on the boundary (that is, with an operator $\Phi(f, g)$ independent of g) was studied in Palamodov (1968).

Chapter 2
Elliptic Systems

§1. Operators with a Constant Defect

We denote by $H^s(E)$ the Sobolev space, regarded as a Hilbert space, of the cross-sections of a bundle E (if Ω is a domain in $\mathbb{R}^n$ and $E = \Omega \times \mathbb{R}^m$, then $H^s(E)$ is the Hilbert space of vector-valued functions defined on Ω with values in $\mathbb{R}^m$, whose generalised derivatives up to order s are square-integrable). If E can be decomposed into a direct sum, that is, $E = \bigoplus_{j=1}^{m'} E_j$, where E_j are bundles over Ω, and $T = (t_1, t_2, \ldots, t_{m'})$ is a multi-index, then we write $H^T(E) = \bigoplus_{j=1}^{m'} H^{t_j}(E_j)$.

Definition 2.1. A differential operator $A_0 : C^\infty(E_0) \to C^\infty(E_1)$ is called an *operator with a constant defect* if for any $x \in \Omega$ and $\xi \in T_x^*\Omega$ ($\xi \neq 0$) the vector ξ is quasi-regular, that is, if $\dim \operatorname{Ker} \sigma A_0(x, \xi)$, $\xi \neq 0$, does not depend on x. A_0 is called an *elliptic operator* if for any $x \in \Omega$ and $\xi \in T_x^*\Omega$ ($\xi \neq 0$) the vector ξ is non-characteristic, that is, if $\dim \operatorname{Ker} \sigma A_0(x, \xi) = 0$.

Example 2.1. Let $\Omega \subset \mathbb{R}^3$ and $A_0 y = \operatorname{curl} y$. The operator A_0 has a constant defect.

Example 2.2. Let $\Omega \subset \mathbb{R}^3$, and let A_0 be the operator defined by

$$A_0(u, v) = (\operatorname{curl} u + \lambda_1 v, \operatorname{curl} v + \lambda_2 u),$$

where $\lambda_1, \lambda_2 \in \mathbb{R}^1$ are constants. The operator A_0 has a constant defect.

Example 2.3. Let $\Omega \subset \mathbb{R}^3$ and $A_0 y = (\operatorname{curl} y, \operatorname{div} y)$. Then A_0 is elliptic.

§2. The Case of Manifolds Without Boundary

First, let Ω be a compact manifold without boundary. The following assertion is well known in the theory of quadratic elliptic systems.

Theorem 2.1 (Taylor (1981)). *If the dimensions of the fibres of the bundles E_0 and E_1 are identical and $A_0 : C^\infty(E_0) \to C^\infty(E_1)$ is an elliptic operator, then the kernel $\operatorname{Ker} A_0$ and cokernel $\operatorname{Coker} A_0 = C^\infty(E_1)/\operatorname{Im} A_0$ of A_0 are finite-dimensional. If k is the order of A_0, then A_0 is a continuous operator from $H^s(E_0)$ to $H^{s-k}(E_1)$ for every s and its kernel and cokernel are finite-dimensional if and only if A_0 is elliptic. The dimensions of the kernel and cokernel do not depend on s.*

Now let $A_0 : C^\infty(E_0) \to C^\infty(E_1)$ be an overdetermined differential operator, and let

$$0 \to C^\infty(E_0) \overset{A_0}{\to} C^\infty(E_1) \overset{A_1}{\to} C^\infty(E_2) \to \cdots \tag{2.1}$$

be its compatibility complex completed on the left with the zero mapping.

Theorem 2.2. *If A_0 is a sufficiently regular elliptic operator, then the cohomologies of the complex (2.1) are finite-dimensional.*

Theorem 2.3. *If A_0 is a formally integrable elliptic operator of order k_0 and A_i in the compatibility complex is of order k_i, then the cohomologies of the complex of Hilbert spaces*

$$0 \to H^s(E_0) \overset{A_0}{\to} H^{s-k_0}(E_1) \overset{A_1}{\to} H^{s-k_0-k_1}(E_2) \overset{A_2}{\to} \cdots$$

are finite-dimensional for every $s \in \mathbb{R}^1$ and their dimensions do not depend on s.

The proof of these theorems can be reduced to Theorem 2.1 in a very simple way, which we now describe. First we consider the general case of a complex of Hilbert spaces.

Statement 2.1. *If*

$$H_1 \overset{\alpha}{\to} H_2 \overset{\beta}{\to} H_3 \tag{2.2}$$

is a complex of Hilbert spaces and linear mappings (that is, $\beta\alpha = 0$), then the cohomology space $\operatorname{Ker}\beta/\operatorname{Im}\alpha$ and the kernel of the operator $\Delta = \alpha\alpha^* + \beta^*\beta :$ $H_2 \to H_2$, where the asterisk denotes the adjoint operator, have the same dimension.

The operator Δ is called the *Laplacian of the complex* (2.2). In the case of manifolds without boundary, the adjoints of the differential operators in Statement 2.1 may be replaced by their formally adjoint operators; at the same time, the Laplacian is a quadratic differential operator. The next assertion indicates a condition for the ellipticity of the Laplacian. Let $H_i = H^{s_i}(E_i)$, and let α and β in (2.2) be differential operators.

Statement 2.2. *If for every $x \in \Omega$ and $\xi \in T_x^*\Omega$ ($\xi \neq 0$) the symbol complex*

$$E_1|_x \overset{\sigma\alpha(x,\xi)}{\to} E_2|_x \overset{\sigma\beta(x,\xi)}{\to} E_3|_x$$

is exact, then the Laplacian $\Delta = \alpha\alpha^* + \beta^*\beta$ is an elliptic operator (α^* and β^* are formally adjoint to α and β, respectively).

Thus, the proofs of Theorems 2.2 and 2.3 reduce to the verifying that the symbol complex generated by the compatibility complex (2.1) is exact. Consequently, the next assertion is fundamental.

Proposition 2.1. *If A_0 is a normalised differential operator (in the sense of Definition 1.17) and $\xi \in T_x^* \Omega$ ($\xi \neq 0$) is a quasi-regular covector, then the symbol complex*

$$E_0|_x \overset{\sigma A_0(x,\xi)}{\to} E_1|_x \overset{\sigma A_1(x,\xi)}{\to} E_2|_x \overset{\sigma A_2(x,\xi)}{\to} E_3|_x \cdots,$$

where A_i is the compatibility operator for A_{i-1} constructed in Chap. 1, Sect. 3, is exact.

The proof of this important assertion follows immediately from the explicit expressions of the compatibility operators (see Proposition 1.5) in a coordinate system $(x_1, x_2, \ldots, x_n)$ where $\xi = dx_n$.

In this way, if A_0 is normalised and elliptic, then from Proposition 2.1 and Statements 2.1 and 2.2 we immediately obtain Theorems 2.2 and 2.3.

Suppose that the operator A_0 is only sufficiently regular. Then, by Theorem 1.2, we can replace it with an equivalent normalised one. In this context, the cohomologies of the compatibility complexes are isomorphic in the corresponding terms. Under this equivalence, the spaces of smooth functions in Theorem 2.2 are mapped again into such spaces; only the bundles change. It is easy to show that in the steps followed when going over from a sufficiently regular operator to an equivalent normalised one (changing to a formally integrable operator, then extending for passing to an involutive one, going over to a first-order operator, and excluding the linear (non-differential) relations), a non-characteristic covector remains non-characteristic. At the same time, sufficiently regular elliptic operators are mapped into equivalent normalised elliptic ones. Hence, Theorem 2.2 is valid for an arbitrary sufficiently regular operator. As regards Theorem 2.3, when going over from a sufficiently regular operator to a formally integrable one (according to Proposition 1.4), the property of spaces being Sobolev spaces may not be preserved. In this case we need to find out what norms we should choose in the spaces of the compatibility complex for A_0 so that the compatibility complex of the equivalent normalised operator contains Sobolev norms. In the general case the following assertion holds, which we formulate only for A_0 and its compatibility operator.

Proposition 2.2. *Let $A_0 : C^\infty(E_0) \to C^\infty(E_1)$ be a sufficiently regular operator and $P : C^\infty(E_1) \to C^\infty(E_1')$ a differential operator such that $A' = (A_0, PA_0)$ is formally integrable. Also, suppose that A' is of order k and elliptic, and that the Banach space $\mathcal{H}^s$ is the completion of $C^\infty(E_1)$ with respect to the norm*

$$\|y\|_{\mathcal{H}^s} = \|(y, Py)\|^{H^{s-k}(E_1 \oplus E_1')}.$$

Then the cohomologies of the complex

$$0 \to H^s(E_0) \overset{A_0}{\to} \mathcal{H}^s \overset{A_1}{\to} H^{s-k'}(E_2)$$

are finite-dimensional and independent of s for some k'.

Of course, if A_0 is elliptic, then so is the operator A' in Proposition 2.2. However, as can be seen from the next example, more interesting and important in applications is the case when A_0 is not elliptic, but A' is.

Example 2.3. Let $\Omega \subset \mathbb{R}^4$ be a sphere in $\mathbb{R}^4$, and consider the operator defined by $A_0 y = \operatorname{curl} y + y$.

A_0 is neither elliptic, not formally integrable, but its equivalent normalised operator

$$A' y = (\operatorname{curl} y + y, \operatorname{div} y) = (f, g)$$

is elliptic. The compatibility operator for A' has the form $A'_1(f, g) = \operatorname{div} f - g$. From Theorem 2.3 it follows that the cohomologies of the complex

$$0 \to H^s(\Lambda^1(T^*\Omega)) \xrightarrow{A'} H^{s-1}(\Lambda^2(T^*\Omega)) \oplus H^{s-1}(\Lambda^3(T^*\Omega))$$

$$\xrightarrow{A'_1} H^{s-2}(\Lambda^3(T^*\Omega)) \to 0,$$

where $\Lambda^i(T^*\Omega)$ is the bundle of i-forms over Ω, are finite-dimensional for all $s \geq 2$. Since the cohomologies of the complex (2.2) are finite-dimensional, from Proposition 2.2 we deduce that so are (for all $s \geq 2$) the cohomologies of the complex

$$0 \to H^s(\Lambda^1(T^*\Omega)) \xrightarrow{A} \mathcal{H}^{s-1} \to 0,$$

where $\mathcal{H}^s$ is the completion of the space $H^2(\Lambda^2(T^*\Omega))$ with respect to the norm

$$\|y\|_{\mathcal{H}^s} = \|y\|_{H^s(\Lambda^2(T^*\Omega))} + \|\operatorname{div} y\|_{H^s(\Lambda^2(T^*\Omega))}$$

(in this case the operator P in Proposition 2.2 coincides with div).

§3. Boundary Value Problems for Operators with a Constant Defect

We go over to boundary value problems. We assume that Ω is a manifold with a smooth boundary $\Gamma = \partial\Omega$. Let $(A, B) : C^\infty(E_0) \to C^\infty(E_1) \times C^\infty(G_1)$ be a boundary value problem operator (here E_0 and E_1 are vector bundles over Ω, and G_1 a vector bundle over Γ). We consider these objects locally, in a sufficiently small neighbourhood U of a point $x \in \Gamma$.

Let $U \subset \mathbb{R}^n$, and suppose that with respect to a coordinate system $(x_1, x_2, \ldots, x_n)$ in $\mathbb{R}^n$ the neighbourhood U is defined locally by the inequality $x_n \geq 0$, so that $U \cap \Gamma$ is given by the equation $x_n = 0$. Suppose that in terms of these coordinates (and in the corresponding trivialisations of the bundles $E_i|_U$ and $G_1|_{U \cap \Gamma}$) the operator (A, B) has the form

$$(Ay)(x) = \sum_{|\alpha| \leq k} a^\alpha(x) \frac{\partial^{|\alpha|} y}{\partial x_1^{\alpha_1} \partial x_2^{\alpha_2} \ldots \partial x_n^{\alpha_n}},$$

$$(By)(x) = \sum_{|\alpha| \leq l} b^\alpha(x) \frac{\partial^{|\alpha|} y}{\partial x_1^{\alpha_1} \partial x_2^{\alpha_2} \ldots \partial x_n^{\alpha_n}} \bigg|_{x_n = 0} \qquad (x \in U \cap \Gamma).$$

We fix a point $x^0 \in U \cap \Gamma$ and a vector $(\eta_1, \eta_2, \ldots, \eta_{n-1}) \in \mathbb{R}^{n-1}$. We replace $\partial/\partial x_j$ by $i\eta_j$ $(j = 1, 2, \ldots, n-1)$ in the expression of $(By)(x)$, fixing the coefficients a^α and b^α at x^0 and discarding the terms of orders lower than k in A and than l in B. We obtain the operator

$$\hat{A}(x^0, \eta)y(x_n) = \sum_{|\alpha|=k} a^\alpha(x^0)(i\eta)^{\alpha'} \frac{d^{\alpha_n} y}{dx_n^{\alpha_n}},$$

$$\hat{B}(x^0, \eta)y(x_n) = \sum_{|\alpha|=l} b^\alpha(x^0)(i\eta)^{\alpha'} \frac{d^{\alpha_n} y}{dx_n^{\alpha_n}} \bigg|_{x_n=0}$$

of a boundary value problem on the semi-axis, in which $\hat{A}$ is an ordinary differential operator with constant coefficients and $\alpha' = (\alpha_1, \ldots, \alpha_{n-1})$.

It is not difficult to describe in invariant terms a construction that associates A and B with operators $\hat{A}$ and $\hat{B}$ with constant coefficients on the semi-axis for every Riemannian manifold Ω with a boundary Γ. This can be done by choosing an orthogonal expansion of the fibre $T_x^*\Omega = T_x^*\Gamma \oplus N$ of the cotangent bundle. We confine ourselves to the above local construction, remarking that the properties of the operators $\hat{A}$ and $\hat{B}$ used below do not depend on the choice of the local coordinate system.

Let $\Phi : C^\infty(E_0) \times C^\infty(G_0) \to C^\infty(E_1) \times C^\infty(G_1)$ be a differential boundary operator such that $\Phi(u, v) = (Au, Bu + Cv)$, and suppose that $G_0 = \bigoplus_{j=1}^{r} G^j$

and $G_1 = \bigoplus_{j=1}^{r'} G'^j$. Then the boundary operator B can be rewritten as the column $(B^j : j = 1, 2, \ldots, r)$ and the differential operator C as the matrix $(c^{ij} : i = 1, 2, \ldots, l; j = 1, 2, \ldots, r')$, so that $B^j : C^\infty(E_0) \to C^\infty(G'^j)$ and $c^{ij} : C^\infty(G^j) \to C^\infty(G'^j)$. Suppose that the integers k and k' and the multi-indices $\beta = (\beta^1, \beta^2, \ldots, \beta^r)$ and $\beta' = (\beta'^1, \beta'^2, \ldots, \beta'^{r'})$, in which the fractional parts of β^j and β'^j are equal to $1/2$, are such that

$$\operatorname{ord} B^j \le k - \beta'^j - \tfrac{1}{2}, \quad \operatorname{ord} c^{ij} = \beta^j - \beta'^i, \quad \operatorname{ord} A \le k - k'$$

for non-zero B^j and c^{ij}.

We fix $x \in \Gamma$ and $\eta \in T_x^*\Gamma$ $(\eta \neq 0)$. The application of the above procedure to each of the A, B^j and c^{ij} regarded as operators of orders $k - k'$, $k - \beta^{ij} - 1/2$ and $\beta^j - \beta'^i$, respectively, leads for fixed $x \in \Gamma$ and $\eta \in T_x^*\Gamma$ $(\eta \neq 0)$ to an ordinary differential boundary operator with constant coefficients in the cross-sections of the bundles over the semi-axis, namely

$$\hat{\Phi}(x, \eta)(u, v) = (\hat{A}(x, \eta)u, \hat{B}(x, \eta)u + \hat{C}(x, \eta)v),$$

where

$$\hat{B}(x, \eta)u = (\hat{B}^1(x, \eta)u, \hat{B}^2(x, \eta)u, \ldots, \hat{B}^{r'}(x, \eta)u),$$
$$\hat{C}(x, \eta)v = ((-i)^{\beta'_s - \beta^j} \sigma c^{sj}(x, \eta)v$$
$$(s = 1, 2, \ldots, r', \quad j = 1, 2, \ldots, r).$$

Let $(A, B) : C^\infty(E_0) \to C^\infty(E_1) \times C^\infty(G_1)$ be a boundary value problem operator, $\Phi_1 : C^\infty(E_1) \times C^\infty(G_1) \to C^\infty(E_2) \times C^\infty(G_2)$ a compatibility operator for it, and $G_1 = \bigoplus_{j=1}^{r_1} G_1^j$ and $G_2 = \bigoplus_{j=1}^{r_2} G_2^j$ the direct sum decompositions of the bundles G_1 and G_2.

The decompositions $G_1 = \oplus G_1^j$ and $G_2 = \oplus G_2^j$ induce a decomposition $(\varphi^{ij} : i = 1, 2, \ldots, r_2; j = 1, 2, \ldots, r_1)$ of Φ_1^{22} (the components of the compatibility operator Φ_1, acting from $C^\infty(G_1)$ to $C^\infty(G_2)$). Let $\operatorname{ord} B^j = \beta_1^j - 1/2$ and $\max_j(\beta_1^j + \operatorname{ord} \varphi^{ij}) = \beta_2^i$.

We denote by $\mathfrak{M}_+$ the set of smooth vector-valued functions on the semi-axis which tend to zero at infinity.

Theorem 2.4. *Let $(A, B) : C^\infty(E_0) \to C^\infty(E_1) \times C^\infty(G_1)$ be a normalised boundary value problem operator, and suppose that A and its tangent part A^τ (Definition 1.20) have a constant defect and that the following coerciveness condition holds: the complex*

$$\operatorname{Ker} \hat{A}(x, \eta) \cap \mathfrak{M}_+ \xrightarrow{\hat{B}(x,\eta)} G_1|_x \xrightarrow{\hat{\Phi}_1^{22}(x,\eta)} G_2|_x$$

is exact for $x \in \Gamma$ and $\eta \in T_x^* \Gamma$ ($\eta \neq 0$).

In this case, if $s - \beta_1 > 1/2$ and the operator Φ_1 is bounded from $H^{s-1}(E_1) \times H^{s-\beta_2}(G_1)$ to $H^{s-2}(E_2) \times H^{s-\beta_2}(G_2)$, then the space $\operatorname{Ker} \Phi_1 / \operatorname{Im}(A, B)$ is finite-dimensional and its dimension remains invariant if s is replaced by $s' > s$.

This theorem can be proved by the same method of going over to the Laplacian of the complex, as in Theorem 2.3. However, in contrast to the case of a manifold Ω without boundary, in the construction of the Laplacian here we leave the set of differential boundary operators. The Laplacians are Boutet de Monvel operators, which form an algebra of operators on a manifold with boundary. Basic details concerning this algebra are given in Sect. 6 of this chapter. We now proceed on the assumption that these details are known.

The ellipticity of the Laplacian follows from Proposition 2.1 and the next assertion.

Proposition 2.3. *If the boundary value problem $(A, B) : C^\infty(E_0) \to C^\infty(E_1) \times C^\infty(G_1)$ satisfies the assumptions in Theorem 2.4, then the complex of DB-operators with constant coefficients on the semi-axis*

$$H^s(\mathbb{R}_+^1, C^{m_0}) \xrightarrow{(\hat{A}, \hat{B})} H^{s-1}(\mathbb{R}_+^1, C^{m_1}) \times C^{r-1} \xrightarrow{\tilde{\Phi}_1} H^{s-2}(\mathbb{R}_+^1, C^{m_2}) \times C^{r_2}),$$

where m_i and r are the dimensions of the fibres in the bundles E_i and G, respectively, is exact for every $x \in \Gamma$ and $\eta \in T_x^* \Gamma$ ($\eta \neq 0$), and $s > l$.

To prove this, we consider the commutative diagram

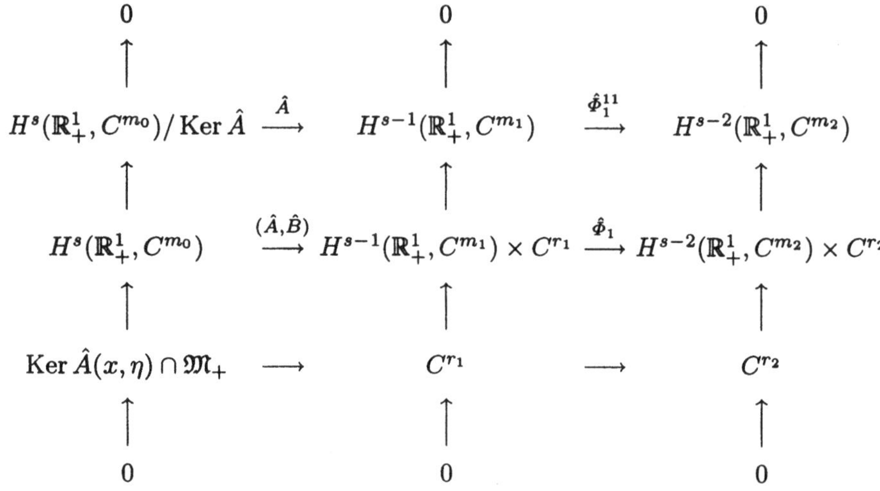

whose columns are exact. From the explicit form of the compatibility operators (Proposition 1.5) and the commutation relations (1.17) it follows that the top row is exact. Then the exactness of the bottom row (the coerciveness condition) yields the exactness of the middle row, which proves Proposition 2.3.

§4. Boundary Value Problems for Elliptic Operators

Let (A, B) be a boundary value problem operator and $L \geq 1$ a number such that in the compatibility complex of (A, B) the bundle G_L is non-zero, while G_{L+1} (consequently, all the G_{L+i}, $i \geq 1$) is zero.

We fix an integer $l \geq 1$ and assume that the multi-indices β^i and the decompositions $G_i = \bigoplus_j G_i^j$ $(i = 1, 2, \ldots, l+1)$ are such that all the mappings in the complex

$$0 \to H^s(E_0) \overset{(A,B)}{\to} H^{s-1}(E_1) \times H^{s-\beta_1}(G_1) \overset{\Phi_1}{\to} \cdots$$
$$\overset{\Phi_l}{\to} H^{s-l+1}(E_{l+1}) \times H^{s-\beta_l+1}(G_{l+1}) \tag{2.3}$$

are bounded (it suffices to assume this for $l \leq L$).

Theorem 2.5. *Let $(A, B) : C^\infty(E_0) \to C^\infty(E_1) \times C^\infty(G_1)$ be a normalised boundary value problem operator, and suppose that A is elliptic and that the following coerciveness condition holds: the complex*

$$0 \to \operatorname{Ker} \hat{A}(x, \eta) \cap \mathfrak{M}_+ \overset{\hat{B}}{\to} G_1|_x \overset{\hat{\Phi}_1^{22}}{\to} \cdots \overset{\hat{\Phi}_l^{22}}{\to} G^{l+1}|_x \tag{2.4}$$

is exact for $x \in \Gamma$ and $\eta \in T_x^* \Gamma$ $(\eta \neq 0)$, and $\dim \operatorname{Ker} \hat{A}(x, \eta)$ is independent of $x \in \Gamma$ and $\eta \in T_x^* \Gamma$ $(\eta \neq 0)$ for $l < L$. Then the cohomologies of the complex

(2.3) are finite-dimensional and their dimensions remain invariant when s is replaced by $s' > s$.

Conversely, suppose that the operator (A, B) is normalised, that the cohomologies of the complex (2.3) are finite-dimensional, and that either

(i) the dimension of the space $\operatorname{Ker} \hat{A}(x, \eta)$ is independent of $x \in \Gamma$ and $\eta \in T_x^* \Gamma$ ($\eta \neq 0$), or

(ii) the tangent part of A has a constant defect.

Then the operator (A, B) satisfies the coerciveness condition (2.4) and A is elliptic.

This theorem reduces to Theorem 2.4 if we remark that in the case when the the boundary value problem operator (A, B) satisfies its conditions, the tangent parts of the A_i (equal to Φ_i^{11} for $i > 0$) have a constant defect.

Corollary 2.1. *If (A, B) is a boundary value problem operator such that its equivalent normalised operator (constructed in finitely many steps by means of the method in Chap. 1, Sect. 4) satisfies the conditions in Theorem 2.4, then the cohomologies of the compatibility complex*

$$0 \to C^\infty(E_0) \overset{(A,B)}{\to} C^\infty(E_1) \times C^\infty(G_1) \overset{\Phi_1}{\to} C^\infty(E_2) \times C^\infty(G_2) \to \cdots$$

are finite-dimensional.

In the above formulation of Theorem 2.5, to verify the coerciveness condition for the normalised operator (A, B) we need to know all the operators Φ_i in its compatibility complex. This is unnecessary. The verification of the exactness of the complex (2.4) can be done in terms of (A, B) alone. Thus, the next assertion holds (Samborski and Fel'dman (1985)).

If (A, B) is a normalised boundary value problem operator, A is elliptic and A^τ has a constant defect, then the cohomologies of the complex (2.4) are finite-dimensional if and only if the following coerciveness condition is satisfied: the operator (A^τ, B) has a constant defect, the mapping $(\hat{A}(x, \eta), \hat{B}(x, \eta))$ is monomorphic on $\mathfrak{M}_+$ for all $x \in \Gamma$ and $\eta \in T_x^* \Gamma$ ($\eta \neq 0$), and

$$\hat{B}(x, \eta)(\operatorname{Ker} \hat{A}^\tau(x, \eta)) = \hat{B}(x, \eta)[\operatorname{Ker} \hat{A}(x, \eta) \cap \mathfrak{M}_+|_{x_n = 0}]$$

for all $x \in \Gamma$ and $\eta \in T_x^* \Gamma$ ($\eta \neq 0$).

§5. Regular DB-Operators

The starting point in the preceding sections was a normalised boundary value problem operator. In applications this is not usually the case, since the order of the differential operator is greater than one, the requirements regarding its formal properties are not satisfied, and the boundary operators are of a more general form. However, as follows from the results in Chap. 1, Sect. 3, every sufficiently regular operator is equivalent to a normalised one, and we can use the results in the preceding section to study the solvability of the given boundary value problem. Furthermore, the passage to an equivalent normalised operator is in many cases an important moment in the construction of compatibility operators. But even for determined operators (that is, whose compatibility operator is zero), the going over to an equivalent normalised one is important for the correct understanding of the type of problems with which we are dealing, as illustrated by the example at the end of the introduction. The distinctive features of problems become clear only when passing to an equivalent normalised operator. However, in this process we may "lose" the coerciveness condition. Here is such an example.

Example 2.4 (the *Dirichlet problem for the Laplace equation*). Let $\Omega = \{(x_1, x_2) \in \mathbb{R}^2 : x_2 \geq 0\}$, and suppose that the operator of the boundary value problem has the form $(A, B)y = (\partial^2 y/\partial x_1^2 + \partial^2 y/\partial x_2^2, y|_{x_2=0})$. The operator (A, B) satisfies the coerciveness condition, but is not normalised. Introducing the new unknown functions $y_1 = \dfrac{\partial y}{\partial x_1}$ and $y_2 = \dfrac{\partial y}{\partial x_2}$, we obtain the normalised operator

$$\tilde{A} = \begin{cases} \partial y/\partial x_1 - y_1, \\ \partial y/\partial x_2 - y_2, \\ \partial y_1/\partial x_1 + \partial y_2/\partial x_2, \\ \partial y_2/\partial x_1 - \partial y_1/\partial x_2, \end{cases} \qquad \tilde{B}(y, y_1, y_2) = y|_{x_2=0}.$$

It is easy to verify that this operator does not satisfy the coerciveness condition (2.4).

We now indicate the construction of a "correct" passage to a normalised operator, involving the adjustment of the boundary conditions, which ensures that the coerciveness condition is satisfied. In the case of Example 2.4, this adjustment consists in adjoining another boundary condition, namely $y_1|_{x_2=0}$, which is a differential consequence of the relations $y|_{x_2=0}$ (the boundary condition) and $(\partial y/\partial x_1 - y_1)|_{x_2=0} = 0$ (the tangent part of $\tilde{A}$).

We describe this "adjustment" procedure locally. Let $U \subset X$, $E_0 = U \times X$ and $E_1 = U \times Y$, where X and Y are Euclidean spaces. Suppose that U is equipped with coordinates $(x_1, x_2, \ldots, x_n)$, and that $(dx_1, dx_2, \ldots, dx_n)$ is the corresponding basis in $T_x^* U = X^*$. Then the space $J^{k-1}(E_0)|_x$ can be represented as a direct product, namely

$$J^{k-1}(E_0)|_x = X \times L(X, Y) \times L_{\mathrm{sym}}^2(X, Y) \times \cdots \times L_{\mathrm{sym}}^{k-1}(X, Y) \qquad (2.5)$$

(see Chap. 1, Sect. 1).

According to what was said in Chap. 1, Sect. 4, starting with an arbitrary sufficiently regular boundary value problem operator (A, B), we can go over to an equivalent operator $(\tilde{A}, \tilde{B})$, where $\tilde{A}$ is a differential, formally integrable, involutive first-order operator and $\tilde{B}$ does not contain differentiation. $(\tilde{A}, \tilde{B})$ is defined on the space $C^\infty(J^{k-1}(E_0))$ of the bundle E_0 for some k. If we rewrite (using the structure of the direct product (2.5)) the elements of $J^{k-1}(E_0)$ in the form $(y, u^1, u^2, \ldots, u^{k-1})$, where $y \in Y$ and $u^i \in L^i_{\mathrm{sym}}(X, Y)$, then the mapping $\tilde{B}$ of order zero can always be rewritten in the form

$$(y, u^1, u^2, \ldots, u^{k-1}) \mapsto (b_0(y), b_1(y, u^1), \ldots, b_{k-1}(y, u^1, \ldots, u^{k-1})),$$

where $b_i : \Gamma \times Y \times L(X, Y) \times \cdots \times L^i_{\mathrm{sym}}(X, Y) \to G^i$ (G^i are some Euclidean spaces), and the mappings $(x, u^i) \mapsto b_i(x, 0, \ldots, 0, u^i)$ from $L^i_{\mathrm{sym}}(X, Y)$ to G^i are surjective. To construct the mapping $\tilde{B}$, we first write the differential operator

$$(x, y, u^1, \ldots, u^{k-1}) \mapsto (b_0(x, y), Db_0(x, y), \ldots, D^{k-1}b_0(x, y),$$
$$b_1(x, y, u^1), Db_1(x, y, u^1), \ldots, D^{k-2}b_1(x, y, u^1),$$

$$\vdots$$

$$b_{k-2}(x, y, \ldots, u^{k-2}), Db_{k-2}(x, y, \ldots, u^{k-2}), b_{k-1}(x, y, \ldots, u^{k-1})$$

(here, as usual in our local considerations, $x \in \Gamma = \partial\Omega = \{x : x_n = 0\}$ and $D = (\partial/\partial x_1, \partial/\partial x_2, \ldots, \partial/\partial x_n)$). Then we transform this differential operator into an operator of order zero by substituting $Dy = u^1$, $Du^1 = u^2, \ldots, Du^{k-1} = u^k$ everywhere in it. The boundary operator of order zero obtained in this way is the desired operator $\tilde{B}$. To go over to a normalised boundary value problem operator $(\tilde{A}, \tilde{B})$ it now remains to exclude the rows in $(A', \tilde{B})$ that "do not contain differentiation".

The above construction (its invariant description can be found in Samborski (1984a)) and the accurate connection with the spaces in which the equivalent mappings in Chap. 1, Sect. 3 act, enable us to use Theorem 2.5 to derive the next assertion, in which, for brevity, we restrict our attention to the study of the kernel and cokernel of a boundary value problem operator.

Theorem 2.6. *Let* $(A, B) : C^\infty(E_0) \to C^\infty(E_1) \times C^\infty(G_1)$ *be a regular boundary value problem operator, and suppose that A is formally integrable and elliptic, and that the following coerciveness condition holds: for any $x \in \Gamma$ and $\eta \in T^*_x \Gamma$ ($\eta \neq 0$) the number* $\dim \mathrm{Ker}\, \hat{A}(x, \eta)$ *does not depend on (x, η) and the complex*

$$0 \to \mathrm{Ker}\, \hat{A}(x, \eta) \cap \mathfrak{M}_+ \overset{\hat{B}(x,\eta)}{\to} G_1|_x \overset{\hat{\Phi}_1^{22}(x,\eta)}{\to} G_2|_x,$$

where Φ_1 is a compatibility operator for (A, B), is exact. In this case, if all the mappings in the complex

$$0 \to H^s(E_0) \overset{(A,B)}{\to} H^{s-k}(E_1) \times H^{s-\beta_1}(G_1) \overset{\Phi_1}{\to} H^{s-k'}(E_2) \times H^{s-\beta_2}(G_2)$$

are bounded, then its cohomologies are finite-dimensional and their dimensions remain invariant when s is replaced by $s' > s$.

If in the compatibility complex of (A, B) the space G_2 is zero, that is, there is no overdetermination on the boundary, then the assumption in Theorem

2.5 on the independence of $\dim \operatorname{Ker} \hat{A}(x, \eta)$ of x and $\eta \neq 0$ is superfluous. This independence follows from the coerciveness condition, which in this case has the same form as for quadratic elliptic systems:

The boundary value problem $\hat{A}(x, \eta)y = 0$, $\hat{B}(x, \eta)y = g$ for a system of ordinary differential equations with constant coefficients on the semi-axis has a unique solution for every g, which tends to zero at infinity.

If A in Theorem 2.5 is not formally integrable, then, in general, we cannot claim that the cohomologies of the complex are finite-dimensional in Sobolev spaces. Instead, the following assertion holds.

Proposition 2.4. *Let $(A, B) : C^\infty(E_0) \to C^\infty(E_1) \times C^\infty(G_1)$ be a regular boundary value problem operator (see Definition 1.22), and let $P : C^\infty(E_1) \to C^\infty(\bar{E}_1)$ be a differential operator such that $A' = (A, PA)$ is formally integrable. If (A', B) satisfies the conditions in Theorem 2.5 and the Banach space $\mathcal{H}^s$ is the completion of $C^\infty(E_1)$ with respect to the norm*

$$\|y\|_{\mathcal{H}^s} = \|(y, Py)\|_{H^{s-k}(E_1 \oplus \bar{E}_1)},$$

then the cohomologies of the compatibility complex

$$0 \to H^s(E_0) \overset{(A,B)}{\to} \mathcal{H}^s \times H^{s-B_1}(G_1) \overset{\Phi_1}{\to} H^{s-k'}(E_2) \times H^{s-\beta_2}(G_2)$$

are finite-dimensional.

Various formulations of the coerciveness condition have been considered in Solonnikov (1969).

Elliptic overdetermined problems have also been investigated in Golovkin spaces in Solonnikov (1975). Dudnikov has studied the smoothness properties of the solutions of elliptic boundary value problems in Gevrey classes, in relation to the smoothness of the right-hand sides and of the boundary conditions (regularity theorems) (see Dudnikov (1985b)).

Krein and L'vin propose an approximate formulation of linear overdetermined boundary value problems (the overdetermination may occur either in the system of equations, or in the boundary conditions). By an approximate solution we understand a vector-valued function which satisfies the boundary conditions exactly and which, when replaced in the system of equations, yields a residual of the right-hand sides that is minimal in some norm. It turns out that finding the approximate solution of an overdetermined boundary value problem reduces to the investigation of a boundary value problem of a higher order, which under definite conditions is already "well posed". (*Author: unfinished sentence?*) The problem of finding compatibility conditions for the right-hand sides is eliminated in this case. Overdetermined elliptic problems in an approximate formulation are studied in Krein and L'vin (1985). In L'vin (1987) this approach is applied to other overdetermined problems, in particular, to parabolic and degenerate elliptic ones.

§6. Additional Comments. Boutet de Monvel Operators

6.1. We denote by H^+ (H^-) the space of complex-valued functions $f(\cdot) \in C^\infty(\mathbb{R}^1)$ that can be extended analytically into the lower (upper) complex half-plane so that the extension is continuous up to the boundary and admits the asymptotic expansion

$$f(\zeta) = \sum_{k \leq -1} a_k \zeta^k \quad \left(f(\zeta) = \sum_{k \leq 0} a_k \zeta^k \right), \quad |\zeta| \to \infty.$$

We denote by P the space of polynomials in the variable t, and by P_d the space of polynomials of degree not exceeding d. Let $H_d^- = H_0^- \oplus P_d$, $H = H^+ \oplus H_0^- \oplus P$ and $H^- = H_0^- \oplus P$.

We denote by $\mathcal{S}_+(\mathbb{R}^1)$ ($\mathcal{S}_-(\mathbb{R}^1)$) the set of functions on $\mathbb{R}^1$ that are equal to zero for $t < 0$ ($t > 0$), are smooth for $t > 0$ ($t < 0$) up to the point $t = 0$, and decrease rapidly as $t \to \infty$ ($t \to -\infty$). Then the functions in H^+ are the Fourier transforms of those in $\mathcal{S}_+(\mathbb{R}^1)$. The functions in H_d^- are the Fourier transforms of those in the space

$$\mathcal{S}_-(\mathbb{R}^1) \oplus \left\{ \sum_{k=0}^{d-1} a_k \delta^{(k)}(t) \right\}.$$

6.2. The Symbols of Boutet de Monvel Operators in a Domain $\Omega' \times \mathbb{R}_+^1$, $\Omega' \subset \mathbb{R}^{n-1}$, with Coordinates (x', x_n), $x' = (x_1, \ldots, x_{n-1})$. A function $a(x, \xi) \in S^m(\Omega' \times \mathbb{R}^1 \times \mathbb{R}^n)$ is called a *symbol* that satisfies the *transmission condition* with respect to the hyperplane $\{x_n = 0\}$ if

$$a(x', 0+, \xi', \langle\xi'\rangle\nu) \in H_\nu^+ \otimes S^m(\Omega' \times \mathbb{R}^{n-1}),$$
$$\partial_\nu^k u(x', 0+, \xi', \langle\xi'\rangle\nu) \in H_\nu^+ \otimes S^m(\Omega' \times \mathbb{R}^{n-1}), \quad k - 1, \ldots,$$

where $\langle\xi'\rangle = (1 + |\xi'|^2)^{1/2}$.

A function $k(x', \xi', \nu) \in C^\infty(\Omega' \times \mathbb{R}^{n-1} \times \mathbb{R}^1)$ is called a *potential symbol of order m* if

$$k(x', \zeta', \langle\xi'\rangle\nu) \in H_\nu^+ \otimes S^m(\Omega' \times \mathbb{R}^{n-1}).$$

A function $t(x', \xi', \nu) \in C^\infty(\Omega' \times \mathbb{R}^{n-1} \times \mathbb{R}^1)$ is called a *trace symbol of order m and type d* if

$$t(x', \xi', \langle\xi'\rangle\nu) \in H_d^- \otimes S^m(\Omega' \times \mathbb{R}^{n-1}).$$

A function $b(x', \xi', \nu, \tau) \in C^\infty(\Omega' \times \mathbb{R}^{n-1} \times \mathbb{R}^1 \times \mathbb{R}^1)$ is called a *Green's symbol of order m and type d* if

$$b(x', \xi', \langle\xi'\rangle\nu, \langle\xi'\rangle\tau) \in H_n^+ \otimes H_{d,\tau}^- \otimes S^m(\Omega' \times \mathbb{R}^{n-1}).$$

6.3. Pseudodifferential Operators on $\Omega' \times \bar{\mathbb{R}}_+^1$. We denote by $j^+ :$ $C^\infty(\Omega' \times \bar{\mathbb{R}}_+^1) \to \mathcal{D}'(\Omega' \times \mathbb{R}^1)$ the operator of extension of functions by zero for $x_n < 0$, and by $r^+ : \mathcal{D}'(\Omega' \times \mathbb{R}^1) \to \mathcal{D}'(\Omega' \times \mathbb{R}^1_+)$ the operator of restriction of a distribution on $\Omega' \times \mathbb{R}^1$ to the open subset $\Omega' \times \mathbb{R}^1_+$.

Let A be a pseudodifferential operator (PDO) in $\Omega' \times \mathbb{R}^1$. We define a pseudodifferential operator $A' : C_0^\infty(\Omega' \times \mathbb{R}^1_+) \to \mathcal{D}'(\Omega' \times \bar{\mathbb{R}}^1_+)$ by $A'u = r^+ A j^+ u$. We say that A satisfies the *transmission condition* if the symbol of A does.

Proposition 2.5. *A PDO with the transmission property in the domain $\Omega' \times \bar{\mathbb{R}}^1_+$ defines a mapping from $C^\infty(\Omega' \times \bar{\mathbb{R}}^1_+)$ to $C^\infty(\Omega' \times \bar{\mathbb{R}}^1_+)$.*

6.4. Potential Operators. An operator $K_0 : C_0^\infty(\Omega') \to C^\infty(\Omega' \times \bar{\mathbb{R}}^1_+)$ acting according to the formula

$$v \to (K_0 v)(x', x_n) = \int_{\Omega'} k_0(x', x_n, y') v(y') dy',$$

where $k_0 \in C^\infty(\Omega' \times \bar{\mathbb{R}}^1_+ \times \Omega')$ and $v \in C_0^\infty(\Omega')$, is called a *smoothing potential operator.*

Let $k(x', \xi', \nu)$ be a potential symbol. An operator $K : C_0^\infty(\Omega') \to C^\infty(\Omega' \times \bar{\mathbb{R}}^1_+)$ of the form

$$(Kv)(x', x_n) = (2\pi)^{-n+1} \mathcal{F}^{-1}_{\nu \to x_n} \int e^{ix'\xi'} k(x', \xi', \nu) \hat{v}(\xi') d\xi' + K_0 v$$

is called a *potential operator.* (Here K_0 is a smoothing potential operator and $\mathcal{F}^{-1} : H^+ \to \mathcal{S}_+(\mathbb{R}^1)$ is the inverse Fourier transformation on the semi-axis.)

We denote by Π^+ and Π^- the projections $H \to H^+$ and $H \to H^-$ that are the Fourier transforms of the operators r^+ and r^- of restriction of functions to the positive and negative semi-axes, respectively; that is, $\Pi^\pm = \mathcal{F} r^\pm \mathcal{F}^{-1}$. We define a linear operator

$$r' : \mathcal{S}_+(\mathbb{R}^1) \oplus \mathcal{S}_-(\mathbb{R}^1) \oplus \left\{ \sum_{k=0}^m c_k \delta^{(k)}(t) \right\} \to \mathbb{C}^1$$

by setting $r'u = \lim_{t \to 0+} u(t)$ if $u \in \mathcal{S}_+(\mathbb{R}^1) \oplus \mathcal{S}_-(\mathbb{R}^1)$, and $r'u = 0$ if $u \in \left\{ \sum_{k=0}^m c_k \delta^{(k)}(t) \right\}$. We define the operator $\Pi' : H \to C$ as the composition $\Pi' = r' \mathcal{F}^{-1}$. Integral formulae for $\Pi^\pm$ and Π' can be found in Boutet de Monvel (1971) and Rempel and Schulze (1982).

6.5. Trace Operators and Green's Operators. By a *smoothing trace operator* we understand an operator $r'T_0 : C_0^\infty(\Omega' \times \mathbb{R}^1_+) \to C^\infty(\Omega')$ defined by

$$u \mapsto r'T_0 u(x') = \sum_{k=0}^{d-1} B_k(r' D_{x_n}^k u)(x') + \int_{\Omega' \times \mathbb{R}^1_+} t_0(x', y', y_n) u(y', y_n) dy' \, dy_n,$$

where $t \in C^\infty(\Omega' \times \Omega' \times \mathbb{R}^1)$, B_k are smoothing operators in Ω' and $u \in C_0^\infty(\Omega' \times \bar{\mathbb{R}}_+^1)$.

Let $t(x', \xi', \xi_n)$ be a trace symbol. By a *trace operator* we understand an operator $T : C_0^\infty(\Omega' \times \bar{\mathbb{R}}_+^1) \to C^\infty(\Omega')$ of the form

$$(Tu)(x') = \int e^{ix'\xi'} \, \Pi_\nu'[t(x', \xi', \nu)(j^{\hat{+}}u)(\xi', v)]d\xi' + T_0 u,$$

where T_0 is a smoothing trace operator.

By a *smoothing Green's operator* we understand an operator $r'B_0 : C_0^\infty(\Omega' \times \bar{\mathbb{R}}_+^1) \to C^\infty(\Omega' \times \bar{\mathbb{R}}_+^1)$ of the form

$$(r'B_0 u)(x', x_n) = \sum_{k=0}^{d-1} K_k(r' D_{x_n}^k u)(x', x_n) + \int_{\Omega' \times \mathbb{R}_+^1} b(x, y)u(y)dy,$$

where K_k are smoothing potential operators, $b(x, y) \in C^\infty(\Omega' \times \bar{\mathbb{R}}_+^1 \times \Omega' \times \bar{\mathbb{R}}_+^1)$ and $u \in C_0^\infty(\Omega' \times \mathbb{R}_+^1)$.

Let $g(x', y', \xi', \nu, \tau)$ be a Green's symbol. By a *Green's operator* we understand an operator $G : C_0^\infty(\Omega' \times \bar{\mathbb{R}}_+^1) \to C^\infty(\Omega' \times \bar{\mathbb{R}}_+^1)$ of the form

$$(Gu)(x', x_n) = \mathcal{F}_{\nu \to \lambda_n}^{-1} \int e^{ix'\xi'} \Pi_\tau[g(x', \xi', \nu, \tau)(j^{\hat{+}}u)(\xi', \tau)]d\xi'$$
$$+ (G_0 u)(x', x_n),$$

where G_0 is a smoothing Green's operator.

6.6. The Algebra of Boutet de Monvel Operators. An operator $\mathcal{A} : C_0^\infty(\Omega' \times \mathbb{R}_+^1) \times C_0^\infty(\Omega') \to C_0^\infty(\Omega' \times \mathbb{R}_+^1) \times C_0^\infty(\Omega')$ of the form

$$(u, v) \mapsto ((A + G)u + Kv, Tu + Qv), \tag{2.6}$$

where A is a pseudodifferential operator on $\Omega' \times \bar{\mathbb{R}}_+^1$ satisfying the transmission condition, K, T and G are a potential, trace and Green's operators, respectively, and Q is a PDO on Ω', is called a *Boutet de Monvel operator*.

These operators form an *algebra*, that is, the composition of such operators, the adjoint and the inverse (if it exists) of an operator of this type are also of the form (2.6).

The symbol of A is called the *interior symbol* of the operator $\mathcal{A}$.

By a *boundary symbol* $\alpha(x', \xi')$ we understand a mapping of the space $H \oplus C^1$ into itself, defined by

$$\alpha(x', \xi')(u(\nu), v) = (\Pi_\nu^+[a(x', \xi', \nu)u(\nu)] + \Pi_\tau'[g(x', \xi', \nu, \tau)u(\tau)]$$
$$+ k(x', \xi', \nu)v\Pi_\nu'[t(x', \xi', \nu)u(\nu)] + q(x', \xi')v),$$

where $a(x', \xi', \nu)$, $k(x', \xi', \nu)$, $t(x', \xi', \nu)$, $g(x', \xi', \nu, \tau)$, and $q(x', \xi')$ are the symbols of the operators A, K, T, G, and Q in (2.6).

Boutet de Monvel operators and their symbols, which act in the cross-sections of vector bundles, are defined by means of trivialisations in the standard way.

If the symbols of A, K, T, G, and Q can be represented as asymptotic sums of homogeneous symbols, then we can define the principal interior and boundary symbols of the operator $\mathcal{A}$.

Operators of the form (2.6) are continuous on Sobolev spaces with indices determined by the their orders.

An operator $\mathcal{A}$ is called *elliptic* if its principal interior and boundary symbols are invertible for all x and $\xi \neq 0$ (x' and $\xi' \neq 0$). Elliptic operators are Fredholm operators on Sobolev spaces (that is, their kernel and cokernel are finite-dimensional).

Differential boundary operators of the form $\Phi(u, v) = (\Phi^{11}u, \Phi^{22}u + \Phi^{22}v)$ belong to the algebra of Boutet de Monvel operators; here Φ^{11}, Φ^{21} and Φ^{22} correspond to A, T and Q in (2.6). The interior symbol of Φ—the symbol of Φ^{11}, the principal boundary symbol—is the Fourier transform on the semi-axis of the ordinary DB-operator $\Phi(x', \eta)$ with constant coefficients defined in the formulation of the coerciveness condition, that is, $\alpha(x', \eta) = \mathcal{F}\Phi(x', \eta)\mathcal{F}^{-1}$.

Chapter 3
Initial Boundary Value Problems
for Parabolic Systems

§1. Parabolic Operators

Let Ω be a domain in $\mathbb{R}^n$ with coordinates $x = (x_1, \ldots, x_n)$, and let Y_1 and Y_2 be Euclidean spaces of dimensions m_1 and m_2, respectively. Also, let $y = (y_1, \ldots, y_{m_1})$ be coordinates in Y_1. We denote by Ω' the set $\Omega \times \mathbb{R}^1$. The coordinates in Ω' are pairs (x, t), where $x \in \Omega$ and $t \in \mathbb{R}^1$. A differential operator $A : C^\infty(\Omega', Y_2) \to C^\infty(\Omega', Y_2)$ is written in the form

$$(Ay)_j(x, t) = \sum_{r, |\alpha| \leq k} \left| \sum_{s=1}^{m_1} a_{r\alpha js}(x, t) \right| \frac{\partial^{|\alpha|+r} y_s}{\partial x^\alpha \partial t^r}(x, t), \qquad (3.1)$$

where $\alpha = (\alpha_1, \ldots, \alpha_n)$ is a multi-index and $a_{r\alpha js}$ are smooth functions. Let b be a natural number. We say that A is an *operator of order (l, b)* (*of order l with weight b with respect to t*) if the functions $a_{r\alpha js}$ in (3.1) are not identically zero only for $|\alpha| + br \leq l$, and there are r, α, j and s such that $|\alpha| + br = l$ and $a_{r\alpha js}(x, t) \not\equiv 0$.

Definition 3.1. Let A be a differential operator of order (l, b) of the form (3.1). By the *principal b-homogeneous symbol of A* at the point $(x, t) \in \mathbb{R}^n \times \mathbb{R}^1$ we understand the mapping $\sigma_A^b(x, t, \xi, \tau) : Y_1 \otimes \mathbb{C}^1 \to Y_2 \otimes \mathbb{C}^1$ defined by

$$(\sigma_A^b(x, t, \xi, \tau)y)_j = \sum_{br+|\alpha|=l} i^{r+|\alpha|} \sum_{s=1}^{m_1} a_{r\alpha js}(x, t)\tau^r \xi^\alpha y_s.$$

Let $\mathbb{C}^-$ be the lower complex half-plane $\{z \in \mathbb{C}^1 : \operatorname{Im} z < 0\}$, and let $\bar{\mathbb{C}}^-$ be the closure of $\mathbb{C}^-$. The principal b-homogeneous symbol of the operator (3.1) is defined by the same formula also for (ξ, τ) with $\tau \in \bar{\mathbb{C}}^-$.

Definition 3.2. A differential operator $A : C^\infty(\Omega', Y_1) \to C^\infty(\Omega', Y_2)$ is called *parabolic* if the mapping $\sigma_A^b(x, t, \xi, \tau) : Y_1 \otimes \mathbb{C}^1 \to Y_2 \otimes \mathbb{C}^1$ is a monomorphism for all $(x, t) \in \Omega'$ and $(\xi, \tau) \in \mathbb{R}^n \times \bar{\mathbb{C}}^-$.

§2. The Formal Theory of Parabolic Systems

In this section we develop a formal theory of differential operators adapted to the parabolic case. We explain why such a special formal theory is needed.

The basic results in the elliptic theory in Chap. 2 were obtained by considering the complex of principal homogeneous symbols generated by the compatibility complex. For parabolic operators it is essential to consider not the homogeneous (as in Chap. 2) symbols, but the principal b-homogeneous symbols defined above. In the formal theory in Chap. 1, the complex of symbols generated by the compatibility complex for a formally integrable operator was exact for quasi-regular covectors. This was the key property in the study of elliptic operators. In the parabolic case, the covector dt is not quasi-regular in general, and we also need to consider complexes of symbols with such a covector. Hence, we need to modify the definition of formal integrability accordingly. Practically, this means that in the derivation of differential consequences we count the order with respect to various weights for x and t. A corollary to this is the exactness of the complex of principal b-homogeneous symbols generated by the compatibility complex for an operator that is formally integrable in this new sense, including for the covector dt. We remark that for a given differential operator there are many compatibility complexes, but such a complex becomes unique if we fix one concept of formal exactness or another. Natuarlly, this meaning is different for the elliptic and parabolic theories, and we make this distinction by introducing anisotropic jet bundles.

Let Ω be a smooth manifold and E a vector bundle over Ω. We write $\Omega' = \Omega \times \mathbb{R}^1$ and denote by E' the inverse image of E with respect to the projection $\Omega' \to \Omega$ defined by $(x, t) \mapsto x$ (locally, if $E = \Omega \times Y$, then $E' = (\Omega \times \mathbb{R}) \times Y)$).

Let $U \subset \Omega$ be a neighbourhood with coordinates $(x_1, \ldots, x_n)$, and suppose that the bundle $E|_U$ is isomorphic to the product $U \times Y$. We denote by U'

a neighbourhood in Ω' of the form $U \times \mathbb{R}^1$. U' is equipped with coordinates of the form $(x_1, \ldots, x_n, t)$. Then $E'|_{U'}$ is isomorphic to $U' \times Y$. Two cross-sections s_1 and s_2 of E' are called (l, b)-equivalent at a point $(x, t) \in U'$ if $(s_1 - s_2)(x, t) = 0$ and $(\partial^{|\alpha|+r}/\partial x_\alpha \partial t^r)(s_1 - s_2)(x, t) = 0$ for all (α, r) such that $|\alpha| + br \leq l$, where $\alpha = (\alpha_1, \ldots, \alpha_n)$ is a multi-index. Obviously, this definition is independent of the choice of trivialisation for the bundle $E'|_{U'}$ and of the coordinate system in U'. We denote by $J^{l,b}_{x,t}(E')$ the set of equivalent classes and write

$$J^{l,b}(E') = \bigcup_{(x,t) \in \Omega'} J^{l,b}_{x,t}(E').$$

Proposition 3.1. *The set $J^{l,b}(E')$ can be equipped with a structure of vector bundle over Ω'.*

We indicate the structure of the bundle $J^{l,b}(E')$ over Ω' in a localisation of E'. Let Ω be a domain in $\mathbb{R}^n$, and suppose that the bundle E is of the form $E = \Omega \times Y$, where Y is an Euclidean space. Then $\Omega' = \Omega \times \mathbb{R}^1$ and $E' = \Omega' \times Y$. The cross-sections of E' have the form $(x, t, f(x, t))$ and can be identified with the functions $(x, t) \mapsto f(x, t)$ from Ω' to Y.

We denote by $D_x^k f$ the derivative of order k of f with respect to x. The derivative $D_x^k \left(\dfrac{\partial^r}{\partial t^r} f \right)$ is a mapping from Ω' to the space $L^k_{\mathrm{sym}}(\mathbb{R}^n, Y)$. We denote by $L^m_{an}(\mathbb{R}^n \times \mathbb{R}^1, Y)$ the direct sum

$$\bigoplus_{r=0}^{[m/b]} L^{m-rb}_{\mathrm{sym}}(\mathbb{R}^n, Y)$$

(here we assume that $L^0_{\mathrm{sym}}(X, Y) = Y$). The bundle $J^{l,b}(\Omega' \times Y)$ has the local trivialisation

$$J^{l,b}(\Omega', Y) = \Omega' \times Y \times L^1_{an}(\mathbb{R}^n \times \mathbb{R}^1, Y) \times \cdots \times L^l_{an}(\mathbb{R}^n \times \mathbb{R}^1, Y).$$

We denote by $j^{l,b}$ the mapping from $C^\infty(E') \to C^\infty(J^{l,b}(E'))$ such that the value of $j^{l,b}s$ for a cross-section $s \in C^\infty(E')$ at a point $(x, t \in \Omega'$ is the (l, b)-equivalence class of s at (x, t). In the localisation, $j^{l,b}$ is a differential operator of the form

$$(j^{l,b}f)(x, t) = \left(f(x, t), \left\{ \frac{\partial^{|\alpha|+r}}{\partial x^\alpha \partial t^r} f : |\alpha| + br \leq l \right\} \right).$$

We define a projection $\pi^k_{l+r,l} : J^{l+r,b}(E') \to J^{l,b}(E')$, $r > 0$. Let s be a cross-section of the bundle E_1, which is a representative of the class $\alpha \in L^{l+r,b}_{x,l}(E')$. Also, let $\beta \in J^{l,b}_{x,t}(E')$ be the (l, b)-equivalence class of s at the point (x, t). Then, by definition, $\pi^b_{l+r}(\alpha) = \beta$. This definition is consistent, since if two cross-sections are $(l + r, b)$-equivalent at (x, t), then they are (l, b)-equivalent at (x, t). In local coordinates, if

$$\alpha = (x, t, y, \{y_{\gamma s} : \gamma = (\gamma_1, \ldots, \gamma_n), |\gamma| + bs \le l + r\}) \in J^{l+r,b}(\Omega' \times Y),$$

where $y, y_{\gamma s} \in Y$, then $\pi_{l+r,l}\alpha = (x, t, y, \{y_{\gamma s} : |\gamma| + bs \le l\})$.

Let $A : C^\infty(E_0') \to C^\infty(E_1')$ be a differential operator. A is called an *operator of order* (l, b) if there is a bundle morphism $p^b(A) : J^{l,b}(E_0') \to E_1'$ such that $A = p^b(A)j^{l,b}$. The definition of an operator of order (l, b) given in Sect. 1 is the localisation of this definition.

For $m \ge 0$ we set

$$R_{l+m,b} = \operatorname{Ker} p^b(j^{m,b}A).$$

The projection

$$\pi^b_{l+m_1, l+m_2} : J^{l+m_1,b}(E_0') \to J^{l+m_2,b}(E_0'),$$

where $m_1 > m_2$, induces a projection $\pi^b_{l+m_1, l+m_2} : R_{l+m_1,b} \to R_{l+m_2,b}$.

Definition 3.3. A differential operator $A : C^\infty(E_0') \to C^\infty(E_1')$ is called *b-regular* if A is sufficiently regular (Definition 1.7) and for all $m \ge 0$ the sets $R_{l+m,b}$ are sub-bundles of the bundles $J^{l+m,b}(E_0')$.

Example 3.1. Every operator with constant coefficients is b-regular.

Example 3.2. Let $\Omega \subset \mathbb{R}^3$, $E_0' = \Omega' \times \mathbb{R}^3$, $E_1' = \Omega' \times \mathbb{R}^6$, and $A : u \mapsto (\partial u/\partial t - a(x,t)\Delta u, \operatorname{curl} u)$, where $a : \Omega' \to \mathbb{R}^1$, $a(x,t) \ge \delta > 0$. If the rank of the mapping $F : \mathbb{R}^3 \to \mathbb{R}^3$ of the form $y \mapsto [\operatorname{grad}_x a(x,t) \times y]$ does not depend on the point $(x,t) \in \Omega'$, then A is 2-regular.

Definition 3.4. A b-regular differential operator $A : C^\infty(E_0') \to C^\infty(E_1')$ of order (l, b) is called *b-formally integrable* if the mappings $\pi^b_{l+mb, l+mb-1} : R_{l+mb,b} \to R_{l+mb-1,b}$ are surjective for all integers $m > 0$.

Example 3.3. The operator A in Example 3.2 is not 2-formally integrable. Adjoining to the relation $Au = 0$ its differential consequences, we can obtain a 2-formally integrable operator. Suppose that the function $a(x,t)$ does not depend on x, that is, $a(x,t) = a(t)$, and let $f, g \in C^\infty(\Omega', \mathbb{R}^3)$. Then we define an operator $P : C^\infty(\Omega' \times \mathbb{R}^6) \to C^\infty(\Omega' \times \mathbb{R}^{15})$ by setting $P(f,g) = (f, g, \{\partial g_k/\partial x_l : k, l = 1, 2, 3\})$. The operator $\tilde{A} = PA$ (equivalent to A in the category $D(\Omega')$) is 2-formally integrable.

Example 3.4. Let A be the operator in Example 3.2, and suppose that the rank of the mapping $F : \mathbb{R}^3 \to \mathbb{R}^3$ of the form $y \mapsto [\operatorname{grad}_x a(x,t) \times y]$ is equal to 2 for all (x,t). We define an operator

$$P_1 : C^\infty(\Omega' \times \mathbb{R}^6) \to C^\infty(\Omega' \times \mathbb{R}^{18})$$

by setting $P_1(f,g) = (f, g, \partial g/\partial t - a\Delta g - \operatorname{curl} r, \{\partial g_k/\partial x_l\})$. The operator $A_1 = P_1 A$, equivalent to A in the category $D(\Omega')$, is 2-formally integrable and can be written in the form

$$(\partial u/\partial t - a(x,t)\Delta u, [\operatorname{grad}_x a(x,t) \times \Delta u], \operatorname{curl} u, \{\partial(\operatorname{curl} u)_k/\partial x_l : k, l = 1, 2, 3\}).$$

Definition 3.5. A complex

$$C^\infty(E_0') \xrightarrow{A} C^\infty(E_1') \xrightarrow{A_1} C^\infty(E_2') \qquad (3.2)$$

is called *b-formally exact* if the complex

$$J^{l+l_1+mb,b}(E_0') \xrightarrow{p^b(j^{l_1+mb}A)} J^{l_1+mb,b}(E_1') \xrightarrow{p^b(j^{mb,b}A_1)} J^{mb,b}(E_2')$$

is exact for every $m \geq 0$.

Proposition 3.2. *If the complex (3.2) is formally exact, then it is a compatibility complex for the operator A (in the category $\mathrm{D}(\Omega')$; see Definition 1.4).*

The compatibility operators that are b-formally exact are called *b-compatibility operators*.

In Sect. 1 we defined the principal b-homogeneous symbol of a differential operator acting on functions $f : \Omega \times \mathbb{R}^1 \to Y$, where $\Omega \subset \mathbb{R}^n$ and Y is a vector space. In the case of an operator $A : C^\infty(E_0') \to C^\infty(E_1')$ we use local trivialisations of bundles and Definition 3.1 to define the principal b-homogeneous symbol of A for a covector $(\xi, \tau) \in T_{x,t}^*\Omega'$ as the mapping

$$\sigma_A^b(x, t, \xi, \tau) : E_0'|_{(x,t)} \otimes \mathbb{C}^1 \to E_1'|_{(x,t)} \otimes \mathbb{C}^1.$$

The symbol can also be defined directly (without a localisation). We denote by $J_\Omega^k(E')$ the inverse image of the bundle $J^k(E)$ over Ω with respect to the projection $\mathrm{Pr} : \Omega' \to \Omega$ defined by $(x, t) \to x$. Then the bundle $J^{l,b}(E')$ is isomorphic to the direct sum $\displaystyle\bigoplus_{r=0}^{[l/b]} J_\Omega^{l-rb}(E')$, as follows from the definitions of the bundles $J^{l,b}$, J^k and J_Ω^k. We denote by $\mathrm{pr}^*S^k(\Omega')$ the inverse image of the bundle $S^k(T^*\Omega)$ with respect to the projection $\mathrm{pr} : \Omega' \to \Omega$, and by $S^{l,b}(\Omega')$ the direct sum $\displaystyle\bigoplus_{r=0}^{[l/b]} \mathrm{pr}^*S^{l-rb}(\Omega')$. The embedding operators $\varepsilon : S^k(T^*\Omega) \otimes E \to J^k(E)$ induce an embedding operator $\varepsilon_b : S^{l,b}(\Omega') \otimes E' \to J^{l,b}(\Omega')$. Let $p^b(A) : J^{lb,b}(E_0') \to E_1'$ be the bundle morphism corresponding to a differential operator A. By the principal *b-homogeneous symbol* of A we understand the mapping $\sigma^b(A) : S^{l,b}(\Omega') \otimes E_0' \to E_1'$ of the form $\sigma^b(A) = p^b(A) \circ \varepsilon_b$. The cotangent bundle $T^*\Omega'$ has the structure of a direct product, namely $T^*\Omega' = \mathrm{pr}^*T^*\Omega \times \mathbb{R}^1$, where $\mathrm{pr}\, T^*\Omega$ is the inverse image of the bundle $T^*\Omega$ with respect to the projection $\mathrm{pr} : \Omega' \to \Omega$. We define an embedding $\varphi : T^*\Omega' \to S^{l,b}(\Omega) \otimes \mathbb{C}^1$ by writing

$$\varphi(\xi, \tau) = \bigoplus_{r=0}^{[l/b]} i^{l-r(b-1)}\tau^r \cdot \xi \otimes \cdots \otimes \xi$$

for an element (ξ, τ) of the bundle $T^*\Omega' = \mathrm{pr}^*T^*\Omega \otimes \mathbb{R}^1$.

By the *principal b-homogeneous symbol of the operator A for a covector* $(\xi,\tau) \in T^*_{(x,t)}\Omega'$ we understand the mapping $\sigma_A^b(x,t,\xi,\tau) : E_0'|_{(x,t)} \otimes \mathbb{C}^1 \to E_1'|_{(x,t)} \otimes \mathbb{C}^1$ of the form $y \mapsto \sigma^b(A)(\varphi(\xi,\tau) \otimes y)$.

We denote by $p^*\Omega'$ the inverse image of the bundle $T^*\Omega \times \bar{\mathbb{C}}^-$ with respect to the projection $\Omega' \to \Omega$. (Here $\bar{\mathbb{C}}^-$ is the closed lower complex half-plane $\{z \in \mathbb{C}^1 : \operatorname{Im} z \leq 0\}$.) In the case when $\Omega \subset \mathbb{R}^n$, the fibre $p^*\Omega'|_{(x,t)}$ consists of a pair (ξ,τ) with $\xi \in \mathbb{R}^n$ and $\tau \in \mathbb{C}^-$. The definition of the principal b-homogeneous symbol for a covector $(\xi,\tau) \in T^*_{(x,t)}\Omega'$ carries over without modification to the case when $(\xi,\tau) \in p^*\Omega'$.

Definition 3.6. A differential operator $A : C^\infty(E_0') \to C^\infty(E_1')$ is called *parabolic with weight b* if the mapping $\sigma_A^b(x,t,\xi,\tau) : E_0'|_{(x,t)} \otimes \mathbb{C}^1 \to E_1'|_{(x,t)} \otimes \mathbb{C}^1$ is a monomorphism for all $(x,t) \in \Omega'$ and $(\xi,\tau) \in p^*\Omega'|_{(x,t)}(\xi) \neq 0$.

Let A be a b-formally integrable differential operator of order (lb,b), and suppose that its principal b-homogeneous symbol for the covector $(0,\tau)$ is a monomorphism (this condition always holds for a parabolic operator). Such an operator, possibly after applying a bundle isomorphism, can be rewritten in the form

$$Ay = \left(\frac{\partial^l y}{\partial t^l} + Ly, My \right), \tag{3.3}$$

where the operators L and M of order (lb,b) do not contain differentiations of order l with respect to t.

We construct a b-compatibilty operator for an operator of the form (3.3). To do this, we use the commutation relations described in the next assertion.

Proposition 3.3. *If an operator A of the form (3.3) is b-formally integrable, then there is an integer $l_1 > 0$ and operators K and R of order $(l_1 b, b)$ not containing differentiations of order l_1 with respect to t and satisfying the commutation relations*

$$\left(\frac{\partial^{l_1}}{\partial t^{l_1}} - K \right) M = R \left(\frac{\partial^l}{\partial t^l} + L \right). \tag{3.4}$$

If $l = 1$ in (3.3), then L and M do not contain $\partial/\partial t$ and depend on t as a parameter. In this case it can be shown that if A satisfies the conditions in Proposition 3.3, then M, regarded as a differential operator with respect to x, is sufficiently regular. The condition $l_1 = 1$ in (3.4) is the analogue of the involutiveness condition. First we consider the case of operators of order (b,b) that satisfy this condition.

Proposition 3.4. *Let A be a b-formally integrable operator of order (b,b) of the form (3.3), and suppose that M in (3.3), regarded as a differential operator with respect to x, is involutive. We also assume that $l_1 = 1$ in the relations (3.4) written for A. Then a b-compatibility operator for A is given by the formula*

$$A_1(f,g) = \left(\frac{\partial g}{\partial t} - Kg - Rf, M'g \right), \qquad (3.5)$$

where K and R are the operators in (3.4) and M' is a compatibility operator of order b for M, which depends on t as a parameter.

In the conditions of Proposition 3.4, R coincides with M and the coefficients of the operator K are found from the commutation relations (3.4). Consequently, (3.4) yields the explicit form of a b-compatibility operator.

We construct a b-compatibility operator of order (lb, b) for (3.3) in the case when l, $l_1 \neq 1$. First we consider an example.

Example 3.5. We construct a b-compatibility operator for $j^{lb,b} : C^\infty(E') \to C^\infty(f^{lb,b}(E'))$. The bundle $J^{lb,b}(E')$ is isomorphic to the direct sum of bundles $\bigoplus_{r=0}^{l} J_\Omega^{rb}(E')$, where $J_\Omega^{rb}(E')$ is the inverse image of $J^{rb}(E)$ with respect to the projection pr $: \Omega' \to \Omega$ and $J_\Omega^0(E')$ is identified with E'. Let $D_{rb}^1 : C^\infty(J^{rb}(E)) \to C^\infty(C_{rb}^1)$ be the compatibility operator for $J^{rb} : C^\infty(E) \to C^\infty(J^{rb}(E))$ described in Example 1.18. We denote by $\hat{C}_{rb}$ the inverse image of the bundle C_{rb}^1 (in the same example) with respect to the projection $\Omega' \to \Omega$. Let $j_\Omega^{rb} : C^\infty(E') \to C^\infty(J_\Omega^{rb}(E'))$ and $\hat{D}_{rb} : C^\infty(J_\Omega^{rb}(E')) \to C^\infty(\hat{C}_{rb})$ be the operators generated by j^{rb} and D_{rb}^1, respectively, in the inverse images. We denote by $C_{lb,b}$ the bundle

$$\left(\bigoplus_{r=1}^{l} J_\Omega^l(\hat{C}_{rb}) \right) \oplus \left(\bigoplus_{r=0}^{l-1} J_\Omega^{rb}(E') \right).$$

Using the isomorphism of $J^{lb,b}(E')$ and $\bigoplus_{r=0}^{l} J_\Omega^{rb}(E')$, we represent an element $f \in C^\infty(J^{lb,b}(E'))$ in the form $(f_0, \ldots, f_l)$, where $f_r \in C^\infty(J_\Omega^{rb}(E'))$, $r = 0, \ldots, l$. Let $\pi'_{m_1,m_2} : J_\Omega^{m_1}(E') \to J_\Omega^{m_2}(E')$, where $m_1 > m_2$, be the morphism induced by the morphism $\pi_{m_1,m_2} : J^{m_1}(E) \to J^{m_2}(E)$. An operator $D_{lb,b} : C^\infty(J^{lb,b}(E')) \to C^\infty(C_{lb,b})$ of the form

$$f \mapsto \left(j_\Omega^{b-1}\hat{D}_b f_{l-1}, \ldots, j_\Omega^{b-1}\hat{D}_l f_0, \frac{\partial}{\partial t} f_0 - \pi'_{b,0} f_1, \ldots, \frac{\partial}{\partial t} f_{l-1} - \pi'_{lb,(l-1)b} f_l \right)$$

is a b-compatibility operator for $j^{lb,b}$. (This is easily verified by rewriting the $j^{lb,b}$ and $D_{lb,b}$ in terms of local coordinates. The localisation of $D_{lb,b}$ is obtained immediately from the form of $D_{lb,b}$ and the localisation of D_k^1.)

To construct a b-compatibility operator for an operator A of the form (3.3), we construct an operator $\tilde{A}$ equivalent to A in the category $\mathrm{D}(\Omega')$ and satisfying the conditions in Proposition 3.4.

We denote by $A^{(k,b)} : C^\infty(E_0') \to C^\infty(J^{kb,b}(E_1'))$ the operator $j^{kb,b}A$.

Proposition 3.5. *$A^{(k,b)}$ is equivalent to A in the category $\mathrm{D}(\Omega')$. If A is a b-formally integrable operator, then so is $A^{(k,b)}$ for all $k \geq 0$. In this case* *there is a number k_0 such that for $k \geq k_0$ the number l in the relations (3.4) written for $A^{(k,b)}$ is equal to one.*

To construct the operator $\tilde{A}$, it remains to choose a sufficiently large number k and lower the order of $A^{(k,b)}$ so that the properties of $A^{(k,b)}$ mentioned in Proposition 3.5 are preserved. To this end, we define an operator

$$\bar{A} : C^\infty(J^{(l+k-1)b,b}(E_0')) \to C^\infty(J^{kb,b}(E_1'))$$

of order (b,b) by setting $\bar{A}j^{(l+k-1)b,b} = A^{(k,b)}$, where l is a number such that (lb,b) is the order of A. We define $\tilde{A} : C^\infty(J^{(l+k-1)b,b}(E_0)) \to C^\infty(J^{kb,b}(E_1') \oplus C_{(l+k-1)b,b})$ by setting $\tilde{A}y = (\bar{A}y, D_{(l+k-1)b,b}y)$, where $C_{(l+k-1)b,b}$ and $D_{(l+k-1)b,b}$ are the bundle and operator constructed in Example 3.5.

Proposition 3.6. *There is a number k such that the above operator $\tilde{A}$ satisfies the conditions in Proposition 3.4. $\tilde{A}$ is equivalent to A in the category* $D(\Omega')$.

From Propositions 3.4 and 3.6 we deduce the following assertion.

Proposition 3.7. *If A is a b-formally integrable operator of the form (3.3), then there exists a b-compatibility operator for A, which can be constructed in finitely many steps.*

The explicit expressions of a b-compatibility operator for A (formula (3.5)) and the method of construction of $\tilde{A}$ yield an explicit procedure for constructing a b-compatibility operator for A.

The following property of a b-compatibility operator is fundamental in the study of parabolic overdetermined operators.

Proposition 3.8. *If A_0 is a parabolic b-formally integrable operator and A_1 its b-compatibility operator, then the complex*

$$0 \to E_0'|_{(x,t)} \otimes \mathbb{C}^1 \xrightarrow{\sigma^b_{A_0}(x,t,\xi,\tau)} E_1'|_{(x,t)} \otimes \mathbb{C}^1 \xrightarrow{\sigma^b_{A_1}(x,t,\xi,\tau)} E_2'|_{(x,t)} \otimes \mathbb{C}^1, \quad (3.6)$$

where $\sigma^b_{A_i}(\cdot)$ is the principal b-homogeneous symbol of A_i, is exact for all $(x,t) \in \Omega'$ and $(\xi,\tau) \in p^*\Omega'|_{(x,t)}, (\xi,\tau) \neq 0$.

Example 3.6. Let A be the operator in Example 3.2, and let $a = a(t)$. Then a b-compatibility operator for A has the form

$$A_1 : (f,g) \mapsto (\partial g/\partial t - a\Delta g - \operatorname{curl} f, \operatorname{div} f, \{(\partial/\partial x_k)\operatorname{div} g : k = 1,2,3\}).$$

A b-compatibility operator for the operator $\tilde{A} = PA$ in Example 3.3 is given by

$$\tilde{A}_1 : (f,g,\{\varphi_{kl} : k,l = 1,2,3\}) \mapsto (\{\partial \varphi_{kl}/\partial t - a\Delta\varphi_{kl} - (\partial/\partial x_l)(\operatorname{pr}_k \operatorname{curl} f)\},$$
$$\partial g/\partial t - a\Delta g - \operatorname{curl} f, D^1(g,\{\varphi_{kl}\}), (\partial/\partial x_k)(D^1(g,\{\varphi_{kl}\}) : k = 1,2,3),$$
$$j_x^1 \operatorname{div} g, j_x^1 \operatorname{div} \varphi_{kl}),$$

where $k, l = 1, 2, 3$, pr_k is the projection of a vector on its kth component, and D^1 is the operator in Example 1.18. $\tilde{A}$ is parabolic and b-formally integrable; hence, the exactness of the complex (3.6) for $\tilde{A}$ and $\tilde{A}_1$ follows from Proposition 3.8. (For these operators, the exactness of the complex (3.6) is also easily verified directly.) We remark that the compatibility operator for $\tilde{A}$ constructed by means of the scheme set forth in Chap. 1 acts according to the formula

$$\tilde{A}_1' : (f, g, \{\varphi_{kl}\}) \mapsto \left(\left\{ \frac{\partial g_k}{\partial t} - a \sum_{l=1} \frac{\partial \varphi_{kl}}{\partial x_l} - \mathrm{pr}_k \, \mathrm{curl} \, f \right\}, \right.$$
$$\left. D^1(g, \{\varphi_{kl}\}), \mathrm{div}\, g, \{\mathrm{div}\, \varphi_l\} \right),$$

and the complex of the b-homogeneous symbols of $\tilde{A}$ and $\tilde{A}_1$ is not exact.

§3. Parabolic Differential Boundary Operators

Let Ω be a smooth manifold with a smooth boundary Γ, $\Omega' = \Omega \times \mathbb{R}^1$, and $\Gamma' = \Gamma \times \mathbb{R}^1$; also, let E_k and G_k be vector bundles over Ω and Γ, respectively, and E_k' and G_k' the vector bundles over Ω' and Γ' that are the inverse images of E_k and G_k with respect to the projections $\Omega' \to \Omega$ and $\Gamma' \to \Gamma$. We consider DB-operators, that is, operators in the category $\mathrm{DB}(\Omega', \Gamma')$. If Φ is such an operator, then $\Phi(f, g) = (\Phi^{11} f, \Phi^{21} f + \Phi^{22} g)$, where $\Phi^{11} : C^\infty(E_0') \to C^\infty(E_1')$ and $\Phi^{22} : C^\infty(G_0') \to C^\infty(G_1')$ are differential operators on Ω' and Γ', respectively, and $\Phi^{21} : C^\infty(E_0') \to C^\infty(G_0')$ is a local boundary operator.

We indicate sufficient conditions for constructing a compatibility operator for a DB-operator Φ with "good" formal properties that take account of different weights for x and t.

Let (lb, b), (k, b) and (k, b) be the orders of the operators Φ^{11}, Φ^{21} and Φ^{22} (the components of Φ), respectively.

We write

$$R_{lb+m}^b = \mathrm{Ker}\, p^b(j^{m,b}\Phi^{11}) \subset J^{lb+m,b}(E_1'),$$
$$\tilde{R}_{k+m}^b = \mathrm{Ker}\, p^b(j^{m,b}(\Phi^{21} + \Phi^{22})) \subseteq J^{k+m,b}(E_0')|_{\Gamma'} \oplus J^{k+m,b}(G_0').$$

Definition 3.7. A DB-operator Φ is called *b-regular* if (i) it is regular (in the sense of Definition 1.22), (ii) Φ^{11} is equivalent in the category $\mathrm{D}(\Omega')$ to an operator of the form (3.3) and is b-formally integrable, (iii) the conormal to Γ' is quasi-regular for Φ^{11} at every point $x \in \Gamma'$, and (iv) the sets

$$(R_{lb+m_1}^b|_{\Gamma'} \oplus J^{k+m_2,b}(G_0')) \cap \tilde{R}_{k+m_2}^b$$

are sub-bundles of the bundles

$$J^{bl+m_1,b}(E_0')|_{\Gamma'} \oplus J^{k+m_2,b}(G_0')$$

for all natural numbers m_1 and m_2 such that $m_1 - m_2 = k - lb$.

If Φ^{11} is a b-formally integrable parabolic operator, then it satisfies the conditions (ii) and (iii) in Definition 3.7.

Example 3.7. Let Ω be a domain in $\mathbb{R}^3$ with boundary Γ, Φ^{11} the operator A in Example 3.2, and $\Phi^{21} : C^\infty(\Omega', \mathbb{R}^3) \to C^\infty(\Omega', \mathbb{R}^3)$ the operator of restriction of a function to Γ'. The operator

$$\Phi : C^\infty(\Omega', \mathbb{R}^3) \to C^\infty(\Omega', \mathbb{R}^6) \times C^\infty(\Gamma', \mathbb{R}^3)$$

that maps a function u into $(\Phi^{11}u, \Phi^{21}u)$, is not b-regular, since the condition (ii) in Definition 3.7 is not satisfied. Let $\tilde{\Phi}^{11}$ be the operator $\tilde{A}$ defined in Example 3.3. The operator

$$\tilde{\Phi} : C^\infty(\Omega', \mathbb{R}^3) \to C^\infty(\Omega', \mathbb{R}^{15}) \times C^\infty(\Gamma', \mathbb{R}^3)$$

defined by $\tilde{\Phi}u = (\tilde{\Phi}^{11}u, \Phi^{21}u)$ is a b-regular DB-operator.

Example 3.8. Let Ω be a domain in $\mathbb{R}^3$ with boundary Γ, $\Phi^{11} : C^\infty(\Omega', \mathbb{R}^2) \to C^\infty(\Omega', \mathbb{R}^2)$ an operator acting according to the formula $u \mapsto \partial u/\partial t - \Delta u$, and $\Phi^{21} : C^\infty(\Omega', \mathbb{R}^2) \to C^\infty(\Gamma', \mathbb{R}^2)$ the operator defined by $(\Phi^{21}u)(x', t) = a_{j1}(x', t)u_1(x', t) + a_{j2}(x', t)u_2(x', t)$, where $j = 1, 2$, $x' \in \Gamma$, and $a_{ij}(x', t)$ are smooth functions on Γ'. Suppose that the determinant of the matrix $A = (a_{ij})$ is equal to zero at some points $(x', t) \in \Gamma'$ but not identically zero. Then the operator $\Phi : C^\infty(\Omega', \mathbb{R}^2) \to C^\infty(\Omega', \mathbb{R}^2) \times C^\infty(\Gamma', \mathbb{R}^2)$ defined by $\Phi u = (\Phi^{11}u, \Phi^{21}u)$ is not b-regular, since the condition (iv) in Definition 3.7 is not satisfied.

We define a b-compatibility operator for a boundary value problem operator. Let $(A, B) : C^\infty(E_0') \to C^\infty(E_1') \times C^\infty(G_1')$ be a b-regular boundary value problem operator. Then (A, B) is regular, hence, it has a compatibility operator (constructed in Chap. 1)

$$\tilde{\Phi} : C^\infty(E_1'') \times C^\infty(G_1'') \to C^\infty(E_2'') \times C^\infty(G_2'')$$

defined by $\tilde{\Phi}(f, g) = (\tilde{\Phi}^{11}f, \tilde{\Phi}^{21}f + \tilde{\Phi}^{22}g)$. Let c be the order of $\tilde{\Phi}^{22}$ and r a natural number such that $d = rbc + k - lb \geq 0$. Since (A, B) is b-regular, it follows that there is a differential operator $D : C^\infty(\tilde{G}_2') \to C^\infty(G_2')$ such that $\operatorname{Ker} p^b(D\tilde{\Phi}^{22}) = p^b(j^{rbc,b}B)(\operatorname{Ker} p^b(j^{d,b}A))$. We denote by Φ the operator

$$(f, g) \mapsto (A_1 f, D\tilde{\Phi}^{21}f + D\tilde{\Phi}^{22}g),$$

where A_1 is a b-compatibility operator for A. Φ is called a *b-compatibility operator* for (A, B).

Proposition 3.9. (i) *Φ is a compatibility operator for (A, B) in the category* DB(Ω', Γ').

(ii) *The complex of vector bundles*

$$R_{lb+d}^{l} \stackrel{p^b(j_{r'}^{rbc}B)}{\longrightarrow} J^{rbc,b}(G_1') \stackrel{p^b(\Phi^{22})}{\longrightarrow} G_2',$$

where $\Phi^{22} = D\tilde{\Phi}^{22}$, *is exact.*

§4. The Solvability of Initial Boundary Value Problems for Parabolic Operators

4.1. The Coerciveness Condition. Let $(A, B) : C^\infty(E_0') \to C^\infty(E_1') \times C^\infty(G_1')$ be a regular DB-operator, A a parabolic operator, and

$$\Phi : C^\infty(E_1') \times C^\infty(G_1') \to C^\infty(E_2') \times C^\infty(G_2')$$

a compatibility operator for (A, B). Also, suppose that $G_j' = \bigoplus_{k=1}^{r_j} G_{jk}'$ $(j = 1, 2)$. Then B and Φ^{22} are of the form $B = (B_1, B_2, \ldots, B_{r_1})$ and $\Phi^{22} = \{\Phi_{km}^{22} : m = 1, \ldots, r_1; \, k = 1, \ldots, r_2\}$, where

$$B_k : C^\infty(E_0') \to C^\infty(G_{1k}'), \quad \Phi_{km}^{22} : C^\infty(G_{1m}') \to C^\infty(G_{2k}') \, .$$

Suppose that the B_k are of order (β_{1k}, b), respectively, and let $\beta_{2k} = \max_{m}\{\alpha_{km}\}$, where α_{km} are numbers such that $(\alpha_{km} - \beta_{1m}, b)$ is the order of Φ_{km}^{22}. We denote the multi-index $(\beta_{11}, \ldots, \beta_{1r_1})$ by β_1 and $(\beta_{21}, \ldots, \beta_{2r_2})$ by β_2.

In the neighbourhood $U' = U \times \mathbb{R}^1$ of a point $\tilde{x} = (x', t) \in \partial\Omega'$ we choose a system of coordinates $(x_1, x_2, \ldots, x_n, t)$ such that the set $\partial\Omega' \cap U'$ is given by the equation $x_n = 0$ and $\Omega' \cap U'$ by the inequality $x_n \geq 0$. In terms of these coordinates (in the corresponding trivialisations of the bundles $E_i'|_{U'}$ and $G_i'|_{U' \cap \partial\Omega'}$) the operators A and B can be rewritten as

$$(Ay)(x, t) = \sum_{|\alpha|+rb \leq lb} a_{\alpha r}(x, t)\left(\frac{\partial^{|\alpha|+r}}{\partial x^\alpha \partial t^r} y\right)(x, t) \, ,$$

$$(B_k y)(x', t) = \sum_{|\alpha|+rb \leq \beta_{1k}} b_{k\alpha r}(x', t)\left(\frac{\partial^{|\alpha|+r}}{\partial x^\alpha \partial t^r} y\right)(x', t) \, ,$$

(3.7)

where $k = 1, \ldots, r_1$, $\alpha = (\alpha_1, \ldots, \alpha_n)$ is a multi-index, and $a_{\alpha r}(\cdot)$ and $\beta_{k\alpha r}(\cdot)$ are matrices. We fix a point $(x^0, t^0) \in U' \cap \partial\Omega'$ and a vector $(\eta_1, \ldots, \eta_{n-1}, \tau) \in \mathbb{R}^{n-1} \times \mathbb{C}^-$. We replace $\partial/\partial x_j$ by $i\eta_j$ $(j = 1, \ldots, n-1)$ and $\partial/\partial t$ by $i\tau$ in (3.7), and fix the coefficients $a_{\alpha r}$ and $\beta_{k\alpha r}$ at (x^0, t^0). Discarding the terms of order less than (lb, b) in A and less than (β_{1k}, b) in B_k, we arrive at a boundary value problem operator with constant coefficients on the semiaxis, namely

$$(A(x^0,t^0,\eta,\tau)y)(\nu) = \sum_{|\alpha|+rb=lb} i^{|\alpha'|+r} a_{\alpha r}(x^0,t^0)\tau^r \eta^{\alpha'} \left(\frac{d^{\alpha_n}}{d\nu^{\alpha_n}}y\right)(\nu)\,,$$

$$B_k(x^0,t^0,\eta,\tau)y = \sum_{|\alpha|+rb=\beta_{1k}} \beta_{k\alpha r}(x^0,t^0)\tau^r \eta^{\alpha'} \left(\frac{d^{\alpha_n}}{d\nu^{\alpha_n}}y\right)(0)\,, \tag{3.8}$$

where $\alpha' = (\alpha_1,\ldots,\alpha_{n-1})$.

We apply the above procedure to the Φ_{km}^{22} regarded as operators of orders $(\alpha_{km}-\beta_{1m},b)$, and obtain the linear fibre mappings

$$\Phi_{km}^{22}(x^0,t^0,\eta,\tau) : G_1'|_{(x^0,t^0)} \to G_2'|_{(x^0,t^0)}\,.$$

Taking into account the form of (3.8) after the change of coordinates, we conclude that (η,τ) is an element of the fibre $p^*\partial\Omega'|_{(x^0,t^0)}$ of the bundle $p^*\partial\Omega'$ (the inverse image of the bundle $T^*\Gamma\times\bar{\mathbb{C}}^-$ with respect to a projection $\partial\Omega'\to\Gamma$ of the form $(x',t)\mapsto x'$).

Definition 3.8. The operator (A,B) satisfies the *coerciveness condition* with multi-indices (β_1,β_2) if

(i) the number $\dim\operatorname{Ker}A(\tilde{x},\tilde{\eta})$ does not depend on the points $\tilde{x}=(x^0,t^0)\in\partial\Omega'$ and $\tilde{\eta}=(\eta',\tau)\in p^*\partial\Omega'|_{\tilde{x}}$, $\tilde{\eta}\neq 0$;

(ii) for all $\tilde{x}\in\partial\Omega'$ and $\tilde{\eta}\in p^*\partial\Omega'|_{\tilde{x}}$, $\tilde{\eta}\neq 0$, the sequence of finite-dimensional spaces

$$0 \to \operatorname{Ker}A(\tilde{x},\tilde{\eta})\cap\mathfrak{M}^+ \stackrel{(A(\tilde{x},\tilde{\eta}),B(\tilde{x},\tilde{\eta}))}{\longrightarrow} G_1'|_{\tilde{x}}\otimes\mathbb{C}^1 \stackrel{\Phi^{22}(\tilde{x},\tilde{\eta})}{\longrightarrow} G_2'|_{\tilde{x}}\otimes\mathbb{C}^1\,,$$

where $\mathfrak{M}^+ = \{f\in C^\infty(\bar{\mathbb{R}}_+^1) : f(x)\to 0 \text{ as } x\to\infty\}$, is exact.

4.2. The Spaces $H^{s,b}$. Let $H^{s,b}(\mathbb{R}^n\times\mathbb{R}^1)$ be the completion of $C_0^\infty(\mathbb{R}^n\times\mathbb{R}^1)$ with respect to the norm

$$\|u\|_{s,b}^2 - \int(1+|\xi|^{2b}+|\tau|^2)^{s/b}|\hat{u}(\xi,\tau)|^2\,d\xi\,d\tau\,,$$

where $\hat{u}(\xi,\tau)$ is the Fourier transform of $u(x,t)$.

We denote by $\overset{+}{H}{}^{s,b}(\mathbb{R}^n\times\mathbb{R}^1)$ the set of functions in $H^{s,b}(\mathbb{R}^n\times\mathbb{R}^1)$ equal to zero for $t<0$. Clearly, $\overset{+}{H}{}^{s,b}(\mathbb{R}^n\times\mathbb{R}^1)$ is a closed subspace of $H^{s,b}(\mathbb{R}^n\times\mathbb{R}^1)$.

Let $\mathbb{R}_+^n = \{(x_1,\ldots,x_n) : x_n>0\}$, and let $\overset{+}{H}{}^{s,b}(\bar{\mathbb{R}}_+^n\times\mathbb{R}^1)$ be the set of all functions f on $\bar{\mathbb{R}}_+^n\times\mathbb{R}^1$ admitting an extension to $\mathbb{R}^n\times\mathbb{R}^1$ which belongs to $\overset{+}{H}{}^{s,b}(\mathbb{R}^n\times\mathbb{R}^1)$. The norm on $\overset{+}{H}{}^{s,b}(\bar{\mathbb{R}}_+^n\times\mathbb{R}^1)$ is defined by

$$\|f\|_{s,b}^+ = \inf_{\varphi\in\overset{+}{H}{}^{s,b}(\mathbb{R}^n\times\mathbb{R}^1),\,\varphi|_{\mathbb{R}_+^n\times\mathbb{R}^1}=f} \|f_{s,b}\|\,,$$

where the norm on the right-hand side is that on $\overset{+}{H}{}^{s,b}(\mathbb{R}^n\times\mathbb{R}^1)$.

Let $\gamma > 0$. We denote by $\overset{+}{H}{}^{s,b,\gamma}(\mathbb{R}^n \times \mathbb{R}^1)$ $(\overset{+}{H}{}^{s,b,\gamma}(\bar{\mathbb{R}}^n_+ \times \mathbb{R}^1))$ the set of all functions $f(x,t)$ such that $e^{-\gamma t}f(x,t)$ belongs to $\overset{+}{H}{}^{s,b}(\mathbb{R}^n \times \mathbb{R}^1)$ $(\overset{+}{H}{}^{s,b}(\bar{\mathbb{R}}^n_+ \times \mathbb{R}^1))$.

The norm on the spaces $\overset{+}{H}{}^{s,b,\gamma}$ is defined by

$$\|f\|_{s,b,\gamma} = \|f_\gamma\|_{s,b}, \quad f_\gamma(x,t) = e^{-\gamma t}f(x,t) \, .$$

Using coverings of the manifolds Ω' and $\partial\Omega'$ by coordinate neighbourhoods U' of the form $U \times \mathbb{R}^1$, and the partition of unity, we define in the usual way the spaces $\overset{+}{H}{}^{s,b}(E')$, $\overset{+}{H}{}^{s,b}(G')$, $\overset{+}{H}{}^{s,b,\gamma}(E')$, and $\overset{+}{H}{}^{s,b,\gamma}(G')$ of the cross-sections of the vector bundles E' over Ω' and G' over $\partial\Omega'$.

Let $T > 0$. We write $\Omega'_T = \Omega \times (-\infty, T)$ and $\partial\Omega'_T = \partial\Omega' \times (-\infty, T)$, and denote by E'_{iT} and G'_{iT} the restrictions of E'_i and G'_i to Ω'_T and $\partial\Omega'_T$, respectively. Also, let $\overset{+}{H}{}^{s,b}(E'_{iT})$ be the set of all the cross-sections of E'_{iT} that can be extended to a cross-section of E'_i belonging to $\overset{+}{H}{}^{s,b}(E'_i)$. The space $\overset{+}{H}{}^{s,b}(G'_{iT})$ is introduced similarly. The norm of an element $f \in \overset{+}{H}{}^{s,b}(E'_{iT})$ is defined by

$$\|f\|_{s,b}^+ = \inf_{\varphi \in \overset{+}{H}{}^{s,b}(E'_i),\, \varphi|_{\Omega'_T} = f} \|\varphi\|_{s,b}^+ \, .$$

The norm on $\overset{+}{H}{}^{s,b}(G'_{iT})$ is defined analogously.

We denote by $H^\beta(G')$ the space $\bigoplus_{j=1}^k H^{\beta_j}(G'_j)$ for a partition $G' = \bigoplus_{j=1}^k G'_j$ and a multi-index $\beta = (\beta_1, \ldots, \beta_k)$.

4.3. Solvability Theorems for Initial Boundary Value Problems for Parabolic Operators. In this case we assume that the manifold Ω (with boundary) is compact. We consider the case of a finite time interval $[0, T]$. The parabolicity of the operator and the coerciveness of the boundary value problem on $[0, T]$ with respect to t mean that in Definitions 3.2 and 3.5 we consider only points (x, t) such that $t \in [0, T]$.

Suppose that all the mappings in the complex

$$0 \to \overset{+}{H}{}^{s,b}(E'_{0T}) \overset{(A,B)}{\to} \overset{+}{H}{}^{s-lb,b}(E'_{1T}) \times \overset{+}{H}{}^{s-\beta_1-1/2,b}(G'_{1T})$$

$$\overset{\Phi}{\to} \overset{+}{H}{}^{s-(l+l_1)b,b}(E'_{2T}) \times \overset{+}{H}{}^{s-\beta_2-1/2,b}(G'_{2T}) \, , \qquad (3.9)$$

where Φ is a compatibility b-operator for (A, B) and (lb, b) and (l_1b, b) are the orders of A and Φ^{11}, respectively, are bounded.

Theorem 3.1. *Let (A, B) be a b-regular boundary value problem operator, and suppose that for $t \in [0, T+\varepsilon]$ the operator A is parabolic with weight b with* *respect to t and (A, B) satisfies the coerciveness condition with multi-indices (β_1, β_2) (in the sense of Definition 3.8). Then the complex (3.9) is exact and*

$$\|u\|_{s,b}^+ \le C(\|Au\|_{s-lb,b}^+ + \|Bu\|_{s-\beta_1-1/2,b}^+) \,, \tag{3.10}$$

where C does not depend on $u \in \overset{+}{H}{}^{s,b}(E'_{0T})$.

We now consider the case of an infinite interval with respect to t. We assume that the coefficients of (A, B) do not depend on t for $t > T$, where T is some positive number. Then neither do the coefficients of Φ (a compatibility b-operator for (A, B)). We assume that all the mappings in the complex

$$0 \to \overset{+}{H}{}^{s,b,\gamma}(E'_0) \overset{(A,B)}{\to} \overset{+}{H}{}^{s-lb,b,\gamma}(E'_1) \times \overset{+}{H}{}^{s-\beta_1-1/2,b,\gamma}(G'_1)$$

$$\overset{\Phi}{\to} \overset{+}{H}{}^{s-(l+l_1)b,b,\gamma}(E'_2) \times \overset{+}{H}{}^{s-\beta_2-1/2,b,\gamma}(G'_2) \tag{3.11}$$

are bounded.

Theorem 3.2. *Let (A, B) be a b-regular boundary value problem operator, and suppose that the coefficients of (A, B) do not depend on t for $t > T$, that A is parabolic with weight b with respect to t, and that (A, B) satisfies the coerciveness condition with multi-indices (β_1, β_2). Then we can find a number $\gamma_0 \ge 0$ such that for any $\gamma \ge \gamma_0$ the complex (3.11) is exact and*

$$\|u\|_{s,b,\gamma}^+ \le C(\|Au\|_{s-lb,b,\gamma}^+ + \|Bu\|_{s-\beta_1-1/2,b,\gamma}) \,,$$

where C is independent of $u \in \overset{+}{H}{}^{s,b,\gamma}(E'_0)$ and $\gamma \ge \gamma_0$.

Theorems 3.1 and 3.2 are proved by applying the theory of Boutet de Monvel parabolic operators, which is analogous to that of Boutet de Monvel elliptic operators set forth in Chap. 2, Sect. 6.

Suppose that the parabolic differential operator A is not b-formally integrable, and that there is a differential operator P such that PA is b-formally integrable (such operators have been considered in Examples 3.2, 3.3 and 3.4). Also, suppose that the boundary value problem operator (PA, B) satisfies the conditions in Theorem 3.1, and that Φ is a b-compatibility operator for (PA, B). Then for the original operator (A, B), the complex

$$0 \to \overset{+}{H}{}^{s,b}(E'_{0T}) \overset{(A,B)}{\to} \mathcal{H}^s \times \overset{+}{H}{}^{s-\beta_1-1/2,b}(G'_{1T})$$

$$\overset{\tilde{\Phi}_1}{\to} \overset{+}{H}{}^{s-(l+l_1)b,b}(E'_{2T}) \times \overset{+}{H}{}^{s-\beta_2-1/2,b}(G'_{2T}) \,,$$

where $\mathcal{H}^s = \{f \in H^{s-lb,b}(\tilde{E}'_1) : Pf \in H^{s-lb,b}(E'_1)\}$ and $\tilde{\Phi}_1$ is defined by $\tilde{\Phi}_1(f, g) = \Phi_1(Pf, g)$, is exact.

Example 3.9 . Let A be the operator in Example 3.2 and $a(x, t) = a(t)$. Also, let $\Phi_0 : C^\infty(\Omega', \mathbb{R}^3) \to C^\infty(\Omega', \mathbb{R}^6)) \times C^\infty(\partial\Omega', \mathbb{R}^1)$ be an operator of the form $u \mapsto (Au, u_n)$, where u_n is the normal component on Γ' of the vector field u.

Let $\Phi_1 : C^\infty(\Omega', \mathbb{R}^6) \times C^\infty(\partial\Omega', \mathbb{R}^1) \to C^\infty(\Omega', \mathbb{R}^4)$ be an operator of the form $(f, g, h) \mapsto (\partial g/\partial t - a\Delta g - \operatorname{curl} f, \operatorname{div} g)$, where $f, g \in C^\infty(\Omega', \mathbb{R}^3)$ and $h \in C^\infty(\partial\Omega', \mathbb{R}^1)$. We denote by $\mathcal{H}^s$ the space $\{(f, g) : f, g, \partial g/\partial x_k \in \overset{+}{H}{}^{s-2,2}(\Omega', \mathbb{R}^3), \ k = 1, 2, 3\}$. Then the complex

$$0 \to \overset{+}{H}{}^{s,2}(\Omega', \mathbb{R}^3) \overset{\Phi_0}{\to} \mathcal{H}^s \times \overset{+}{H}{}^{s-1/2,2}(\partial\Omega', \mathbb{R}^1) \overset{\Phi_1}{\to} \overset{+}{H}{}^{s-4,2}(\Omega', \mathbb{R}^4)$$

is exact.

In conclusion, we remark that the a priori estimates (3.10) and (3.11) can also be obtained in a more general case, without having to develop a complicated formal theory. The finite-dimensional nature of the kernel and the closed character of the image of a parabolic boundary value problem were first noted in Khachatryan (1977). Theorems 3.1 and 3.2 establish the exactness of the compatibility complexes, which is a considerably sharper result.

Chapter 4
Initial Boundary Value Problems
for Hyperbolic Systems

§1. Strictly Hyperbolic Operators

Let Ω be a smooth manifold and $\Omega' = \mathbb{R}^1 \times \Omega$; also, let E_i $(i = 0, 1)$ be smooth real finite-dimensional vector bundles over Ω, E_i' $(i = 0, 1)$ the inverse image of E_i with respect to the natural projection $\Omega' \to \Omega$ (see the localisation in Chap. 3, Sect. 1), and $A_0 : C^\infty(E_0') \to C^\infty(E_1')$ a normalised operator (in the sense of Definition 1.17). Denoting the points of the manifold Ω' by (t, x) $(t \in \mathbb{R}^1, x \in \Omega)$ and assuming that the covector dt is non-characteristic for A_0, we can rewrite A_0 in the form

$$A_0 y = \left(\frac{\partial y}{\partial t} + L_0\left(t, x, \frac{\partial}{\partial x}\right) y, M_0\left(t, x, \frac{\partial}{\partial x}\right) y \right), \qquad (4.1)$$

where the differential operators L_0 and M_0 do not contain differentiation with respect to t. Let $\sigma L_0(t, x, \xi) : Y_0 \to Y_1$ and $\sigma M_0(t, x, \xi) : Y_0 \to \bar{Y}_1$ (Y_0 and $Y_1 = Y_0 \oplus \bar{Y}_1$ are the fibres of the bundles E_0' and E_1', respectively) be the symbols of L_0 and M_0 at the point (t, x) for the covector $(0, \xi) \in T^*_{(t,x)}\Omega' = \mathbb{R}^1 \times T^*_x\Omega$. Since A_0 is normalised, from the commutation relations in Chap. 1 it follows that the operator $\sigma L_0(t, x, \xi)$ maps the space

$F_{t,x,\xi} = \operatorname{Ker} \sigma M_0(t, x, \xi)$ into itself. We denote by $\overline{\sigma L}_0(t, x, \xi)$ the restriction of $\sigma L_0(t, x, \xi)$ to the subspace $F_{t,x,\xi}$.

Definition 4.1. A normalised operator A_0 is called *strictly hyperbolic* if for all $(t, x) \in \Omega'$ and covectors $\xi \in T_x^* \Omega$ $(\xi \neq 0)$

(i) the dimension d of the space $F_{t,x,\xi}$ is independent of $(t, x) \in \Omega'$ and ξ;

(ii) $F_{t,x,\xi}$ has a basis consisting of eigenvectors of the operator $\overline{\sigma L}_0(t, x, \xi)$;

(iii) all the eigenvalues of $\overline{\sigma L}_0(t, x, \xi)$ are real and their multiplicities are independent of $(t, x) \in \Omega'$ and ξ.

Example 4.1. Let $A_0 : C^\infty(E_0') \to C^\infty(E_1')$ be a normalised operator of the form (4.1). If M_0 is elliptic with respect to the space variable, then A_0 is strictly hyperbolic.

Example 4.2. The operator that defines the left-hand side of Maxwell's equations

$$-\varepsilon_0 \varepsilon \frac{\partial E}{\partial t} + \operatorname{curl} H = j, \, \rho \mu_0 \mu \frac{\partial H}{\partial t} + \operatorname{curl} E = 0,$$

$$\operatorname{div} H = 0,$$

$$\varepsilon_0 \varepsilon \operatorname{div} E = \rho$$

is strictly hyperbolic.

For quadratic operators (that is, for which $\dim Y_0 = \dim Y_1$) we adopt the following definition of strict hyperbolicity (Agranovich (1971)).

Definition 4.2. An operator $A : C^\infty(E_0') \to C^\infty(E_1')$ of order k is called *strictly hyperbolic* if

(i) the covector dt is non-characteristic for A at every point $(t, x) \in \Omega'$;

(ii) all the roots λ of the equation

$$\det \sigma A(t, x, \lambda, \xi) = 0,$$

where $\sigma A(t, x, \lambda, \xi) : Y_0 \to Y_1$ is the symbol of A at $(t, x) \in \Omega'$ for the covector $(\lambda, \xi) \in T^*_{(t,x)} \Omega' = \mathbb{R}^1 \times T_x^* \Omega$, are real and distinct at every $(t, x) \in \Omega'$ for every non-zero $\xi \in T_x^* \Omega$.

The next assertion describes the connection between Definitions 4.1 and 4.2.

Proposition 4.1. *If* $A : C^\infty(E_0') \to C^\infty(E_1')$ *is a quadratic operator of order k, strictly hyperbolic in the sense of Definition 4.2, then there is a normalised operator A_0 equivalent to A (in the category* $\mathrm{D}(\Omega')$*) and strictly hyperbolic in the sense of Definition 4.1.*

§2. The Solvability of Initial Boundary Value Problems for Strictly Hyperbolic Operators

2.1. The Uniform Lopatinskij Condition. We now assume that the manifold Ω is smooth and has a smooth boundary Γ. We denote by $E|_{\partial\Omega'}$ the restriction of a bundle E to the manifold $\partial\Omega' = \mathbb{R}^1 \times \Gamma$ ($\partial\Omega'$ is the boundary of the manifold Ω'). In addition to the bundles E_0' and E_1' over Ω', we consider a finite-dimensional bundle G' over $\partial\Omega'$, which is the inverse image of some bundle G over Γ with respect to the projection $\partial\Omega' = \mathbb{R}^1 \times \Gamma \to \Gamma$; also, in addition to the operator $A_0 : C^\infty(E_0') \to C^\infty(E_1')$, we consider a boundary operator $B : C^\infty(E_0') \to C^\infty(G')$ of order zero and such that it coincides with the composition of the operator of restriction of the cross-sections of E_0' to $\partial\Omega'$ and the bundle morphism $\beta : E_0'|_{\partial\Omega'} \to G'$.

Let $A_0(\lambda, \pi, t, x, d/dx_n)$ be the ordinary differential operator with constant coefficients obtained from A_0 (in some coordinate system in the neighbourhood of the point $(t, x) \in \partial\Omega'$ in which dx_n is the conormal and the covectors dx_i ($i < n$) are cotangent to Γ) by removing the terms of order zero, fixing the coefficients at (t, x) and replacing $\partial/\partial x_j$ by $i\eta_j$ for $j = 1, 2, \ldots, n-1$ and $\partial/\partial t$ by $\lambda = \gamma + i\tau$. Let $S_+ = \{(\lambda, \eta) \in C^1 \times \mathbb{R}^{n-1} : |\lambda|^2 + |\eta|^2 = 1, \operatorname{Re}\lambda > 0\}$, and let $\bar{S}_+$ be the closure of S_+ in $C^1 \times \mathbb{R}^{n-1}$.

Definition 4.3. We say that the operator $(A_0, B) : C^\infty(E_0') \to C^\infty(E_1') \times C^\infty(G')$ satisfies the *homogeneous Lopatinskij condition* if
(i) the dimension of the space $\operatorname{Ker} A_0(\lambda, \eta, t, x, d/dx_n)$ is independent of the points $(t, x) \in \Omega'$ and $(\lambda, \mu) \in \bar{S}_+$;
(ii) the boundary value problem on the semi-axis

$$A_0(\lambda, \eta, t, x, d/dx_n)y(x_n) = 0,$$
$$(By)|_{x_n=0} = g$$

is uniquely solvable in the space $\mathfrak{M}_+$ of functions that tend to zero as $x_n \to \infty$, for all points $(t, x) \in \partial\Omega'$, $(\lambda, \eta) \in S_+$ and vectors $g \in C^r$;
(iii) there is a constant $c > 0$ independent of $(t, x) \in \partial\Omega'$ and $(\lambda, \eta) \in S_+$ such that

$$|y(0)| \leq C|(By)|_{x_n=0}$$

for all functions $y(x_n) \in \operatorname{Ker} A_0(\lambda, \eta, t, x, d/dx_n) \cap \mathfrak{M}_+$.

2.2. The Spaces $H_\gamma^{q,s}(E')$ and $H_\gamma^s(G')$. Let $\mathbb{R}^{n+1}$ and Y be Euclidean spaces. We write the points in $\mathbb{R}^{n+1}$ in the form $(t, x) = (t, x_1, x_2, \ldots, x_n)$, where $t \in \mathbb{R}^1$ and $x \in \mathbb{R}^n$. We write the "dual" variables with respect to the Laplace-Fourier transformation in the form $(-i\lambda, \xi) = (-i\lambda, \eta, \xi_n)$, where $\lambda \in C^1$, $\xi \in \mathbb{R}^n$ and $\eta \in \mathbb{R}^{n-1}$. We denote by $\hat{y}(\lambda, \eta, \xi_n)$ and $\hat{y}(\lambda, \eta, x_n)$, respectively, the Fourier transform and the partial Fourier transform with respect to the variable $x' = (x_1, x_2, \ldots, x_{n-1})$ of the function $y(t, x)\exp(-\gamma t)$.

We denote by $H_\gamma^s(\mathbb{R}^{n+1}, Y)$ $(s, \gamma \in \mathbb{R}^1)$ the completion of the space $C_0^\infty(\mathbb{R}^{n+1}, Y)$ of smooth functions with compact support in $\mathbb{R}^{n+1}$ and taking values in Y, with respect to the norm

$$\|y\|_{s,\gamma}^2 = \int\limits_{\mathbb{R}^{n+1}} (\rho^2 + |\xi|^2)^s |\hat{y}(\lambda, \xi)|^2 d\xi\, d\tau,$$

where $\rho^2 = |\lambda|^2 = \tau^2 + \gamma^2$.

Let $\mathbb{R}_+^{n+1}$ be the half-space $\{(t, x) \in \mathbb{R}^{n+1} : x_n > 0\}$, and $\bar{\mathbb{R}}_+^{n+1}$ the closure of $\mathbb{R}_+^{n+1}$ in $\mathbb{R}^{n+1}$. For $q = 0, 1, 2, \ldots$ and $\gamma, s \in \mathbb{R}^1$, we denote by $H_\gamma^{q,s}(\mathbb{R}^{n+1}, Y)$ the completion of the space $C^\infty(\bar{\mathbb{R}}_+^{n+1}, Y)$ of smooth functions with compact support in $\bar{\mathbb{R}}_+^{n+1}$ and taking values in Y, with respect to the norm

$$\|y\|_{q,s,Y}^2 = \sum_{k=0}^{q} \int\limits_{x_n > 0} \left\| \frac{\partial^{q-k} y}{\partial x_n^{q-k}} \right\|_{s+k,\gamma}^2 dx_n.$$

We now choose a Euclidean structure on each of the bundles E_i' and G'. Then the spaces $H_\gamma^{q,s}(E_1')$ and $H_\gamma^s(G')$ are defined in the standard way, by means of a partition of unity (Agranovich (1971)).

2.3. The Solvability of Initial Boundary Value Problems for Strictly Hyperbolic Operators. Let

$$C^\infty(E_0') \xrightarrow{A_0} C^\infty(E_1') \xrightarrow{A_1} C^\infty(E_2') \xrightarrow{A_2} \cdots \tag{4.2}$$

be a compatibility complex for the nornalised operator A_0 constructed in Chap. 1. We set

$$\begin{aligned}
\mathcal{H}_{q,s,\gamma}^0 &= H_\gamma^{q,s}(E_0') \times H_\gamma^{s+q}(E_0'|_{\partial\Omega'}), \\
\mathcal{H}_{q,s,\gamma}^1 &= H_\gamma^{q,s}(E_1') \times H_\gamma^{q+s}(G'), \\
\mathcal{H}_{q,s,\gamma}^j &= H_\gamma^{q,s}(E_j') \quad (j = 1, 2, \ldots, m+1).
\end{aligned}$$

We define a family of unbounded operators $U_{q,s,\gamma}^j : \mathcal{H}_{q,s,\gamma}^j \to \mathcal{H}_{q,s,\gamma}^{j+1}$ $(j = 0, 1, \ldots, m)$ as follows.

(i) The domain $D(U_{q,s,\gamma}^0)$ of $U_{q,s,\gamma}^0$ consists of pairs of functions

$$(u, v) \in H_\gamma^{q+1,s}(E_0') \times H_\gamma^{q+s}(E_0'|_{\partial\Omega'}) \subset \mathcal{H}_{q,s,\gamma}^0$$

such that $u|_{\partial\Omega'} = V$. The operator $U_{q,s,\gamma}^0$ is defined on $D(U_{q,s,\gamma}^0)$ by $U_{q,s,\gamma}^0(u, v) = (A_0 U, \beta v)$.

(ii) The domain $D(U_{q,s,\gamma}^1)$ of $U_{q,s,\gamma}^1$ consists of pairs of functions $(f, g) \in H_\gamma^{q+1,s}(E_1') \times H_\gamma^{q+s}(G') \subset \mathcal{H}_{q,s,\gamma}^1$, and $U_{q,s,\gamma}^1$ is defined on it by $U_{q,s,\gamma}^1(f, g) = A_1 f$.

(iii) For $j > 1$, the domain $D(U_{q,s,\gamma}^j)$ of $U_{q,s,\gamma}^j$ coincides with the space $H_\gamma^{q+1,s}(E_j)$, and $U_{q,s,\gamma}^j$ is defined on it by $U_{q,s,\gamma}^j f = A_j f$, where $A_j :$

$C^\infty(E_j') \to C^\infty(E_{j+1}')$ are the operators in the compatibility complex (4.2) for A_0.

Proposition 4.2. (i) *The operator* $U_{q,s,\gamma}^j$ *has a closure* $\bar{U}_{q,s,\gamma}^j$ *for all* $j \geq 0$.
(ii) *The image of* $\bar{U}_{q,s,\gamma}^{j-1}$ *belongs to the domain* $D(\bar{U}_{q,s,\gamma}^j)$ *of* $\bar{U}_{q,s,\gamma}^j$.
(iii) *The equality*

$$\bar{U}_{q,s,\gamma}^k \circ \bar{U}_{q,s,\gamma}^{k-1} y = 0$$

holds for all $y \in D(\bar{U}_{q,s,\gamma}^{k-1})$.

Thus, the complex of closed linear unbounded operators

$$0 \to \mathcal{H}_{q,s,\gamma}^0 \overset{\bar{U}_{q,s,\gamma}^0}{\to} \mathcal{H}_{q,s,\gamma}^1 \overset{\bar{U}_{q,s,\gamma}^1}{\to} \mathcal{H}_{q,s,\gamma}^2 \overset{\bar{U}_{q,s,\gamma}^2}{\to} \ldots \tag{4.3}$$

is well defined.

Theorem 4.1. *Let* $A_0 : C^\infty(E_0') \to C^\infty(E_1')$ *be a normalised strictly hyperbolic operator and* $B : C^\infty(E_0') \to C^\infty(G')$ *a boundary operator of order zero. Also, suppose that*
(i) *the coefficients of* A_0 *and* B *are independent of* t;
(ii) *the conormal to the boundary* $\partial\Omega'$ *of the manifold* Ω' *is a noncharacteristic covector for* A_0;
(iii) *the operator* $(A_0, B) : C^\infty(E_0') \to C^\infty(E_1') \times C^\infty(G')$ *satisfies the homogeneous Lopatinskij condition.*
Then for any $q = 0, 1, 2, \ldots$ *and* $s \in \mathbb{R}^1$ *there is a number* $\gamma_0(q, s) > 0$ *such that the complex (4.3) is exact for all* $\gamma > \gamma_0(q, s)$.

Theorem 4.1 remains valid if the $\mathcal{H}_{q,s,\gamma}^k$ are replaced by the spaces $\overset{0}{\mathcal{H}}_{q,s,\gamma}^k$ of functions that are equal to zero for $t > 0$. These spaces correspond to the Cauchy problem with homogeneous initial conditions.

The first condition in Theorem 4.1, which requires the coefficients of A_0 and B to be independent of t, may be relaxed without detriment to the statement of the theorem if we assume that these coefficients are independent of t for t large.

Additional Comments. Coupled Systems

5.1. Along with physical processes whose mathematical description leads to typical systems of partial differential equations (such as elliptic, parabolic and hyperbolic ones), situations have long been investigated where two (or more) typical processes interacting with each other lead to mathematical models that do not fit the above classification. For example, the interaction of heat conduction with elastic waves leads to the system of thermoelasticity, where the rather complex behaviour of the unknown functions does not permit us to apply the technique of typical systems. Fixing our attention on one of the intervening processes leads to the consideration of overdetermined systems.

Indeed, let $A : C^\infty(\Omega, Y) \to C^\infty(\Omega, W)$ be a linear differential operator, and suppose that $Y = Y_1 + Y_2$. Then $Ay = A_1 y_1 + A_2 y_2$ (where $A_i : C^\infty(\Omega, Y_i) \to C^\infty(\Omega, W)$, $i = 1, 2$). If A is a quadratic operator (that is, $\dim Y = \dim W$), then the operators A_1 and A_2 are obviously overdetermined. The study of the problem $A_1 y_1 = f_1$ may be useful for the original one $Ay = f$. This also applies to boundary value problems. We consider examples.

Example 5.1. In the domain $\Omega' = \Omega \times \mathbb{R}^1$, where $\Omega \subset \mathbb{R}^3$, we consider the system of equations

$$
A(u, v) = \begin{cases}
\dfrac{\partial^2 v}{\partial t^2} - \alpha \Delta v - \beta \operatorname{grad} \operatorname{div} v + \gamma \operatorname{grad} u = f_1, \\[2mm]
\dfrac{\partial u}{\partial t} - a \Delta u + b \dfrac{\partial}{\partial t} \operatorname{div} v = f_2,
\end{cases}
\tag{5.1}
$$

where v, $f_1 \in C^\infty(\Omega', \mathbb{R}^3)$, u, $f_2 \in (\Omega', \mathbb{R}^1)$, and α, β, γ, a and b are positive constants. The equations of thermoelasticity, for example, reduce to such a system. The system (5.1) has the form $A_1 u + A_2 v = f$, where $f = (f_1, f_2)$ and the operators $A_1 : C^\infty(\Omega', \mathbb{R}^1) \to C^\infty(\Omega', \mathbb{R}^4)$ and $A_2 : C^\infty(\Omega', \mathbb{R}^3) \to C^\infty(\Omega', \mathbb{R}^4)$ are, respectively, defined by

$$
A_1 : u \mapsto \left(\gamma \operatorname{grad} u, \frac{\partial u}{\partial t} - a \Delta u \right),
$$

$$
A_2 : v \mapsto \left(\frac{\partial^2 v}{\partial t^2} - \alpha \Delta v - \beta \operatorname{grad} \operatorname{div} v, b \frac{\partial}{\partial t} \operatorname{div} v \right).
$$

A_1 is an overdetermined parabolic operator; A_2 is also overdetermined.

Overdetermined systems may also occur in a more complex situation.

Example 5.2. The system of equations of magnetohydrodynamics can be written in the following form. Let Ω be a domain in $\mathbb{R}^3$, S the boundary of Ω, and $\Omega_T = \Omega \times [0, T]$. Also, let H, v, f and j be functions from Ω_T to $\mathbb{R}^3$, p a function from Ω_T to $\mathbb{R}^1$, and a a function from $S \times [0, T]$ to $\mathbb{R}^3$. We consider the initial boundary value problem

$$
\frac{\partial v}{\partial t} - \nu \Delta v = \sum_{k=1}^{3} v_k \frac{\partial v}{\partial x_k} - \frac{\mu}{\rho} \sum_{k=1}^{3} H_k \frac{\partial H}{\partial x_k} + \frac{1}{\rho} \operatorname{grad} \left(p + \frac{\mu H^2}{2} \right) = f,
$$

$$
\frac{\partial H}{\partial t} - \frac{1}{\sigma \mu} \Delta H - \operatorname{curl}[v \times H] = \frac{1}{\sigma \mu} \operatorname{curl} j,
$$

$$
\operatorname{div} v = 0, \quad \operatorname{div} H = 0,
\tag{5.2}
$$

$$
v|_S = a, \quad H_n|_S = 0, \quad \operatorname{curl}_\tau H_S = j\tau|_S,
$$

$$
v|_{t=0} = v_0(x), \quad H_{t=0} = H_0(x)
$$

for the unknown functions H, v and p. The investigation of (5.2) in Sakhaev (1975) involves the initial boundary value problem for a linear overdetermined parabolic system

$$\sigma\mu\frac{\partial H}{\partial t} - \Delta H = \operatorname{curl} j, \quad \operatorname{div} H = 0,$$

$$H_n|_S = 0, \quad \operatorname{curl}_\tau H|_S = j\tau|_S, \tag{5.3}$$

$$H|_{t=0} = H_0(x).$$

It is shown that (5.3) is solvable in the spaces W_p^{kl}, $p \geq 3$, and estimates for the norm of the solution are derived, which are similar to (3.9) (but in the norms of W_p^{kl}, $p \geq 3$). These results are used in Sakhaev (1975) to prove that the problem (5.2) is solvable.

In a situation like that in Example 5.1 the use of the theory of overdetermined systems can be extended. First we consider an abstract algebraic situation.

5.2. Let $\mathfrak{A} = \{\operatorname{Ob}\mathfrak{A}, \operatorname{Mor}\mathfrak{A}\}$ be an Abelian subcategory of the category of vector spaces and linear mappings, and let H_i $(i = 1, 2, 3)$ be objects and $A_i : H_i \to H_3$ $(i = 1, 2)$ morphisms in $\mathfrak{R}$.

Let

$$Ay = A_1 y_1 + A_2 y_2 = f \tag{5.4}$$

be a linear equation, where $y = (y_1, y_2) \in H_1 \oplus H_2$ and $f \in H_3$. In order to describe the conditions required for the unique solvability of (5.4), we also consider the equation

$$A_1 y_1 = \varphi. \tag{5.5}$$

Generally speaking, the morphism A_1 is overdetermined in $\mathfrak{A}$; therefore, (5.5) is solvable only if φ satisfies the solvability conditions $A_1' \varphi = 0$, where A_1' is some morphism in $\mathfrak{A}$. Consequently, the solvability of the equation

$$A_1' A_2 y_2 = A_1' f \tag{5.6}$$

is a necessary condition for the solvability of (5.4).

We denote by $\tilde{A}_2$ the morphism $A_1' A_2$ which, in general, is also overdetermined, so that for the existence of a y_2 satisfying (5.6), $A_2' f$ must satisfy the compatibility conditions $\tilde{A}_2' A_2' f = 0$.

Proposition 5.1. *If the complexes*

$$0 \to H_1 \xrightarrow{A_1} H_2 \xrightarrow{A_1'} H_4, \tag{5.7}$$

$$0 \to H_2 \xrightarrow{\tilde{A}_2} H_4 \xrightarrow{\tilde{A}_2'} H_5, \tag{5.8}$$

where $\tilde{A}_2 = A_1' A_2$ and H_i are some objects in the category $\mathfrak{A}$, are exact in $\mathfrak{A}$, then the equation (5.4) has a unique solution $y = (y_1, y_2)$ if and only if $\tilde{A}_2' A_1' f = 0$. In particular, if the morphism A is determined (in $\mathfrak{A}$), then (5.4) is uniquely solvable for any $f \in H_3$.

5.3. We now go over to differential equations. To illustrate the problems arising when the scheme in Sect. 5.2 is applied, we consider the case of a manifold without boundary. In this case, the category $\mathfrak{A}$ in Sect. 5.2 becomes the category $D(\Omega'_+)$ (in Example 1.1, where $\Omega'_+ = \Omega \times \mathbb{R}^1_+$, Ω is a compact manifold without boundary, and $\overset{0}{C}(E') = \{f \in C^\infty(E') : (\partial^k f/\partial t^k)|_{t=0} = 0,\ k = 0, 1, \ldots\}$.

If A is a parabolic or hyperbolic operator, then the compatibility complex

$$0 \to \overset{0}{C}(E'_0) \overset{A}{\to} \overset{0}{C}(E'_1) \overset{A_1}{\to} \overset{0}{C}(E'_2) \tag{5.9}$$

is exact in the category $D(\Omega')$; this follows from the results in Chap. 3 and Chap. 4 as a special case for manifolds without boundary.

We introduce another class of differential operators for which the compatibility complex is exact. First we remark that if C is a prabolic operator, then it has a left inverse of the form $L + R$, where L is a pseudodifferential operator and R an operator with a small norm (in the spaces $\overset{+}{H}{}^{a,b}$ defined in Chap. 3). The principal b-homogeneous symbol of L is equal to the left inverse of the principal b-homogeneous symbol of C, and in local coordinates it has the form $p^{-1}(\xi,\tau)L(\xi,\tau)$, where $p(\xi,\tau)$ is a polynomial that does not vanish for $\xi \in \mathbb{R}^n$ and $\tau \in \mathbb{C}^-$, and $L(\xi,\tau)$ is a matrix of polynomials (such an operator is constructed in Khachatryan (1977)). This symbol, which we denote by $\sigma_c^{-1}(x,t,\xi,\tau)$, belongs to the class of symbols N_ρ^{-m}, $m = \operatorname{ord} C$, $\rho = b^{-1}$, defined in Taylor (1981).

Definition 5.1. A symbol $a(x,t,\xi,\tau)$ is of *class N_ρ^m* if for $|\xi| < |\tau|$

$$|D_{x,t}^\beta D_\xi^\gamma D_\tau^r a(x,t,\xi,\tau)| \le C_{r,\beta,\gamma}\tau^{m-r}(|\tau|^\rho + |\xi|)^{-|\gamma|}$$

and $a \in S_{1,0}^m$ on any set $\Psi \subset T^*\Omega'$ that does not intersect the set $\{(x,t,\xi,\tau) \in T^*\Omega' \setminus 0 : \xi = 0\}$.

Definition 5.2. A differential operator A is of *class $\mathfrak{B}$* if $A = MC + K$, where M, C and K are differential operators and

(i) C is a parabolic operator (with compatibility operator C');

(ii) the operator $u \mapsto (Mu, C'u)$ is strictly hyperbolic;

(iii) K is a term of "lower order" in the sense that the symbol $K(x,t,\xi,\tau)\sigma_c^{-1}(x,t,\xi,\tau)$ is of class N_p^r, where $r \le \operatorname{ord} B - 1$ and $K(\cdot)$ is the full symbol of K.

Proposition 5.2. *If A is a regular differential operator of class $\mathfrak{B}$, then the compatibility complex (5.9) is exact.*

We consider a system of differential equations of the form (5.4).

Definition 5.3. A system of the form (5.4) is called a *coupled parabolic-hyperbolic system (PH-system)* if A_1 is a parabolic operator and $\tilde{A}_2 = A'_1 A_2$ is an operator of class $\mathfrak{B}$ (A'_1 is a compatibility operator for A_1).

The next assertion follows from the exactness of the complex (5.9) for parabolic, hyperbolic and class $\mathfrak{B}$ operators, and Proposition 5.1.

Proposition 5.3. *If* (5.4) *is a determined coupled PH-system, then it is uniquely solvable in* $\overset{0}{C}$*-crosss-sections for any* $\overset{0}{C}$*-cross-section* f.

5.4. Now let Ω be a compact smooth manifold with a smooth boundary Γ. We denote by Ω'_+ and Γ'_+ the manifolds $\Omega \times \mathbb{R}^1_+$ and $\Gamma \times \mathbb{R}^1_+$, respectively. We also denote by $\overset{0}{C}(E')$ and $\overset{0}{C}(G')$ the subspaces of $C^\infty(E')$ and $C^\infty(G')$ consisting of the cross-sections equal to zero over $\Omega \times \mathbb{R}^1_-$ and $\Gamma \times \mathbb{R}^1_-$, where $\mathbb{R}^1_- = (-\infty, 0]$.

Let $(A, B) : \overset{0}{C}(E'_0) \to \overset{0}{C}(E'_1) \times \overset{0}{C}(G'_1)$ be a boundary value problem operator and

$$0 \to \overset{0}{C}(E'_0) \overset{(A,B)}{\to} \overset{0}{C}(E'_1) \times \overset{0}{C}(G'_1) \overset{\Phi_1}{\to} \overset{0}{C}(E'_2) \times \overset{0}{C}(G'_2) \tag{5.10}$$

its compatibility complex. If (A, B) satisfies the conditions in the theorems in Chap. 3 or Chap. 4, then the complex (5.10) is exact. In the parabolic case, the left inverse of (A, B) is an operator of the form $L + R$, where L is a Boutet de Monvel operator and R an operator with a small norm in the spaces $\overset{+}{H}{}^{s,b}$ defined in Chap. 3, Sect. 5.

The principal b-homogeneous interior and boundary symbols of L are constructed explicitly in terms of those of (A, B) (Fel'dman (1987)). When the boundary operator is itself a trace operator in the Boutet de Monvel algebra, the procedure for constructing the symbol of L remains unchanged. This technique can also be applied if the coerciveness condition is relaxed: the complex in Definition 3.8 needs to be exact only in the term $\operatorname{Ker} A(\tilde{x}, \eta) \cap \mathfrak{M}^+$. A boundary value problem (A, B) with a parabolic operator A, which satisfies this weakened coerciveness condition is called a *parabolic boundary value problem*.

We need some additional notation. Let $L : (f, g) \mapsto L_1 f + L_2 g$ be the left inverse of a parabolic boundary value problem operator (C, E) (f is a cross-section of a bundle over Ω' and g of one over Γ'), and C' a b-compatibility operator for C.

Definition 5.4. The boundary value problem operator (A, E) is of *class* $\mathfrak{B}$ if

(i) $A = MC + K$ is of class $\mathfrak{B}$ in the sense of Definition 5.3;
(ii) (C, E) is a parabolic boundary value problem operator;
(iii) the hyperbolic operator $N : u \mapsto (Mu, C'u)$ is determined and its corresponding boundary value problem $u \mapsto EL_2 u|_{T \times \mathbb{R}}$ is uniquely solvable in the space of smooth cross-sections.

We remark that condition (iii) is satisfied, for example, if the boundary value problem mentioned in it satisfies the condition of uniform coerciveness. To verify this condition we need to know only the principal symbol of L_1, which can be written out explicitly.

We consider the boundary value problem

$$Ay = A_1 y_1 + A_2 y_2 = f, \quad By = B_1 y_1 + B_2 y_2 = g, \qquad (5.11)$$

where A_i are differential and B_i boundary operators. We denote by Φ_1 a b-compatibility operator for (A_1, B_1); Φ_1 is defined by $(f, g) \mapsto (A_1' f, \Phi_1^{21} f + \Phi_1^{22} g)$.

Definition 5.5. The problem (5.11) is called a *parabolic-hyperbolic boundary value problem* (*PH-problem*) if A_1 is a parabolic operator, the boundary value problem operator (A_1, B_1) is b-regular and satisfies the coerciveness condition (Definition 3.8), and the boundary value problem operator $y_2 \mapsto (A_1' A_2 y_2, [(\Phi_1^{21} A_2 y_2)|_{\Gamma'} + \Phi_1^{22}(B_2 y_2)|_{\Gamma'}])$ is of class $\mathfrak{B}$.

Theorem 5.1. *If (A, B) is a regular boundary value problem operator that is determined in the category* $\mathrm{DB}(\Omega_+', \Gamma_+')$*, and (5.11) is a PH-problem, then (5.11) has a unique solution for any f and g in spaces of $\overset{0}{C}$-cross-sections corresponding to bundles over Ω' and Γ'.*

Consequently, the above scheme enables us to reduce the investigation of the solvability of a PH-problem to that of a boundary value problem for a hyperbolic operator.

Apart from applying it to prove solvability theorems, we can also use this technique of splitting the original non-typical problem into overdetermined "typical" ones to describe the wave fronts of the solutions of PH-problems (Samborski and Fel'dman (1987)), in particular, in the case of boundary value problems for the equation of thermoelasticity with non-tangent bicharacteristics.

Example 5.3. Let Ω be a compact domain in $\mathbb{R}^3$ with a smooth boundary Γ', and A the operator of the system (5.1) in Example 5.1. For A we consider the boundary value problem

$$A(u, v) = f,$$
$$u|_{\Gamma'} = g_1, \quad v|_{\Gamma'} = g_2, \qquad (5.12)$$

where $f = (f_1, f_2)$, $g_1 \in \overset{0}{C}(\Gamma', \mathbb{R}^1)$ and $g_2 \in \overset{0}{C}(\Gamma', \mathbb{R}^3)$. Also, let A_1 be the parabolic operator defined in Example 5.1. This operator has the 2-compatibility operator $\hat{A}_1' : (\varphi_1, \varphi_2) \mapsto (j_\Omega^1 \operatorname{curl} \varphi_1, \partial \varphi_1 / \partial t - a \Delta \varphi_1 - \gamma \operatorname{grad} \varphi_2)$, where $\varphi_1 \in \overset{0}{C}(\Omega_+', \mathbb{R}^3)$ and $\varphi_2 \in \overset{0}{C}(\Omega_+', \mathbb{R}^1)$. We show that the conditions

$$\frac{\partial \varphi_1}{\partial t} - a \Delta \varphi_1 - \gamma \operatorname{grad} \varphi_2 = 0,$$
$$(\operatorname{curl} \varphi_1)_\tau|_{\Gamma'} = 0 \qquad (5.13)$$

yield $\operatorname{curl} \varphi_1 = 0$ in Ω'. We denote $\operatorname{curl} \varphi_1$ by F. Then F belongs to $\overset{0}{C}(\Omega_+', \mathbb{R}^3)$. From the definition of F it follows that

$$\frac{\partial F}{\partial t} - a\Delta F = 0, \tag{5.14}$$
$$\operatorname{div} F = 0.$$

This is an overdetermined parabolic system. Adjoining the boundary condition $F_\tau|_{\Gamma'} = 0$ to it, we arrive at a coercive boundary value problem. The unique solution of this problem is $F \equiv 0$, that is, $\operatorname{curl}\varphi_1 \equiv 0$. Hence, the equation $\hat{A}_1(\varphi_1, \varphi_2) = 0$ in Ω' is equivalent to the boundary value problem $\hat{\Phi}(\varphi_1, \varphi_2) = 0$, where

$$\hat{\Phi}(\varphi_1, \varphi_2) = \left(\frac{\partial \varphi_1}{\partial t} - a\Delta\varphi_1 - \gamma \operatorname{grad} \varphi_2, (\operatorname{curl}\varphi_2)_\tau|_\Gamma \right).$$

Consequently, when we apply the scheme described in Sect. 5.3 and Sect. 5.4, we may assume that the differential boundary operator $\hat{\Phi}$ is a compatibility operator for A_1. It can be verified directly that (5.12) is a PH-problem (the condition (iii) in Definition 5.4 may be replaced by a weaker formulation, namely, $(N, EL_2)u \in C^\infty$ implies that $u \in C^\infty$). The operators M, C and K in Definition 5.2 are, respectively, $M : v \mapsto (\partial^2 v/\partial t^2 - \alpha\Delta v - \beta \operatorname{grad}\operatorname{div} v)$, $C : y \mapsto (\partial y/\partial t - a\Delta y)$, where $y \in \overset{0}{C}(\Omega', \mathbb{R}^3)$, and $K : v \mapsto (\partial/\partial t) \operatorname{grad}\operatorname{div} v$. A compatibilty operator for (A_1, B_1) has the form

$$(\varphi_1, \varphi_2, g_1, g_2) \mapsto \left(\frac{\partial \varphi_1}{\partial t} - a\Delta\varphi_1 - \gamma \operatorname{grad} \varphi_2, g_2, \operatorname{grad}_\Gamma g_1 - \varphi_1\tau, \right.$$
$$\left. \frac{\partial g_1}{\partial t} - a\Delta_{\Gamma'} g_1 - \frac{a}{\gamma} \frac{\partial \varphi_{1n}}{\partial n}\bigg|_{\Gamma'} - \varphi_2|_{\Gamma'} \right).$$

Comments on the References

General systems of quasilinear partial differential equations were studied by Riquier and Cartan (Cartan (1945)). Cartan formulated the property of involutiveness in the language of Pfaffian systems, which, as shown by Kuranishi (1967), is equivalent to Definition 1.13 introduced by Kuranishi himself. Cartan also assumed that the operation of extension of systems in a sufficiently large class leads to involutive systems. This was later proved by Rashevskij (1947) in a particular case, and by Kuranishi (1967) in the general case. In the 1960s, the work of Spencer, Quillen and Goldschmidt recast the formal theory into an elegant mould. The definition of involutiveness given in Chap. 1, Sect. 3.4 is due to Spencer. The equivalence of Definitions 1.11 and 1.13 was shown by J-P. Serre. Definition 1.14 was given by Dudnikov and is equivalent to those of Cartan, Kuranishi and Spencer (although the proof of this is non-trivial).

The importance of the commutation relations was indicated by Guillemin (1968). The generalisations of these relations to quasi-regular covectors, and their use in the explicit procedure for constructing compatibility operators and for proving the existence theorems discussed in Chap. 1, Sect. 3.6–3.8, have been taken from Samborski (1981a,b,1983).

Techniques for constructing compatibility operators in the case of differential operators with constant coefficients can be found in Gudovich and Krein (1974) and Solonnikov (1971,1974).

The construction of Spencer complexes has been taken from Spencer (1969), where it is explained in more detail (see also Pommaret (1978)).

Various proofs of the local exactness of a compatibility complex in the real-analytic case (Spencer's result) can be found in Goldschmidt (1967), Pommaret (1978) and Spencer (1969). The proof given in this survey is a trivial corollary to the commutation relations (Samborski (1981a,b)).

The equivalence relation in Definition 1.5, Chap. 1, in the form applied constantly throughout the survey to differential and differential boundary operators, has been taken from Samborski (1983,1984a), although cochain equivalence (see MacLane (1963)) was used extensively in various situations in the work of Spencer's group.

The results in Chap. 1, Sect. 4 and the DB-operators have been taken from Samborski (1981b); the fundamental theorems in Chap. 2 and the elliptic theory (Theorems 2.5 and 2.6)—from Samborski (1984a) (a particular case without overdetermination on the boundary can be found in Dudnikov and Samborski (1981)); the results in Chap. 3 (the parabolic theory)—from Samborski (1983,1984b) and Samborski and Fel'dman (1987) (Sect. 6 in Chap. 2—the additional comments—contains information from Boutet de Monvel (1971); see also the monograph Rempel and Schulze (1982) and the survey Brenner and Shagorodsky (in press)). The results in Chap. 4 (the hyperbolic theory) belong to Dudnikov (1985a), and the additional comments to Samborski and Fel'dman (1987).

References*

Agranovich, M.S. (1971): Boundary value problems for systems with a parameter. Mat. Sb., Nov. Ser. *84*, 27–65. English transl.: Math. USSR, Sb. *13*, 25–64 (1971). Zbl. 207,108

Atiyah, M., Bott, R. (1967): A Lefschetz fixed point formula for elliptic complexes. I. Ann. Math., II. Ser. *86*, 374–407. Zbl. 161,432

Bourbaki, N. (1971): Variétés Différentielles et Analytiques. Hermann, Paris. Zbl. 217,204

Boutet de Monvel, L. (1971): Boundary problems for pseudodifferential operators. Acta Math. *126*, 11–51. Zbl. 206,394

---

* For the convenience of the reader, references to reviews in Zentralblatt für Mathematik (Zbl.), compiled by means of the MATH database, have, as far as possible, been included in this bibliography.

Brenner, A.V., Shagorodsky, E.M. (in press): Boundary Value Problems for Elliptic Pseudodifferential Operators. In: Itogi Nauki Tekh., Sovr. Probl. Mat., Fundam. Napr. *79*, VINITI, Moscow. English transl. in: Encycl. Math. Sci. *79*, Springer, Berlin Heidelberg New York (in preparation)

Cartan, E. (1945): Les Systèmes Différentiels Extérieurs et Leurs Applications Géometriques. Paris

Dudnikov, P.I. (1985a): On the regularity of the solutions of boundary value problems for overdetermined systems of equations of elliptic type. Preprint no. 2678-Uk85. Kiev Politekh. Inst., Kiev

Dudnikov, P.I. (1985b): The solvability of initial boundary value problems for overdetermined systems of partial differential equations of hyperbolic type. Preprint no. 2679-Uk85. Kiev Politekh. Inst., Kiev

Dudnikov, P.I., Samborski, S.N. (1981): On Noetherian boundary value problems for overdetermined systems of partial differential equations. Preprint no. 81. Inst. Mat. Akad. Nauk UkrSSR, Kiev (see also Dokl. Akad. Nauk SSSR *258*, 283–288 (1981)). English transl.: Sov. Math., Dokl. *23*, 508–512 (1981). Zbl. 548.35091

Dudnikov, P.I., Samborski, S.N. (1986): Boundary value problems for systems with curl in their dominant part. Ukr. Mat. Zh. *38*, 782–785. English transl.: Ukr. Math. J. *38*, 662–665 (1986). Zbl. 628.35032

Egorov, Yu.V., Shubin, M.A. (1987): Linear Partial Differential Equations. (Itogi Nauki Tekh., Sovr. Probl. Mat. *30*.) VINITI, Moscow. English transl.: Encycl. Math. Sci. *30*, Springer, Berlin Heidelberg New York (1992). Zbl. 657.35002

Fel'dman, M.A. (1987): Coercive boundary value problems for parabolic overdetermined systems with variable coefficients. Ukr. Mat. Zh. *39*, 493–500. English transl.: Ukr. Math. J. *39*, 398–404 (1987). Zbl. 649.35066

Goldschmidt, H. (1967): Existence theorems for analytic linear partial differential equations. Ann. Math., II. Ser. *86*, 246–270. Zbl. 154,351

Gudovich, I.S., Krein, S.G. (1974): Boundary Value Problems for Overdetermined Systems of Partial Differential Equations. Inst. Fiz. Mat. Akad. LitSSR, Vilnyus. Zbl. 344.35066

Guillemin, V. (1968): Some algebraic results concerning the characteristics of overdetermined partial differential equations. Am. J. Math. *90*, 270–284. Zbl. 162,406

Hörmander, L. (1963): Linear Partial Differential Operators. Springer, Berlin Göttingen Heidelberg. Zbl. 108,93

Khachatryan, A.G. (1977): Overdetermined parabolic boundary value problems. Zap. Nauchn. Semin. Leningr. Otd. Mat. Inst. Steklova *69*, 240–272. English transl.: J. Sov. Math. *8*, 171–193 (1978). Zbl. 355.35047

Krein, S.G., L'vin, S.Ya. (1985): Overdetermined and Sub-definite Elliptic Problems. Functional Analysis and Mathematical Physics. Novosibirsk, 106–116. Zbl. 584.35074

Krein, S.G., Shikhvatov, A.M. (1970): Linear differential equations on a Lie group. Funkts. Anal. Prilozh. *4*, 52–61. English transl.: Funct. Anal. Appl. *4*, 46–54 (1970). Zbl. 198,187

Kuranishi, M. (1967): Lectures on Involutive Systems of Partial Differential Equations. Publ. Soc. Math., Sao Paulo. Zbl. 163,120

Lewy, H. (1957): An example of a smooth linear partial differential equation without solution. Ann. Math., II. Ser. *66*, 155–158. Zbl. 78,81

L'vin, S.Ya. (1987): Overdetermined boundary value problems. Differ. Uravn. *23*, 1081–1084. Zbl. 662.35029

MacLane, S. (1963): Homology. Springer, Berlin Göttingen Heidelberg. Zbl. 133,265

Palamodov, V.P. (1967): Linear Differential Operators with Constant Coefficients. Nauka, Moscow. Zbl. 191,434. English transl.: Springer, Berlin Heidelberg New York 1970, Grundlehren Math. Wiss. *168*. Zbl. 191,434

Palamodov, V.P. (1968): Differential operators in the class of convergent power series, and Weierstrass's auxiliary lemma. Funkts. Anal. Prilozh. *2*, 58–69. English transl.: Funct. Anal. Appl. *2*, 235–244 (1968). Zbl. 176,457

Palamodov. V.P. (1969): Systems of Linear Differential Equations. (Itogi Nauki Tekh., Mat. Anal. 1968, 5-37.) VINITI, Moscow. English transl.: Progress Math. *10*, 1–35 (1971). Zbl. 197,364

Pommaret, J.F. (1978): Systems of Partial Differential Equations and Lie Pseudogroups. Gordon and Breach, New York London Paris. Zbl. 401.58006

Rashevskij, P.K. (1947): Geometric Theory of Partial Differential Equations. Nauka, Moscow. Zbl. 36,64

Rempel, S., Schulze, B.-W. (1982): Index Theory of Elliptic Boundary Problems. Akademie-Verlag, Berlin. Zbl. 504.35002

Sakhaev, Sh. (1975): An estimate for the solutions of a boundary value problem of magnetohydrodynamics. Tr. Mat. Inst. Steklova *127*, 58–75. English transl.: Proc. Steklov Inst. Math. *127*, 67–86 (1977). Zbl. 318.35041

Samborski, S.N. (1977): On the weak solvability of systems of nonlinear equations on Banach spaces. Ukr. Mat. Zh. *25*, 685–689. English transl.: Ukr. Math. J. *29*, 525–528 (1977). Zbl. 369.34027

Samborski, S.N. (1980): On the Cauchy problem for involutive systems of partial differential equations. Differ. Uravn. *16*, 516–521. English transl.: Differ. Equations *16*, 330–333 (1980). Zbl. 429.35007

Samborski, S.N. (1981a): On the formal properties of boundary value problems for overdetermined systems of partial differential equations. Dokl. Akad. Nauk UkrSSR *12*, 18–22. Zbl. 477.35067

Samborski, S.N. (1981b): Formal properties of boundary value problems for overdetermined systems of partial differential equations. Zap. Nauchn. Semin. Leningr. Otd. Mat. Inst. Steklova *110*, 203–216. English transl.: J. Sov. Math. *25*, 949–959 (1984). Zbl. 497.35068

Samborski, S.N. (1983): On overdetermined elliptic boundary value problems with a parameter, and parabolic boundary value problems. Dokl. Akad. Nauk SSSR *271*, 544-549. English transl.: Sov. Math., Dokl. *28*, 128–133 (1983). Zbl. 554.35089

Samborski, S.N. (1984a): Coercive boundary value problems for overdetermined systems. Elliptic problems. Ukr. Mat. Zh. *36*, 340–346. English transl.: Ukr. Math. J. *36*, 305–310 (1984). Zbl. 595.35081

Samborski, S.N. (1984b):Coercive boundary value problems for overdetermined systems. Parabolic systems. Ukr. Mat. Zh. *36*, 473–479. English transl.: Ukr. Math. J. *36*, 385–390 (1984). Zbl. 587.35070

Samborski, S.N., Fel'dman, M.A. (1985): On the coerciveness condition for overdetermined boundary value problems. Ukr. Mat. Zh. *37*, 616–622. English transl.: Ukr. Math. J. *37*, 500-505 (1985). Zbl. 625.35067

Samborski, S.N., Fel'dman, M.A. (1987): Coupled systems of partial differential equations (initial boundary value problems and the distribution of the singularities). Izv. Vyssh. Uchebn. Zaved., Mat. 1987, No. 10, 14–23. English transl.: Sov. Math. J. *31*, No. 10, 16–28 (1987). Zbl. 639.35056

Solonnikov, V.A. (1969): On the condition of being complementary for overdetermined systems. Zap. Nauchn. Semin. Leningr. Otd. Mat. Inst. Steklova *14*, 237–255. English transl.: Semin. Math. Steklov Math. Inst. *14*, 122–131 (1971). Zbl. 213,104

Solonnikov, V.A. (1971): Overdetermined elliptic boundary value problems. Zap. Nauchn. Semin. Leningr. Otd. Mat. Inst. Steklova *21*, 112–158. English transl.: J. Sov. Math. *1*, 477–512 (1973). Zbl. 251.35033

Solonnikov, V.A. (1974): On a class of Noetherian overdetermined elliptic boundary value probems. Zap. Nauchn. Semin. Leningr. Otd. Mat. Inst. Steklova *47*, 138–154. English transl.: J. Sov. Math. *9*, 244–259 (1978). Zbl. 354.35037

Solonnikov, V.A. (1975): The study of overdetermined elliptic boundary value problems in Golovkin fractional spaces. Tr. Mat. Inst. Steklova *127*, 93–114. English transl.: Proc. Steklov Inst. Math. *127*, 109–134 (1977). Zbl. 319.35034
Spencer, D.C. (1969): Overdetermined systems of linear partial differential equations. Bull. Am. Math. Soc., New Ser. *75*, 179–239. Zbl. 185,338
Sysoev, Yu.S. (1974): Linear non-homogeneous equations on a Lie group. Differ. Uravn. *10*, 364–366. English transl.: Differ. Equations *10*, 273–275 (1975). Zbl. 272.35006
Taylor, M.E. (1981): Pseudodifferential Operators. Princeton Univ. Press, Princeton, NJ. Zbl. 453.47026
Vinogradov, A.M., Krasil'shchik, I.S., Lychagin, V.V. (1986): Introduction to the Geometry of Nonlinear Differential Equations. Nauka, Moscow. English transl.: Gordon & Breach, New York 1986. Zbl. 592.35002
Wells, R.O. (1973): Differential Analysis on Complex Manifolds. Prentice-Hall. Zbl. 262.32005

II. Spectral Analysis of a Dissipative Singular Schrödinger Operator in Terms of a Functional Model

B. S. Pavlov

Translated from the Russian
by C. Constanda

Contents

Acknowledgement

The author is indebted to N. K. Nikolskij and V. Vasyunin for reading the English version of the text and suggesting some improvements.

§1. Introduction

1.1. Spectral Analysis in Terms of the Resolvent. Historically, the first general method in the spectral analysis of non-selfadjoint differential operators was the Riesz integral, complemented by the refined technique of estimating the resolvent on the contours that divide the spectrum. Using this method, Lidskij (1962) proved the summation over groups ("with brackets") of the spectral resolution of a general regular second order differential operator. Since then, the so-called "bases with brackets" have been studied extensively by his successors (see the references in Sadovnichij (1973)). Unfortunately, the arrangement of the "brackets", that is, the combination into one group of the sets of eigenvectors and root vectors corresponding to some neighbouring points of the spectrum, is defined non-uniquely and, to a large extent, non-constructively. Hence, as a rule, the assertions concerning bases with brackets have the character of existence theorems.

In the case of singular differential operators the situation is even worse: beside the continuous spectrum, here one may encounter points of accummulation of the eigenvalues at a finite distance. An attempt to estimate the resolvent of the operator on the contours that divide the spectrum near the set of such points leads to tremendous analytic difficulties. We can get a good idea of what these difficulties are if we recall that for regular operators, the corresponding estimates can be based on theorems describing the behaviour of the modulus of an entire function of a given order outside a fixed neighbourhood of the set of the roots (Lidskij (1962)). Analogous results for general analytic functions near the points of the boundary of the holomorphy domain are not available.

Obviously, this is the main reason why in earlier studies of the spectrum of the non-selfadjoint singular Schrödinger operator the investigators concentrated on the case of finitely many eigenvalues (Gel'fand (1952), Lyantse (1964a,b), Martirosyan (1957), and Naimark (1954)). However, even the exclusion of the accumulation points of the eigenvalues does not on its own guarantee that the standard technique of the Riesz integral can be applied to obtain spectral resolutions; this technique works only if the resolvent has linear growth near the spectrum σ, that is,

$$\|R_1\| \leq \{\mathrm{dist}(\lambda, \sigma)\}^{-1}\mathrm{const.} \tag{1}$$

Even in the simplest cases (in particular, for the one-dimensional Schrödinger operator with a complex potential with compact support) the continuous spectrum contains points—so-called *spectral singularities*—where the condition (1) is not satisfied. This does not allow us to employ the usual Riesz integral technique to construct spectral projections on the parts of the spectrum that contain such points.

In earlier work (see Martirosyan (1957) and Naimark (1954)) is was shown that coarse conditions which ensure that there are only finitely many eigenvalues also guarantee that there are only finitely many spectral singularities, and even that the growth of the resolvent at each of the latter is of finite order.

Apart from the smallness of the potential, no other effective conditions have been found so far under which the non-selfadjoint Schrödinger operator has no spectral singularities.

Methods were subsequently developed (see Lyantse (1964a,b)) for constructing unbounded spectral projections on the intervals of the spectrum containing spectral singularities of the multiple pole type. These methods reduce, in essence, to the isolation of spurious "eigenfunctions" and "root functions" at the points of the spectral singularities, which are not square-integrable and should therefore occur in a complete system in L_2 only as part of a package and not individually. Clearly, their artificial isolation means that the remaining part of the resolution loses its spectral character.

1.2. Spectral Analysis and the Characteristic Function. The futility of the attempts to construct a spectral resolution for a non-selfadjoint differential operator in the same way as for a selfadjoint one became clear when Marchenko (1960) showed that a spectral function in such a construction is a distribution, and that, by the same argument, the class of elements in the space which admit a spectral representation is automatically very restricted. The inflexibility of the exact conditions found soon after that in Pavlov (1961,1962) under which there are only finitely many eigenvalues and spectral singularities, and the examples constructed there of one-dimensional singular Schrödinger operators with a rich spectral structure signalled that the time had come to seek other approaches to the problem of spectral analysis.

By that time it had already become obvious that the analytic "tool" basis of the spectral analysis of non-selfadjoint operators was inadequate, reducing practically to the Cauchy integral. In his address to the International Congress of Mathematicians in Moscow in 1966, M.G. Krein predicted that any further progress in the spectral theory of non-selfadjoint operators would make use of the very latest developments in the theory of analytic functions.

And this is exactly what happened.

Even before the Moscow congress, the Hungarian mathematician Szökefalvi-Nagy and the Romanian mathematician Foiaş had published papers (see Szökefalvi-Nagy and Foiaş (1970)) in which they studied the spectral properties of an operator of a very simple form, which is in fact the universal model of a linear contraction in a Hilbert space. The basic parameter of this model, which carries full information about its properties, is not the resolvent of the operator, but a much simpler object, namely, its characteristic function, introduced earlier by Livšic (see Lidskij (1962)). This is a contracting function analytic in the unit disk in the plane of the spectral parameter. To the authors of the model it was extremely important that the properties of such functions had already been fully investigated. Their systematic study began at the turn of the century with the work of Nevanlinna. In 1926 Smirnov obtained a decisive result on factorisation, valid for a significantly larger class of functions with an integrable logarithm of the modulus (see Smirnov (1932)); every such function is obtained by multiplying three factors: a Blaschke factor, a singular inner function, and an outer function with the same modulus on the circle as the original one:

$$f(\zeta) = \prod_{l=1}^{\infty} \frac{\zeta_l - \zeta}{1 - \bar{\zeta}_l \zeta} \frac{\bar{\zeta}_l}{|\zeta_l|} \exp\left\{ - \int \frac{e^{i\varphi} + \zeta}{e^{i\varphi} - \zeta} d\mu \right\}$$

$$\times \exp\left\{ \frac{1}{2\pi} \int \frac{e^{i\varphi} + \zeta}{e^{i\varphi} - \zeta} [\lim_{r \to 1} \ln |f(re^{i\varphi})|] d\varphi \right\}, \quad f(\xi_l) = 0. \qquad (2)$$

This theorem was proved again later by Beurling, who used it as a basis for describing the invariant subspaces (Beurling (1949)) of the shift operator in the space of square-integrable functions on a circle. Later Potapov (1955) found a many-dimensional analogue of the factorisation (2), which Szökefalvi-Nagy and Foiaş (1970) used as a basis for the description of the spectral properties of a multiple shift.

1.3. Spectral Analysis and Scattering Theory. At the same time as the papers by Szökefalvi-Nagy and Foiaş, the work of the American mathematicians Lax and Phillips came to light, on the energy decay in a bounded domain near a compact obstacle $\mathbb{R}^3 \setminus \Omega$ in acoustic scattering in an unbounded domain $\Omega \subset \mathbb{R}^3$ (Lax and Phillips (1967)). These authors describe the evolution of the wave field near the obstacle by means of a contraction semigroup $Z_t = \exp\{iBt\}$ in the Hilbert space of the Cauchy data $U = \{u_0, u_1\}$ of the acoustic problem with the *energy metric*

$$|U|_E^2 = \frac{1}{2} \int_{\Omega} (|\nabla u_0|^2 + |u_1|^2) d^3x.$$

It turns out that the corresponding scattering matrix S is an analytic contracting operator-function in the upper half-plane of the spectral parameter, which is the wave number.

The complex roots of the S-matrix—the resonances—coincide with the eigenvalues of the generator of the contraction semigroup Z_t, which is a dissipative operator B.

Almost immediately after this, Krein's students Adamyan and Arov (1966) decoded the non-standard definition of the S-matrix given by Lax and Phillips and wrote it in the usual way in terms of the wave operators. This led to the discovery of the remarkable fact that the scattering matrix is the characteristic function of the semigroup describing the evolution of the wave field near the obstacle. At the same time, a direct link was observed between the asymptotic properties of the wave field and its behaviour near the obstacle.

This abstract result was established in the general case, solely under the condition of orthogonality of the incoming and outgoing invariant subspaces $D_{\mp}$ of the given unitary group U_t:

$$U_t D_- \subset D_-, \ \ t < 0; \quad U_t D_- \to 0, \ \ t \to -\infty,$$
$$U_t D_+ \subset D_+, \ \ t > 0; \quad U_t D_+ \to 0, \ \ t \to \infty. \tag{3}$$

This makes it possible to use it not only in the case of a compact obstacle, but also in the general situation where the spectrum of the dissipative operator has a continuous component of Lebesgue type. The only requirement is that the dissipative operator in question should be the generator of some contraction semigroup obtained by compressing some unitary group on its *semi-invariant subspace K*, which is the orthogonal complement of the pair of mutually orthogonal incoming and outgoing subspaces, that is, $K = H \ominus \{D_- \oplus D_+\}$; in other words,

$$Z_t = e^{iBt} = P_K U_t | K.$$

This condition is satisfied in many interesting problems of mathematical and theoretical physics. Using the results of Szökefalvi-Nagy and Foiaş, Lax and Phillips, and Adamyan and Arov, we can reduce the dissipative operators arising in such problems to model ones by rewriting them in the spectral representation of the original unitary group (see Sect. 5). It is remarkable that in a number of cases we can, starting from the given dissipative operator B, construct a unitary group in a larger space, from which B can be obtained by means of the procedure described above. In this way, we succeed in finding an adequate solution to the problem of spectral analysis for a well-known classical object, namely the dissipative Schrödinger operator with an absorbent complex potential q such that $\operatorname{Im} q(x) = 2\rho^2(x) \geq 0$. It turns out that there is a canonical set of eigenfunctions of the absolutely continuous spectrum of this operator, obtained by compressing on the original space the solutions of the scattering problem for the generator of the unitary group constructed in this way—a so-called *dilation* of the original operator, which is a selfadjoint operator acting in a larger space. We can obtain various forms of functional models simply by writing the given operator in terms of different spectral representations of the dilation. Thus, if we use Joost's solutions as the eigenfunctions of the absolutely continuous spectrum of the dilation, then we obtain a symmetric form of model, described in Pavlov (1972, 1975b). This form was found to be convenient in many problems of spectral analysis (see Pavlov (1975c,1976,1977,1979b)) and was thoroughly investigated in the following years (Nikol'skij and Khrushchev (1987)). In particular, it was used as a basis for constructing a new universal model for a non-dissipative non-selfadjoint operator in a space with a positive metric (Naboko (1980)).

1.4. Spectral Analysis in Terms of a Functional Model—a Survey.
The view that the functional model of an operator is an adequate replacement for its spectral representation, which was widely held in St. Petersburg fifteen years ago (see, for example, Pavlov (1976)), is generally accepted these days. Conversely, now it has also become usual for the spectral representation of a selfadjoint (or unitary) operator to be called its functional model. The advantages of such a viewpoint for analysis are indisputable: passing to a

representation in which the operator has a model form, we discover analytic mechanisms that specify the geometric features of the operator as Hilbert space objects.

In this context, the most important operator-theoretic problems reduce to questions of the theory of analytic functions in the *Hardy classes* $H^2_\pm$, which consist of the square-integrable functions on the real axis admitting an analytic extension to the upper (lower) half-plane with a bounded norm on every line parallel to the real axis. Thus, the problem of expansion in the eigenfunctions of the discrete spectrum is replaced by an interpolation problem, that of completeness by one of factorisation, the question of joint completeness of the eigenfunctions of the operator and its adjoint by that of the positiveness of the angle between the corresponding subspaces in the Hardy class H^2_+, the problem of separability of the spectral components by that of bounded invertibility of a Toeplitz operator, which is the compression of the operator of multiplication by an analytic function (see Pavlov (1971b,1975b,1976), Khrushchev, Nikol'skij and Pavlov (1981), and Nikol'skij and Khrushchev (1987)). A remarkable consequence of the existence of such connections between the geometric (operator) problems and those of function theory was the opening of the latter to ideas and observations from the theory of differential operators, and even from theoretical physics. Thus, the use in Pavlov (1973c) of the condition of skew connection of pairs of subspaces in $L_2(0, \infty)$ in the study of exponential bases coincides in essence with the condition of completeness of the resonance states found in one of the early works on scattering theory by Regge (1958). Further analysis of this condition led to the discovery of a test for the basis property of exponentials on a closed interval (Pavlov (1979a)) and of a large set of corresponding factors for general systems of reproducing kernels (Khrushchev, Nikol'skij and Pavlov (1981)).

The aim of this survey is to introduce the reader to a number of ideas and methods growing on the fertile patch between the spectral theory of singular dissipative differential operators, the theory of functional models, the theory of analytic functions, and mathematical physics. The survey is structured as follows. In Sect. 2 we describe briefly a symmetric functional model of a dissipative operator. In Sect. 3 we discuss theorems on the separability of the spectral components, on eigenfunction expansion, and on the summation of spectral resolutions in the presence of spectral singularities, as well as results on the joint completeness of the eigenfunctions of an operator and its adjoint. Sect. 4 is devoted to the analysis of examples of dissipative operators arising in problems of resonance scattering. In Sect. 5 we describe the localisation of the spectrum of a one-dimensional singular Schrödinger operator, and use functional model methods to analyse a classical example of such an operator with a real potential and a complex boundary condition. In the last part of this section we go beyond the circle of problems with a finite non-selfadjointness defect and use the functional model methods to perform the spectral analysis of a three-dimensional dissipative Schrödinger operator.

Given the limited space, we do not touch on results obtained in the analysis of other interesting operators encountered in mathematical physics. In the specific domain of spectral analysis that forms the object of our study, these results are similar to those described below, and are obtained by means of analogous methods.

§2. The Functional Model

2.1. The Semigroup. Let S be a contracting operator-function, holomorphic in the upper half-plane $\mathbb{C}_+ = \{k : \operatorname{Im} k > 0\}$ and with values $S(k)$, $S(k) : E \to E$, in the ring of bounded operators over a complex Euclidean space[1] E, $\dim E < \infty$. We assume that this function is invertible at least at one point $k_0 = s_0 + i\tau_0$, $k_0 \in \mathbb{C}_+$. Then S is invertible everywhere on $\mathbb{C}_+$ except, possibly, on some discrete set $\sigma_p(S)$. We extend S to the lower half-space $\mathbb{C}_-$ by means of the symmetry principle, setting $S(k) = \{S^+(\bar{k})\}^{-1}$, $\operatorname{Im} k > 0$, $\bar{k} \notin \sigma_p$; the analytic function in $\mathbb{C}_- \cup \mathbb{C}_+$ constructed in this way is again denoted by S. We denote by $\sigma(S)$ the set of all the singularities of S^{-1} and call it the *spectrum of S*. Clearly, $\sigma_p(S) \subset \sigma(S)$; in general, the spectrum lies in the closed upper half-plane. According to well-known theorems, S has finite normal upper limit values almost everywhere on the real axis, which define a measurable bounded function. We denote these values by $S(s)$, that is,

$$S(s) = \lim_{\tau \to 0+} S(s + i\tau).$$

We consider the lineal of all measurable, complex-valued, two-component vector-functions $f = (f_1, f_2)$ for which the integral

$$\langle\!\langle f \rangle\!\rangle^2 = \frac{1}{2\pi} \int\limits_{-\infty}^{\infty} \left\langle \begin{pmatrix} I & S^+(s) \\ S(s) & I \end{pmatrix} \begin{pmatrix} f_0 \\ f_1 \end{pmatrix}, \begin{pmatrix} f_0 \\ f_1 \end{pmatrix} \right\rangle ds$$

is finite. The metric defined by this integral on the above lineal is degenerate; we denote by $H = L_2 \begin{pmatrix} I & S^+ \\ S & I \end{pmatrix}$ the quotient space with respect to the set of elements of zero norm. H is a separable Hilbert space. In what follows we identify the classes of H with its two-component representatives and write $f = (f_1, f_2)$, as before.

In H we consider the unitary *translation group* $U_t : f(s) \to \exp(ist)f(s)$, whose generator is the operator of multiplcation by the independent variable. The next assertion is easily verified if we take the analyticity of S into account.

---

[1] The majority of the results below also remain valid in the infinite-dimensional case, that is, when $\dim E = \infty$. We will remark on this at the appropriate places in the text.

Theorem 1. *The translation group has an incoming subspace $D_- = (0, H_-^2)$ and an outgoing subspace $D_+ = (H_+^2, 0)$ which are orthogonal and generate the whole of H, that is,*

$$\overline{\bigcup_{-\infty < t < \infty} U_t(D_- \oplus D_+)} = H.$$

The following assertion plays a fundamental role in our constructions.

Theorem 2 (Nagy and Foiaş (1970)). *The family of operators Z_t obtained by compressing the translation group on the shift-invariant subspace K, that is,*

$$K = H \ominus (D_- \oplus D_+),$$
$$Z_t = P_K U_t | K, \quad t \geq 0,$$

is a strongly continuous, completely non-unitary semigroup with a dissipative generator $B = \lim_{t \to 0}(it)^{-1}\{Z_t - I\}$.

The proof of the semigroup properties follows from the basic properties (3) of the incoming and outgoing subspaces. Indeed, the definition of K yields $P_K U_{t_1} | D_+ = 0$ for $t_1 \geq 0$ and $D_- U_{t_2} | K = \{P_K U_{-t_2} | D_-\}^+ = 0$ for $t_2 \geq 0$, since $U_{t_1} D_+ \subset D_+$, $U_{-t_2} D_- \subset D_-$, and $D_\pm \perp K$. Thus,

$$Z_{t_1} Z_{t_2} = Z_{t_1 + t_2} - P_K U_{t_1} P_{D_+} U_{t_2} | K - P_K U_{t_1} P_{D_-} U_{t_2} | K = Z_{t_1 + t_2}.$$

The full non-unitary character follows from the degeneracy condition, and the strong continuity is implied by the continuity of a unitary group. In turn, the strong continuity of Z_t implies that this semigroup has a dissipative generator B, that is, $Z_t = \exp(iBt)$, $t > 0$.

Along with the continuous group $\{U_t\}$, sometimes it is convenient to consider the discrete group $\{V^n\}$ generated by multiplication by the fraction $\zeta = (k - i)(k + 1)^{-1}$. This group is unitary in the corresponding space of two-component functions on a circle weight function $\hat{S}(\zeta) = S(i(1 + \zeta)(1 - \zeta)^{-1})$.

The compression of V^n on the corresponding shift-invariant subspace $\hat{K}$ is a semigroup $\hat{T}^n = P_{\hat{K}} V^n | \hat{K}$ whose generating element $\hat{T} = P_{\hat{K}} V | \hat{K}$ is the Cayley transform of the generator $T = (B - iI)(B + iI)^{-1}$ of the continuous semigroup.

2.2. The Contracting Operator-Function.

The fundamental parameter that yields the properties of both semigroups is a contracting function S in the half-plane, or, equally, the corresponding contracting function in a disk.

We denote by Δ_T and Δ_{T^+} the defect operators of the contraction T; their images $\Delta_T \hat{H}$ and $\Delta_{T^+} \hat{H}$ are called *defect subspaces*, and the dimensions of the images are called the *non-selfadjointness defects*. In our case, the index of the contraction T is zero and the defects are both equal to d.

Fundamental Theorem (Livšic-Szökefalvi-Nagy-Foiaş). *The character-istic function θ_T of the contraction T, defined by*

$$\theta_T(\zeta) = \Delta_{T+}^{-1}(1 - \zeta T^+)^{-1}(\zeta - T)\Delta_T,$$

coincides with the fundamental parameter $\hat{S}(\zeta)$ up to constant right and left isometric factors mapping the defect spaces on to $\mathbb{C}^d$. Every contracting operator T' whose characteristic function coincides with $\hat{S}$ up to constant isometric factors is unitarily equivalent to T. Conversely, if a contracting operator is unitarily equivalent to T, then its characteristic function coincides with $S(\zeta)$ up to constant isometric factors.

The proof of this classical theorem can be found in Szökefalvi-Nagy and Foiaş (1970), where other spectral representations of the translation group are used, namely

$$\begin{pmatrix} f_0 \\ f_1 \end{pmatrix} \xrightarrow{\mathcal{T}_-} \begin{pmatrix} S & I \\ I & 0 \end{pmatrix} \begin{pmatrix} f_0 \\ f_1 \end{pmatrix}, \quad L_2 \begin{pmatrix} I & S^+ \\ S & I \end{pmatrix} \xrightarrow{\mathcal{T}_-} L_2 \begin{pmatrix} I & 0 \\ 0 & I - S^+S \end{pmatrix},$$

$$D_- = \begin{pmatrix} 0 \\ H_-^2 \end{pmatrix} \xrightarrow{\mathcal{T}_-} \begin{pmatrix} H_-^2 \\ 0 \end{pmatrix}, \quad D_+ = \begin{pmatrix} H_+^2 \\ 0 \end{pmatrix} \xrightarrow{\mathcal{T}_-} \left\{ \begin{pmatrix} Sg \\ g \end{pmatrix}, g \in H_+^2 \right\} \tag{4}$$

for the incoming one $\mathcal{T}_-$, and

$$\begin{pmatrix} f_0 \\ f_1 \end{pmatrix} \xrightarrow{\mathcal{T}_+} \begin{pmatrix} I & S^+ \\ 0 & I \end{pmatrix} \begin{pmatrix} f_0 \\ f_1 \end{pmatrix}$$

for the outgoing one $\mathcal{T}_+$.

An analogous assertion also holds for the dissipative generator B of the semigroup $\{Z_t\}$ and all the operators unitarily equivalent to it.

We mention that the requirement stated at the very beginning that the operator-function S should be finite-dimensional and invertible, that is, S should be finite-dimensional and dim $\Delta_T H = d = $ dim Δ_{T_+} (the non-unitarity defects should be equal), is superfluous. All the constructions and the fundamental theorem remain fully valid in a considerably larger class of situations, for example, when the defect operators Δ_T and Δ_{T_+} are trace class operators. This, however, does not exhaust all possible conditions under which the above construction is preserved; in essence, these conditions consist only of the invertibility of S almost everywhere in its regularity domain and the existence of finite limit values for it on the boundary of this domain (see Szökefalvi-Nagy and Foiaş (1970)). An update of the state of this problem can be found in Nikol'skij and Khrushchev (1987).

The universal character of the operators B and T prompted Szökefalvi-Nagy and Foiaş to consider them to be *functional models* for dissipative operators and contractions, respectively. A detailed analysis of the various methods of description of a model can also be found in Nikol'skij and Khrushchev (1987). In what follows we make use of the symmetric form introduced above.

2.3. Factorisations. In the remaining part of this section we prepare the ground for the spectral analysis of a dissipative model operator. The analytic basis of this study consists of factorisation theorems, analogous to Smirnov's theorem.

An analytic operator-function S_i is called an *inner function in the unit disk* (*upper half-plane*) if it is contracting there and unitary almost everywhere on the circle (real axis).

Example 1. For a given countable set of points ζ_i such that $|\zeta_i| < 1$ and $\sum(1 - |\zeta_i|) < \infty$, and an ordered family of orthogonal projections $\{P_i\}$ in E, in the disk $|\zeta| < 1$ we can construct the *Blaschke-Potapov product*

$$\Pi(\zeta) = \prod_{l=1}^{\infty} \left\{ \frac{\zeta_l - \zeta}{1 - \bar{\zeta}_l \zeta} \cdot \frac{\bar{\zeta}_l}{|\zeta_l|} P_l + (I - P_l) \right\},$$

which is the multi-dimensional analogue of the Blaschke product. The analogous object in the half-plane $\operatorname{Im} z > 0$ is also defined for a set of points z_l such that $\operatorname{Im} z_l > 0$ and $\sum \operatorname{Im} z_l (1 + |z_l|^2)^{-1} < \infty$, and an ordered family of orthogonal projections P_l:

$$\Pi(z) = \prod_{l=1}^{\infty} \left\{ \frac{z - z_l}{z - \bar{z}_l} \cdot \frac{i - \bar{z}_l}{i - z_l} \cdot \frac{|i - z_l|}{|i - \bar{z}_l|} P_l + (I - P_l) \right\}.$$

Obviously, the Blaschke-Potapov products constructed above are inner functions.

We mention that in the infinite-dimensional case, that is, when $\dim E = \infty$, the convergence of similar products is not necessarily connected with the condition $\sum(1 - |\zeta_l|) < \infty$; we can easily convince ourselves of this by means of very simple examples.

Example 2. We consider a non-negative, singular, operator-valued measure μ_φ on the unit circle and construct the *multiplicative integral*

$$\Theta(\zeta) = \int \exp\left\{ -\frac{e^{i\varphi} + \zeta}{e^{i\varphi} - \zeta} \, d\mu_\varphi \right\} \equiv \lim_{\Delta_l \varphi \to 0} \prod_l \exp\left\{ -\frac{e^{i\varphi_l} + \zeta}{e^{i\varphi_l} - \zeta} \mu(\Delta_l \varphi) \right\}; \quad (5)$$

if, similarly, μ_s is defined on the real axis, then the integral is

$$\Theta(z) = \int \exp\left\{ i\frac{1 + sz}{s - z} \, d\mu_s \right\} = \lim_{\Delta_l s \to 0} \prod \exp\left\{ i\frac{1 + zs_l}{s_l - z} \mu(\Delta_l s) \right\}. \quad (6)$$

For specific conditions on the measure μ (see Potapov (1955)), the above functions exist and are inner ones. They are called *singular inner functions*. General singular inner functions are obtained as products of factors of the form (5).

A contracting analytic function in a disk is called an *outer function* if the operator of multiplication by it is invertible in H_+^2 in the sense that $\overline{S_e H_+^2} = H_+^2$.

Potapov (1955) studied in detail the factorisation of contracting analytic matrix-functions. In particular, he showed that every finite-dimensional contracting analytic function S in a disk admits a *canonical factorisation*, that is, it can be represented as a product of inner and outer functions, with the inner one in turn obtained by multiplying a Blaschke product and a singular function. The specific form of the elementary factors depends on the choice of the sequence of roots ζ_l and corresponding projections (for the Blaschke product) and of the singular measure (for a singular function).

A contracting outer function can also be represented as a multiplicative integral (5) in a disk, or (6) in a half-plane, but in the latter case with an absolutely continuous measure connected with the function Θ by the equality

$$(1 + s^2)\frac{d}{ds}\operatorname{tr}\mu_s = \frac{1}{\pi}\ln|\det\Theta|.$$

The outer function is unitary-valued on the complement of the support of the measure.

In spite of the fact that the specific form of the factors occurring in the canonical factorisation of our contracting operator-function depends, generally speaking, on the order in which they are arranged, that is,

$$S = S_i\tilde{S}_e = S_e\tilde{S}_i, \quad S_{i(e)} \neq \tilde{S}_{i(e)},$$

and that the elementary factors themselves from which these factors are chosen depend on their selected order, nevertheless their determinants in the finite-dimensional case are found uniquely up to a non-essential constant factor Θ_0 of modulus 1:

$$\Theta_0 \cdot \det S_e = \det\tilde{S}_e, \quad \det S_i = \det\tilde{S}_i \cdot \Theta_0.$$

Here the determinant of the outer factor is an outer function, while the determinant of the inner one is an inner function. Furthermore, for every representation of the inner factor as a product of a Blaschke factor and a singular one we have

$$S_i = \Pi\hat{\Theta} = \Theta\hat{\Pi}, \quad \tilde{S}_i = \tilde{\Pi}\hat{\tilde{\Theta}} = \tilde{\Theta}\hat{\tilde{\Pi}}.$$

The determinants of the Blaschke factors coincide up to a unitary constant and are Blaschke scalar products. A similar statement holds for the singular factors. Moreover, all these elementary facts also remain valid in the infinite-dimensional case when the given function has a determinant (see Szökefalvi-Nagy and Foiaş (1970)). At the same time, the scalar contracting function that is the determinant of the characteristic function is the instrument with which we study the spectral properties of an operator.

A more refined scalar characteristic contracting function is the *scalar multiple*.

We say that a contracting analytic function S has a *scalar multiple* if there are a "complementary" contracting analytic function $\tilde{S}$ and a scalar contracting analytic function s such that

$$\tilde{S}S = sI$$

identically in the unit disk (upper half-plane). In this case s is called a *scalar multiple* of S.

The largest common divisor of all the scalar multiples of a given contracting function is called the *minimal scalar multiple*. This object is very useful in spectral analysis because of its "economical" nature. Thus, it is clear that the determinant of a Blaschke product is a scalar multiple, but it is minimal only when the projections P_i are one-dimensional. In particular, a *scalar Blaschke factor*

$$\frac{\zeta_l - \zeta}{1 - \bar{\zeta}_l \zeta} \cdot \frac{\zeta_l}{|\zeta_l|}$$

is a scalar multiple of the *elementary Blaschke factor*

$$\frac{\zeta_l - \zeta}{1 - \bar{\zeta}_l \zeta} \cdot \frac{\bar{\zeta}_l}{|\zeta_l|} P_l + (I - P_l),$$

since the role of the complementary function is played by the elementary factor with re-arranged projections $P_l \leftrightarrow I - P_l$. At the same time, the determinant of this elementary factor is the *multiple Blaschke factor*

$$\frac{\zeta_l - \zeta}{1 - \bar{\zeta}_l \zeta} \cdot \frac{\bar{\zeta}_l}{|\zeta_l|}.$$

The concept of scalar multiple and the properties of infinite-dimensional contracting analytic functions having a scalar multiple were investigated in detail by Szőkefalvi-Nagy and Foiaş (1970). In particular, they proved that the inner and outer factors of a contracting function that has a minimal scalar multiple s also have scalar multiples, and that the corresponding minimal scalar multiples s_i and s_e are, respectively, an inner factor and an outer factor of s. An analogous assertion also holds separately for an inner factor: its minimal scalar multiple s_i is the product of minimal scalar multiples of the corresponding Blaschke factor and singular factor, which are a scalar Blaschke product and a scalar singular function, respectively.

In Szőkefalvi-Nagy and Foiaş (1970) a remarkable class C_0 of contracting operators was identified and studied in detail, whose characteristic functions have a scalar multiple. These operators admit a full spectral description. Consequently, in the analysis of specific dfferential operators it is important to be able to recognize those that belong to C_0 (see Sect. 5.3 below).

§3. Spectral Analysis in Terms of the Functional Model

3.1. The Scattering Problem. The only parameter of the functional model of a dissipative operator described in the preceding section is its characteristic function S, which is a contracting operator-function analytic in the upper half-plane of the spectral parameter. In view of the results of Adamyan and Arov (1966), this function coincides with the scattering matrix for the dilation of the given dissipative operator. This remarkable result is easily verified for a symmetric model.

Along with the fundamental group $\{U_t\} = \{e^{ikt}\}$ of translations in H, we also consider the group of translations in the orthogonal sum $D_- \oplus D_+ = H_0$, given by

$$\overset{0}{U}_t \begin{pmatrix} f_+ \\ f_- \end{pmatrix} = \begin{pmatrix} P_+ e^{ikt} f_+ + P_+ e^{ikt} f_- \\ P_- e^{ikt} f_- + P_- e^{ikt} f_+ \end{pmatrix}, \quad -\infty < t < \infty .$$

It is clear that this group is isomorphic to the translation group in L_2; hence, it can play the role of the non-perturbed group in the scattering problem for the pair $(U_t, \overset{0}{U}_t)$. Following Adamyan and Arov, we define the *wave operators* $W_\pm$ as the strong limits

$$W_+ = s\text{-}\lim_{t \to \infty} \overset{0}{U}_{-t} P_{D_+} U_t, \quad W_- = s\text{-}\lim_{t \to -\infty} \overset{0}{U}_{-t} P_{D_-} U_t . \tag{7}$$

Theorem 3. *The wave operators* (7) *exist, map respectively the subspaces* $H_+ = \begin{pmatrix} L_2 \\ 0 \end{pmatrix}$ *and* $H_- = \begin{pmatrix} 0 \\ L_2 \end{pmatrix}$ *of H on to H_0 isometrically, and annihilate their orthogonal complements in H. The scattering operator $W_- W_+^+$ reduces in H_0 to multiplication by the characteristic function.*

The proof can be reduced to direct computation by means of the formulae for the projections on $D_\pm$. Denoting by $P_\pm$ the orthogonal projections in L_2 on the Hardy classes $H_\pm^2$, for example,

$$P_+ f(z) = \frac{1}{2\pi i} \int_{-\infty}^{\infty} (\zeta - z)^{-1} f(\zeta) \, d\zeta, \quad \operatorname{Im} z > 0 ,$$

we can rewrite $P_{D_\pm}$ in the form

$$P_{D_+} \begin{pmatrix} f_0 \\ f_1 \end{pmatrix} = \left\{ \begin{matrix} P_+(f_0 + S^+ f_1) \\ 0 \end{matrix} \right\} ,$$

$$P_{D_-} \begin{pmatrix} f_0 \\ f_1 \end{pmatrix} = \left\{ \begin{matrix} 0 \\ P_-(S f_0 + f_1) \end{matrix} \right\} ,$$

from which we find immediately that

$$W_+ = s\text{-}\lim_{t\to\infty} e^{-ikt} P_+ e^{ikt}(f_0 + S^+ f_1) = f_0 + S^+ f_1 \,,$$

$$W_- = s\text{-}\lim_{t\to\infty} e^{-ikt} P_- e^{ikt}(S f_0 + f_1) = S f_0 + f_1 \,.$$

We see directly from the definition that W_+ maps $(L_2, 0)$ into L_2 and W_- maps $(0, L_2)$ into L_2. Then the scattering operator $W^- W_+^+$ defined on the pre-image of W_+ acts as multiplication by S, and the operator $W_+ W_-^+$ on the pre-image of W_- as multiplication by S^+.

3.2. Spectral Analysis of the Dilation. The eigenfunctions

$$\psi_s^+(e) = \begin{pmatrix} \delta(k-2)e \\ 0 \end{pmatrix}, \quad e \in E \,,$$

of the translation group, which form a complete orthonormal system of functions in the invariant subspace H_+, play the role of *outgoing scattered waves*. Similarly, the eigenfunctions

$$\psi_s^-(e) = \begin{pmatrix} 0 \\ \delta(k-s)e \end{pmatrix}, \quad e \in E \,,$$

which form a complete system in the invariant subspace H_-, play the role of *incoming scattered waves*. The spectral representations of the translation group connected with these waves are called *outgoing* and *incoming*, respectively. To construct these spectral representations, we need to complete each of the two systems of scattered waves by adjoining the corresponding system of eigenfunctions in the subspace orthogonal to H_+ or H_-, defined by

$$H \ominus H_+ \equiv H^> = \left\{ \begin{pmatrix} -S^+ g \\ g \end{pmatrix}, \; \langle (1 - SS^+)g, g\rangle_{L_2(E)} < \infty \right\},$$

$$H \ominus H_- \equiv H^< = \left\{ \begin{pmatrix} h \\ -Sh \end{pmatrix}, \; \langle (1 - S^+ S)h, h\rangle_{L_2(E)} < \infty \right\}.$$

In Szökefalvi-Nagy and Foiaş (1970) the subspaces $H^<$ and $H^>$ were called, respectively, *residual* and **-residual*. In view of the physical significance of the translation eigenfunctions in these subspaces, we call $H^<$ the *radiating* subspace and $H^>$ the *absorbing* subspace. As a complete orthonormal system of translation eigenfunctions in $H^>$ we can use the family of distributions

$$\psi^>(s) = \begin{pmatrix} -S^+(k)\delta(k-s)\pi_S \\ \delta(k-s)\pi_S \end{pmatrix} (\Delta_s^+)^{-1} \,, \tag{8}$$

where π_S is an eigenvector corresponding to the non-zero eigenvalue Δ_S^+ of the defect operator $\Delta^+ = I - SS^+$, that is,

$$(I - SS^+)\pi_S = \Delta_S^+ \pi_S, \quad \Delta_S^+ > 0, \quad |\pi_S|_E = 1 \,.$$

The corresponding system of eigenfunctions in $H^<$ can be chosen to be of the form

102 B. S. Pavlov

$$\psi^<(s) = \begin{pmatrix} \delta(k-s)\nu_S \\ -S(k)\delta(k-s)\nu_S \end{pmatrix}(\Delta_S)^{-1} , \qquad (9)$$

where ν_S is an eigenvector corresponding to the non-zero eigenvalue Δ_S of the defect operator $\Delta = I - S^+S$, that is,

$$(I - S^+S)\nu_S = \Delta_S\nu_S, \quad \Delta_S > 0, \quad |\nu_S|_E = 1 .$$

We call $\psi^<$ *radiating* eigenfunctions and $\psi^>$ *absorbing* eigenfunctions of the translation group in H.

Direct computation shows that the above systems are complete and orthogonal in $H^>$ and $H^<$, respectively, and normalised "with respect to the δ-function" with a suitable weight; that is, arbitrary elements $G^> \in H^>$ and $G^< \in H^<$ admit in $H^>$ and $H^<$ orthogonal representations of the form

$$\begin{aligned}
G^> &\equiv \begin{pmatrix} -S^+g \\ g \end{pmatrix} = \int \sum_{e_S^+} (I - SS^+)\psi^> \langle G^>, \psi^> \rangle_H dS , \\
G^< &\equiv \begin{pmatrix} g \\ -Sg \end{pmatrix} = \int \sum_{e_S} (I - S^+S)\psi^< \langle G^<, \psi^< \rangle_H dS .
\end{aligned} \qquad (10)$$

Fully analogous formulae also hold in the infinite-dimensional case (when $d = \infty$), with natural modifications corresponding to the fact that here the spectrum of the defect operators Δ and Δ^+ may have a more complicated structure. As a complete orthogonal and normalized ("with respect to the δ-function") set of eigenfunctions of the translation group in H we can use either $\{\psi^-, \psi^<\}$ with weight $\mathrm{diag}(I, \Delta)$, or $\{\psi^+, \psi^>\}$ with weight $\mathrm{diag}(I, \Delta^+)$. The radiating eigenfunctions can be used as blocks for constructing a canonical set of eigenfunctions of the absolutely continuous spectrum of the model operator B. Similarly, from the absorbing eigenfunctions we can construct a canonical set of eigenfunctions of the absolutely continuous spectrum of the adjoint operator B^+.

3.3. The Spectrum and the Resolvent of the Semigroup Generator. The unitary translation group $\{U_t\} = \{e^{ikt}\}$ in H, compressed on the shift-invariant subspace $K = H \ominus \{D_+ \oplus D_-\}$ by means of the orthogonal projection

$$P_K \begin{pmatrix} f_0 \\ f_1 \end{pmatrix} = \begin{pmatrix} f_0 - P_+(f_0 + S^+f_1) \\ f_1 - P_-(Sf_0 + f_1) \end{pmatrix} , \qquad (11)$$

yields two strongly continuous semigroups

$$\begin{aligned}
Z_t &= \{P_k U_t : K = \exp(iBt), \ t \geq 0\} , \\
Z_t &= \{P_k U_t^+ : K = \exp(-iB^+t), \ t \geq 0\} ,
\end{aligned}$$

which play the role of model semigroups in what follows. The characteristic function of the dissipative operator B coincides with S, and that of $-B^+$

coincides with $S^+(-\bar\lambda)$. It is clear that the canonical factorisation of the characteristic function determines the corresponding canonical factorisation of the characteristic function of the adjoint operator $-B^+$; in particular,

$$S^+(-\bar\lambda) = \tilde S_i^+(-\bar\lambda) \cdot S_e^+(-\bar\lambda) = \tilde S_e^+(-\bar\lambda) \cdot S_i^+(-\bar\lambda) \ .$$

Solving the non-homogeneous equation in K

$$(B - \lambda I)u = f, \quad f \in K \ ,$$

we can construct the resolvent of the model operator in the explicit form

$$u = (B - \lambda I)^{-1} f = \begin{pmatrix} u_0 \\ u_1 \end{pmatrix} ,$$

where

$$\begin{pmatrix} u_0 \\ u_1 \end{pmatrix} = \begin{cases} \dfrac{1}{z - \lambda}\begin{pmatrix} f_0(z) - (f_0 + S^+ f_1)(\lambda) \\ f_1(z) \end{pmatrix}, & \operatorname{Im}\lambda < 0, \\[2ex] \dfrac{1}{z - \lambda}\begin{pmatrix} f_0(z) - S^{-1}(\lambda)(Sf_0 + f_1)(\lambda) \\ f_1(z) \end{pmatrix}, & \operatorname{Im}\lambda > 0. \end{cases} \tag{12}$$

From these formulae it is clear that the spectrum of B coincides with the set of the singularities of S^{-1}, that is, with its spectrum in the function-theoretic sense. Solving the homogeneous equation in K

$$(B - \lambda I)u_\lambda = 0 \ ,$$

we find the eigenfunctions of the model operator: if $\tilde e \in \operatorname{Ker} S(\lambda)$, $\operatorname{Ker} S(\lambda) \neq 0$, $\operatorname{Im}\lambda > 0$, and $\tilde e \neq 0$, then $u_\lambda = ((z - \lambda)^{-1}\tilde e, 0)$ is an eigenvector. If the characteristic function has a multiple root, that is, S contains the elementary Blaschke factor $\dfrac{z - \lambda}{z - \bar\lambda}\tilde P + (1 - \tilde P)$ to the power r, $r > 1$, then together with the eigenvectors we can also construct explicitly the root vectors, which are the solutions of the equations $(B - \lambda I)^l u_\lambda^l = 0$, $l \leq r$. They have the form $((z - \lambda)^{-l}\tilde e, 0)$, $\tilde e \in \tilde P E$. The homogeneous adjoint equation $(B^+ - \bar\lambda I)v_\lambda = 0$ has non-trivial solutions $v_\lambda = (0, (z - \bar\lambda)^{-1}e)$, $e \in \operatorname{Ker} S^+(\lambda)$, at the complex-conjugate points of the spectral parameter. In the finite-dimensional case, the dimensions of the eigenspaces of the operators B and B^+ at complex-conjugate points are the same, and the systems of eigenvectors can be chosen to be biorthogonal (if the algebraic multiplicity r of the eigenvalue λ is equal to one). To show this, we need to separate the elementary Blaschke factor on the right

$$S(z) = S_\lambda(z)\left(\frac{z - \lambda}{z - \bar\lambda}\tilde P + (1 - \tilde P)\right) ,$$

connect the basis $\{\tilde e_\lambda^{(k)}\}$ for $\operatorname{Ker} S(\lambda)$ with the basis $\{\tilde e_\lambda^{(k)}\}$ for $\operatorname{Ker} S^+(\lambda)$ by means of the invertible operator $S_\lambda(\lambda)$, and construct a basis $\{e_\lambda^{(l)}\}$ biorthogonal to it in $\operatorname{Ker} S^+$. Thus, we arrive at

$$\frac{1}{2\pi i}\int_{-\infty}^{\infty}\left\langle\left(\begin{matrix} I & S^+(z) \\ S(z) & I \end{matrix}\right)\left(\begin{matrix} \dfrac{\tilde{e}_\lambda^{(k)}}{z-\lambda} \\ 0 \end{matrix}\right),\left(\begin{matrix} 0 \\ \dfrac{e_\lambda^{(l)}}{z-\bar{\lambda}} \end{matrix}\right)\right\rangle dz$$

$$=\frac{1}{2\pi i}\int_{-\infty}^{\infty}\left\langle\frac{S(z)\tilde{e}_\lambda^{(k)}}{z-\lambda},\frac{e_\lambda^{(l)}}{z-\bar{\lambda}}\right\rangle dz=\frac{1}{2\pi i}\int_{-\infty}^{\infty}\left\langle\frac{e_\lambda'^{(k)}}{z-\lambda},e_\lambda^{(l)}\right\rangle dz$$

$$=\langle e_\lambda'^{(k)},e_\lambda^{(l)}\rangle=\delta_{kl}\,.$$

The spectral analysis of the model operator (see Pavlov (1975c, 1976)) can now be performed on the basis of these explicit formulae and their analogues for the eigenfunctions of the continuous spectrum (see below).

3.4. The Spectral Components. Proceeding with the spectral analysis of the model operator, we start from the fact that a number of problems have already been solved concerning the factorisation of the characteristic function S into relatively prime factors invertible almost everywhere, given in the finite-dimensional case by contracting square matrix-functions depending analytically on the spectral parameter:

$$S = S_i\tilde{S}_e = S_e\tilde{S}_i,\quad S_i = \Pi\hat{\theta} = \theta\hat{\Pi},\quad \tilde{S}_i = \tilde{\Pi}\hat{\tilde{\theta}} = \tilde{\theta}\hat{\tilde{\Pi}}.$$

These problems are fairly difficult (see, for example, Szökefalvi-Nagy and Foiaş (1970)). Fortunately, all the factors occurring in the finite-dimensional case are two-sided: if $S_i(z)$ is an inner function in the upper half-plane, then $S_i^+(z)$ is an inner function in the lower half-plane, and so on. In the infinite-dimensional case we must especially make sure that the above factorisation properties are satisfied (see Szökefalvi-Nagy and Foiaş (1970)). By the *inner subspace N_i* of the dissipative operator B we understand the maximal of its invariant subspaces in K in which the corresponding part of B has an inner characteristic function. In this context, the restriction $B|_{N_i} = B_i$ is called *the inner part of the operator B.*

Similarly, by the *outer subspace N_e* of B we understand the maximal of its invariant subspaces in which the corresponding parts of B have outer characteristic functions. The restriction $B|_{N_e} = B_e$ of B to N_e is called *the outer part of the operator.* Sakhnovich (1968) suggested that a dissipative operator coinciding with its outer part should be called *absolutely continuous* (see Sakhnovich (1968)). In what follows we will see that there are good reasons for using this term. Following Sakhnovich, sometimes we call the outer part of a dissipative operator its *absolutely continuous part.* The next assertion establishes the form and the main property of the inner and outer subspaces of model operators.

Theorem 4 (Pavlov (1975b)). *The inner subspace N_i and outer subspace N_e of the model operator B have the form*

$$N_i = \left\{ \begin{pmatrix} f_0 \\ 0 \end{pmatrix}, \ f_0 \in H^2_-(E) \ominus \tilde{S}^+_i H^2_-(E) \right\},$$

$$N_e = \bar{\tilde{N}}_e, \quad \tilde{N}_e = P_K H^<.$$

These subspaces satisfy the linear independence and completeness relations

$$N_i \cap N_e = 0, \quad \overline{N_i + N_e} = K.$$

Analogously, the inner subspace N_i^+ and outer subspace N_e^+ of the operator $-B^+$ have the form

$$N_i^+ = \left\{ \begin{pmatrix} 0 \\ g_1 \end{pmatrix}, \ g_1 \in H^2_+(E) \ominus S_i H^2_+(E) \right\},$$

$$N_e^+ = \bar{\tilde{N}}_e^+, \quad \tilde{N}_e^+ = P_K H^>.$$

These subspaces satisfy the linear independence and completeness relations

$$N_i^+ \cap N_e^+ = 0, \quad \overline{N_i^+ + N_e^+} = K,$$

and the biorthogonality relations

$$N_i^+ \perp N_e, \quad N_i \perp N_e^+, \quad N_i^+ \oplus N_e = N_i \oplus N_e^+ = K.$$

The *spectrum σ_i of the inner component*, whose characteristic function is the inner factor $\tilde{S}_i$, coincides with the set of points where this factor is not invertible, and the *spectrum σ_e of the outer component* coincides with the closure of the set of points on the real axis across which the characteristic function S_e cannot be extended continuously by means of the symmetry principle, that is, this function is not unitary, or, in other words, the defect operators $\Delta = I - \tilde{S}^+_e \tilde{S}_e$ and $\Delta^+ = I - S_e S^+_e$ do not vanish.

Following Sakhnovich (1968), we call the spectrum of the outer component the *absolutely continuous spectrum of the operator B*.

3.5. Spectral Singularities and the Separation of the Inner and Outer Components. The spectral components of a dissipative operator cannot always be separated by means of a bounded projection $\mathcal{P}_{ie}$ on N_i parallel to N_e. The problem is that the linear independence condition $N_i \cap N_e = 0$ does not guarantee that the angle between the subspaces N_i and N_e is positive, that is, it does not ensure the existence of the corresponding bounded projection. This projection exists only under additional conditions, whose formulation requires the isolation of a certain singular subset of the absolutely continuous spectrum σ_e.

We say that a point $\lambda \in \sigma_e$, $|\lambda| < \infty$, is a *regular point* of the absolutely continuous spectrum of the model operator if it has a neighbourhood $\omega^\varepsilon_\lambda = \{|k - \lambda| < \varepsilon, \ \mathrm{Im}\, k > 0\}$ where the characteristic function of the outer component $\tilde{S}_e(\lambda)$ is uniformly boundedly-invertible, that is,

$$\sup_{k \in \omega_\lambda^\varepsilon} |\tilde{S}_e^{-1}(k)| < \infty.$$

If $\lambda \in \sigma_e$, is not regular, then it is called a *point of spectral singularity*. For the point at infinity the concept is modified in a natural way by introducing the corresponding system of neighbourhoods. We denote by σ_0 the set of all spectral singularities.

From the uniqueness theorem for scalar functions of class H^∞ it follows that in the most interesting practical case, when the characteristic function has a scalar multiple, the set σ_0 is closed and its Lebesgue measure is zero, while near its complement (the set of regular points), by the well-known estimate (see Szökefalvi-Nagy and Foiaş (1970), Chap. IV)

$$|(B_e - \lambda I)^{-1}| \leq (\operatorname{Im} \lambda)^{-1} |\tilde{S}_e^{-1}(\lambda)|, \quad \operatorname{Im} \lambda > 0,$$

the resolvent of the outer component B_e has a first power order of growth. The importance of the spectral singularities in the theory of singular differential non-selfadjoint operators was noted by Naimark (1954). Lyantse (1964a,b) proposed the method of renormalisation of the spectral resolution to eliminate the unpleasant effects occurring at these points.

The next example shows that the spectral singularities play an important role in the separation of the spectral components.

Example 3. Let k_α be the eigenvalues of the operator B and

$$e_\alpha \in \operatorname{Ker} S(k_\alpha) = \operatorname{Ker} \tilde{S}_i(k_\alpha), \quad |e_\alpha| = 1.$$

Then the angle β between the normalised eigenvector

$$f_\alpha = ((k - k_\alpha)^{-1} e_\alpha \sqrt{\operatorname{Im} k_\alpha}, 0)$$

of B and the outer ("absolutely continuous") subspace is computed by means of the formula

$$\sin \beta = |PSf_\alpha| = |\tilde{S}_e(k_\alpha) e_\alpha|$$

and tends to zero if there is a sequence of eigenvalues k_α accumulating at a point of spectral singularity, that is, $k_\alpha \to k_0$, so that $\tilde{S}_e(k_\alpha) e_\alpha \to 0$. From this it is obvious that in the one-dimensional case, when the characteristic function is scalar and continuous, the spectral components are separable if and only if σ_i and σ_e are disjoint.

It is convenient to formulate sufficient conditions for the separability of the spectral components in the terms introduced by Carleson in the study of the equivalence of various Hilbert norms of classes of analytic functions.

We say that a rectifiable contour γ in the upper half-space $\operatorname{Im} k \geq 0$ is a *Carleson contour* if the Lebesgue measure $|d\gamma|$ concentrated on it (that is, its length) is such that

$$\sup_{-\infty<\kappa<\infty}\ \sup_{\varepsilon}\ \int\limits_{|\kappa-k|<\varepsilon}|d\gamma(k)| = C_0(\gamma) < \infty. \tag{13}$$

From the results in Carleson (1962) it follows that under the condition (13) the scalar analytic functions in H_+^2 are square-integrable over γ and

$$\int\limits_\gamma |f(k)|^2|d\gamma(k)| \le C(\gamma)\int\limits_{-\infty}^\infty |f(k)|^2 dk.$$

Theorem 5 (Pavlov (1975b)). *If there is a Carleson contour that is homotopic to the real axis, separates σ_0 and σ_i, and is such that*

$$\sup_{k\in\gamma}|S^{-1}(k)| \le C < \infty$$

almost everywhere on it, then there exists a bounded spectral projection $\mathcal{P}_{ie}$ on N_i parallel to N_e, given by

$$\mathcal{P}_{ie}\begin{pmatrix} f_0 \\ f_1 \end{pmatrix} = \begin{pmatrix} P_{\tilde{K}_-}S^{-1}P_{K_+}(Sf_0 + f_1) \\ 0 \end{pmatrix},$$

where $P_{\tilde{K}_-}$ and P_{K_+} are the orthogonal projections on $H_-^2 \ominus \tilde{S}_i^+ H_-^2$ and $H_+^2 \ominus S_i H_+^2$, respectively. In this case, $|\mathcal{P}_{ie}| \le C$.

A separability test for the spectral components in terms of the level lines of S_i was formulated in Nikol'skij and Khrushchev (1987).

Separability conditions for the spectral components of model operators of class C_0 can also be given in terms of a scalar multiple s_e of the outer factor S_e.

Theorem 6. *If $B_i = B|_{N_i}$, then*

$$\sin(N_e, N_i) \ge |s_e^{-1}(B_i)|^{-1}.$$

Corollary . *If γ is a Carleson contour separating σ_i and σ_0 and s is a scalar multiple of the characteristic function, then*

$$\sin(N_e, N_i) \ge \inf_{k\in\gamma}|s(k)|.$$

2-36

3.6. The Decomposition of the Inner Component. We now turn to the problem of further decomposing the inner component of the operator. We assume that canonical factorisations

$$\tilde{S}_i = \Theta\hat{\Pi} = \Pi\hat{\Theta}$$

for $\tilde{S}_i$ have already been constructed. Here Π and $\hat{\Pi}$ are Blaschke products and Θ and $\hat{\Theta}$ are singular functions.

By the *singular subspace* N_s and *discrete subspace* N_d of a dissipative operator we understand the maximal invariant subspaces of its inner component where the characteristic function of the corresponding restriction of the operator is a Blaschke product and a singular inner function, respectively. These subspaces are easily computed in terms of the canonical factorisations, being given by

$$N_s = H_-^2 \ominus \hat{\Theta}^+ H_-^2, \quad N_d = H_-^2 \ominus \hat{\Pi}^+ H_-^2.$$

In addition, $\overline{N_s + N_d} = N_i$ and $N_d \cap N_s = 0$. The orthogonal complements of N_s and N_d in N_i are

$$N_s^\perp = \hat{\Theta}^+(H_-^2 \ominus \Pi^+ H_-^2), \quad N_d^\perp = \hat{\Pi}^+(H_-^2 \ominus \Theta^+ H_-^2).$$

Denoting by P_s, P_d, $P_s^\perp$, and $N_d^\perp$ the orthogonal projections on N_s, N_d, $N_s^\perp$, and $N_d^\perp$, respectively, we can formulate a separability test for the singular and discrete components of the model operator B.

Theorem 7 (Pavlov (1975b)). *The angle β between N_s and N_d is positive if and only if at least one of the operators*

$$\Delta_{ds} = P_d^\perp|_{N_s} \to N_d^\perp,$$
$$\Delta_{sd} = P_s^\perp|_{N_d} \to N_s^\perp$$

is invertible. Then the other operator is also invertible and $|\sin \beta| \geq (\inf |\Delta_{ds}^{-1}|,$ $|\Delta_{sd}^{-1}|)^{-1}$, and the projections $\mathcal{P}_{ds}$ on N_d parallel to N_s and $\mathcal{P}_{sd}$ on N_s parallel to N_d are given by the formulae

$$\mathcal{P}_{sd} = \Delta_{ds}^{-1} P_d^\perp, \quad \mathcal{P}_{ds} = \Delta_{sd}^{-1} P_s^\perp.$$

These operators are easily rewritten in terms of the canonical factorisations. For example,

$$\Delta_{ds} = \hat{\Pi}^+ P_{H_-^2 \ominus \Theta^+ H_-^2} \Pi P_{H_-^2 \ominus \hat{\Theta}^+ H_-^2},$$

since the invertibility of Δ_{ds} is equivalent to that of the operator $P_{H_-^2 \ominus \Theta^+ H_-^2}$ $\times \Pi P_{H_-^2 \ominus \hat{\Theta}^+ H_-^2}$ from $H_-^2 \ominus \hat{\Theta}^+ H_-^2$ to $H_-^2 \ominus \Theta^+ H_-^2$.

Theorem 8 (Pavlov (1975b)). *If the Blaschke product Π has a scalar multiple π and the operator $\pi(B_\Theta) = P_{H_+^2 \ominus \Theta H_+^2} \pi|H_+^2 \ominus \Theta H_+^2$ is invertible, then $\sin \beta \geq |\pi^{-1}(B_\Theta)|^{-1}$.*

In particular, the above inequality yields the one-dimensional estimate

$$|\sin \beta| \geq \prod_i |\Theta(\lambda_i)|^{m_i}$$

derived by Vasyunin (see Nikol'skij and Khrushchev (1987)). Further discussion of separability questions for the spectral components can be found in Nikol'skij and Khrushchev (1987).

3.7. Spectral Analysis of the Discrete Spectrum. Having separated the spectral components, we may now proceed with the analysis of each of them. Here we restrict our attention to the spectral analysis of the discrete and absolutely continuous components. The spectral analysis of the singular component can also be conducted in function-theoretic terms, as follows from recent papers by Nikol'skij (see Nikol'skij and Khrushchev (1987)).

We perform the spectral analysis of the discrete component of the model operator by assuming that the algebraic multiplicity of the eigenvalues $\{\lambda\}$ is equal to one, that is, when after separating the elementary Blaschke factor at the point λ, the remaining part of the characteristic function is invertible:

$$S = S_e\tilde{\Theta}\tilde{\Pi} = S_e\tilde{\Theta}\tilde{\Pi}_\lambda\left[\frac{z-\lambda}{z-\bar{\lambda}}\tilde{P}_\lambda + (I - \tilde{P}_\lambda)\right]$$

$$= \left[\frac{z-\lambda}{z-\bar{\lambda}}P_\lambda + (1 - P_\lambda)\right]\Pi_\lambda\Theta\tilde{S}_e = \Pi\Theta\tilde{S}_e,$$

where $S_e\tilde{\Theta}\tilde{\Pi}_\lambda$ and $\Pi_\lambda\Theta\tilde{S}_e$ are invertible operators and

$$S_e\tilde{\Theta}\tilde{\Pi}_\lambda : \operatorname{Ker} S(\lambda) = \tilde{P}_\lambda E \to P_\lambda E = \operatorname{Ker} S^+(\lambda),$$
$$\Pi_\lambda\Theta\tilde{S}_e : \operatorname{Ker} S^+(\lambda) = \tilde{P}_\lambda E \to P_\lambda E = \operatorname{Ker} S(\lambda).$$

It is convenient to take the vectors $u_\lambda = ((z - \lambda)^{-1}\tilde{e}_\lambda^{(k)}, 0)$, where $\tilde{e}_\lambda^{(k)}$ is some basis for $\operatorname{Ker} S(\lambda)$, as the eigenfunctions of the discrete spectrum of the operator B at the point λ, and the vectors $v_\lambda = (0, (z - \bar{\lambda})^{-1}e_\lambda^{(l)})$, where $e_\lambda^{(l)}$ is a basis for $\operatorname{Ker} S^+(\lambda)$, as the eigenfunctions of B^+ at the complex-conjugate point. For the eigenfunctions of the discrete spectra of B and B^+ to be biorthogonal, that is,

$$\left\langle\left(\begin{array}{c}(z-\lambda)^{-1}\tilde{e}_\lambda^{(k)}\\0\end{array}\right), \left(\begin{array}{c}0\\(z-\bar{\lambda}')^{-1}e_{\lambda'}^{(l)}\end{array}\right)\right\rangle$$

$$= \begin{cases} \langle S(\lambda)\tilde{e}_\lambda^{(k)}, e_{\lambda'}^{(l)}\rangle_E \dfrac{2\pi i}{\lambda - \lambda'} = 0, & \lambda \neq \lambda', \\[2ex] \langle S_e(\lambda)\tilde{\Theta}(\lambda)\tilde{\Pi}_\lambda\tilde{e}_\lambda^{(k)}, \tilde{e}_{\lambda'}^{(l)}\rangle\dfrac{\pi}{\operatorname{Im}\lambda} = \delta_{kl}, & \lambda = \lambda', \end{cases}$$

we need to choose the basis $\{e_\lambda^{(l)}\}$ for $\operatorname{Ker} S^+(\lambda)$ to be biorthogonal to the basis $\{S_e(\lambda)\tilde{\Theta}(\lambda)\tilde{\Pi}_\lambda e_\lambda^{(k)}\}$ for $\operatorname{Ker} S(\lambda)$ with the factor $\pi(\operatorname{Im}\lambda)^{-1}$.

By analogy with the expansion in the eigenfunctions of B in the case of a purely discrete spectrum, here we have a biorthogonal expansion. Thus, if $F = (f_0, 0) \in N_d$ and $f_0 \in H^2_- \ominus \tilde{\Pi}^+ H^2_-$, then

$$F \sim \sum_{k,\lambda} \begin{pmatrix} (z-\lambda)^{-1}\tilde{e}_\lambda^{(k)} \\ 0 \end{pmatrix} \left\langle \begin{pmatrix} f_0 \\ 0 \end{pmatrix}, \begin{pmatrix} 0 \\ (z-\bar{\lambda})^{-1}e_\lambda^{(k)} \end{pmatrix} \right\rangle_H .$$

This is an interpolation series in the generalised sense that its coefficients are determined by the values of the expanded function at the points of the spectrum (more precisely, by their projections on $\operatorname{Ker} S^+(\lambda)$), that is,

$$\left\langle \begin{pmatrix} f_0 \\ o \end{pmatrix}, \begin{pmatrix} 0 \\ (z-\bar{\lambda})^{-1}e_\lambda^{(k)} \end{pmatrix} \right\rangle_H = \langle Sf_0(\lambda), e_\lambda^{(k)} \rangle_E ,$$

and the series consisting of the first components multiplied by S

$$\sum_{k,\lambda} (z-\lambda)^{-1} S\tilde{e}_\lambda^{(k)} \langle Sf_0(\lambda), e_\lambda^{(k)} \rangle \tag{14}$$

is the interpolation series for the function Sf_0. The convergence of interpolation series of the form (14) has been studied by various authors, starting with Carleson (1962). The spectral character of these objects has been investigated in detail in Katsnel'son (1967) and Nikol'skij and Pavlov (1970) (see also Nikol'skij and Khrushchev (1987)). In the scalar case, when the auxiliary space is one-dimensional, the series (14) for a function $f_0 \in S_-^2 \ominus \tilde{\Pi} H_-^2$ becomes

$$f_0(z) = \sum_\lambda \frac{1}{z-\lambda} \frac{(\tilde{\Pi} f_0)(\lambda)}{\tilde{\Pi}_\lambda(\lambda)} \cdot 2\operatorname{Im}\lambda .$$

The unconditional convergence of this series for an arbitrary function f_0 in $H_-^2 \ominus \tilde{\Pi} H_-^2$ is ensured by the *Carleson condition*

$$|\tilde{\Pi}_\lambda(\lambda)| = \prod_{\mu \neq \lambda} \left| \frac{\lambda-\mu}{\lambda-\bar{\mu}} \right| \geq \delta > 0 .$$

This coincides with the *condition of uniform minimality* of the family of normalised eigenfunctions $\sqrt{2\operatorname{Im}\lambda}\,(z-\lambda)^{-1} = u_\lambda$ of the model operator, which can be written as

$$\sin[u_\lambda, \bigvee_{\mu \neq \lambda} \{u_\mu\}] = |\tilde{\Pi}_\lambda(\lambda)| \geq \delta > 0 ;$$

therefore, it is a necessary and sufficient condition for a basis consisting of the eigenvectors of a model dissipative operator whose characteristic function is a Blaschke product to be equivalent to an orthonormalised one (see Nikol'skij and Pavlov (1970)).

In the infinite-dimensional case $\dim E = \infty$, the condition of uniform minimality does not guarantee the *basis property*. In Nikol'skij and Pavlov (1970) a case of "serial" bases was studied, for which the family of *directional vectors* $\{\tilde{e}_\lambda\}$ can be decomposed into several sequences $\tilde{e}_{\lambda_s^{(l)}}^{(l)} \to \tilde{e}_\infty^{(l)}$, $l = 1, 2, \ldots, d$.

If the vectors $\tilde{e}_\infty^{(l)}$, $l = 1, 2, \ldots, d$, are linearly independent and each of the sequences of corresponding eigenvalues $\{\lambda_s^{(l)}\}$, $s \to \infty$, satisfies the Carleson condition, then the eigenvectors $\{u_s^{(l)}\}$ of the corresponding model operator form a basis equivalent to an orthonormalised one. The connection between uniform minimality and the basis property of the families of eigenvectors of model operators with a discrete spectrum was fully investigated by Trejl' (1986), who proved the following general result.

Theorem 9. *If the family of directional vectors $\{e_\lambda\}$ corresponding to the eigenfunctions $u_\lambda = (z - \lambda)^{-1} e_\lambda \sqrt{2 \operatorname{Im} \lambda}$ of a model operator with a discrete spectrum is compact in the auxiliary space, then the family $\{u_\lambda\}$ forms an unconditional basis (equivalent to an orthonormalised one) if an only if it is uniformly minimal.*

In particular, from this it follows that the uniform minimality of the family of eigenvectors of a dissipative operator with a finite-dimensional non-selfadjointness defect (that is, with a finite-dimensional characteristic function) is always equivalent to the unconditional basis property.

Unfortunately, for differential operators the uniform minimality condition is satisfied only in exceptional circumstances; consequently, the methods of summation of spectral resolutions become extremely important.

The following assertion for discrete operators is a simple example of this kind of statement.

If an element u in N_d is such that the series $\sum_\lambda e^{-\operatorname{Im} \lambda t} \|u_\lambda\| |\langle u, v_\lambda \rangle|$ converges for every positive t, then the spectral resolution of this element is summable in the sense of Abel, that is,

$$\lim_{t \to 0+} \sum_\lambda e^{i\lambda t} u_\lambda \langle u, v_\lambda \rangle = u.$$

The proof follows immediately from the strong continuity of the model semigroup.

3.8. Spectral Analysis of the Absolutely Continuous Spectrum.

According to a well-known result of Szökefalvi-Nagy and Foiaş, the absolutely continuous component B_e of the model operator is "quasi-equivalent" to the remaining (radiating) component of the dilation. This is directly connected with Theorem 4. Indeed, using the fact that $\bar{N}_e = P_K H^<$ is dense in the absolutely continuous (outer) subspace N_e, we remark that every element $f^< = (f, -Sf)$, $f^< \in H^<$, such that $(1 - S_e^+ S)f = \Delta f \neq 0$, satisfies

$$P_K e^{ikt} f' - e^{iBt} P_K f^< = P_K e^{ikt} \begin{pmatrix} P_+ \Delta f \\ 0 \end{pmatrix} = 0, \quad t \geq 0, \qquad (15)$$

since $D_+ = (H_+^2, 0)$ is invariant with respect to the translation group in $K \perp D_+$. From Theorem 4 we see immediately that the quasi-equivalence of the radiating component of the dilation and the outer (absolutely continuous)

112   B. S. Pavlov

component of the model operator is realised by the orthogonal projection operator

$$P_K : \begin{pmatrix} f \\ -Sf \end{pmatrix} \mapsto \begin{pmatrix} f - P_+ \Delta f \\ -Sf \end{pmatrix}.$$

In general, this operator is not boundedly invertible; however, it is invertible in a weak sense. Remarking that $S^+ S = \tilde{S}_e^+ \tilde{S}_e$ and $|(f, -Sf)|_H^2 = \langle \Delta f, f \rangle_{L_2}$, we see that

$$\langle \tilde{S}_e^+ \tilde{S}_e \Delta f, f \rangle_{L_2} \le \left| P_K \begin{pmatrix} f \\ -Sf \end{pmatrix} \right|_H^2 \le \langle \Delta f, f \rangle_{L_2},$$

that is, invertibility occurs on the set of elements for which the operator-function $\tilde{S}_e^+ \tilde{S}_e$ is invertible on the support of the corresponding function f. This set is dense in $\tilde{N}_e$, hence, also in N_e, since the analytic matrix of functions $\tilde{S}_e$ in the upper half-plane cannot be equal to zero on a set of positive measure.

The absolutely continuous spectrum of the model operator B coincides with the closure of the set of eigenvalues λ for which the defect operator $\Delta(\lambda) \equiv 1 - S^+ S(\lambda)$ or $\Delta^+(\lambda) \equiv 1 - SS^+(\lambda)$ is positive. The spectral projections and eigenfunctions are computed by means of formulae that contain the eigenvectors of the defect operators π and ν, that is, $\Delta^+(\lambda)\pi = \Delta_\pi^+ \pi$ and $\Delta(\lambda)\nu = \Delta_\nu \nu$, corresponding to the positive eigenvalues Δ_π^+ and Δ_ν. Using the polar representation of the characteristic function, we can connect the families $\{\pi\}$ and $\{\nu\}$ for a given ν by means of the equalities $S^+(\lambda)\pi = s_\pi \cdot \nu$ and $S(\lambda)\nu = s_\nu \pi$, where $s_\nu = s_\pi = \sqrt{1 - \Delta_\pi} = \sqrt{1 - \Delta_\nu^+}$.

Theorem 10 (Pavlov (1975c)). *If χ_ω is the indicator function of a closed interval ω on the real axis which does not contain spectral singularities, that is, $\omega \cap \sigma_0 = \emptyset$, then the spectral projection of the model operator on the invariant subspace corresponding to the part $\sigma_e(\omega) = \sigma_e \cap \omega$ of the absolutely continuous spectrum contained in ω is bounded, and is given by the formula*

$$\mathcal{E}_\omega^+ \begin{pmatrix} f - P_+ \Delta f \\ -Sf \end{pmatrix} = \begin{pmatrix} \chi_\omega f - P_+ \chi_\omega \Delta f \\ -S\chi_\omega f \end{pmatrix}$$

for any elements $P_k\{f, -Sf\}$ in $\tilde{N}_e$.

The eigenfunctions φ_ν of the absolutely continuous spectrum of the model operator B are the distributions

$$\varphi_\nu(k, \lambda) = P_K \psi_\nu^< = \Delta_\nu^{-1}(\lambda) \begin{pmatrix} \delta(k - \lambda)\nu + \dfrac{1}{2\pi i} \dfrac{\Delta(\lambda)\nu}{k - \lambda + i0} \\ -S(\lambda)\delta(k - \lambda)\nu \end{pmatrix}.$$

Similarly, the spectral projection on the invariant subspace of the adjoint operator B^+ corresponding to the same part of the spectrum is bounded, and is given by the formula

$$\mathcal{E}_\omega^+ \begin{pmatrix} -S^+ f \\ f - P_- \Delta^+ f \end{pmatrix} = \begin{pmatrix} -S^+ \chi_\omega f \\ \chi_\omega f - P_- \chi_\omega \Delta^+ f \end{pmatrix}$$

for any elements $P_K\{-S^+ f, f\}$ in $\tilde{N}_e^+$.

The eigenfunctions ψ_π of the absolutely continuous spectrum of the operator B^+ are

$$\psi_\pi(k,\lambda) = P_K \psi_\pi^> = \Delta_\pi^{-1} \begin{pmatrix} -S^+(\lambda)\delta(k-\lambda)\pi \\ \delta(k-\lambda)\pi - \dfrac{1}{2\pi i}\dfrac{\tilde{\Delta}(\lambda)\pi}{k-\lambda-i0} \end{pmatrix}.$$

If the orthonormal systems of vectors $\{\pi\}$ and $\{\nu\}$ are connected for every λ by the polar representation of the characteristic function $S^+(\lambda)\pi = s_\pi\nu$, $S\nu = s_\nu\pi$, $\Delta_\pi = 1 - s_\pi^2$, then the families φ and ψ are biorthogonal, that is,

$$\langle \varphi_\nu(\cdot,\lambda), \psi_\pi(\cdot,\lambda)\rangle_H = -\delta(\lambda-\lambda')s_{\pi/(1-2/(s\pi))}\langle \pi,\pi'\rangle_E,$$

and the spectral projection on the invariant subspace corresponding to the part $\sigma_e(\omega)$ of the spectrum can be rewritten in terms of the eigenfunctions as an integral operator with kernel:

$$\mathcal{E}(k,k') = -\int\limits_\omega \sum_{\nu \perp \mathrm{Ker}\,\Delta} (1-s_\pi^2)s_\pi^{-1}\varphi_\nu(k,\lambda)\bar{\psi}_\pi(k',\lambda)d\lambda.$$

The presence of spectral singularities causes a substantial deterioration in the convergence of spectral resolutions. Nevertheless, various procedures can be indicated which also yield spectral representations in this case.

Let $\{\sigma_\varepsilon\}$ be a family of subsets of the real axis $\mathbb{R}$ which does not contain spectral singularities and is such that

1) $|S_e^{-1}(\lambda)| < 1/\varepsilon$, $\lambda \in \sigma_\varepsilon$;
2) $\lim\limits_{\varepsilon\to 0} \sigma_\varepsilon = \mathbb{R}$.

In addition, let $\mathcal{E}(\sigma_\varepsilon)$ be the spectral projections on the invariant subspaces corresponding to the parts $\sigma_e \cap \sigma_\varepsilon$ of the absolutely continuous spectrum. Then the following assertion holds.

Theorem 11 (Pavlov (1975b,1976)). *If $\mathcal{F} \in \tilde{N}_e$, then $s\!-\!\lim\limits_{\varepsilon\to 0} \mathcal{E}(\sigma_\varepsilon)\mathcal{F} = \mathcal{F}$.*

For the entire set N_e, the spectral resolution converges in the weaker metric

$$\left| P_K \begin{pmatrix} f \\ -Sf \end{pmatrix} \right|_W = |S_e \Delta^{1/2} f|_{L_2},$$

that is,

$$|\mathcal{E}(\sigma_\varepsilon)\mathcal{F} - \mathcal{F}|_W \to 0, \quad \varepsilon \to 0.$$

When spectral singularities are present, we can also indicate procedures for the summation of spectral resolutions with respect to the absolutely continuous spectrum. We describe such a technique, which makes use of the scalar multiple s_e of the outer factor of the characteristic function. We construct a family of outer functions $\{s_e^\delta\}$, $\delta > 0$, such that

1) s_e is a divisor of s_e^δ, $\delta > 0$, in the class of bounded functions;
2) $s_e^\delta \to 1$ as $\delta \to 0$ almost everywhere on the real axis.

114 B. S. Pavlov

We can, for example, take outer functions whose modulus is bounded above by the modulus of the scalar multiple near spectral singularities; thus, using a smooth cut-off function η near σ_0, such as

$$\eta(x) = \begin{cases} 1, & |x| > 2, \\ 0, & |x| < 1, \end{cases}$$

we set

$$s_e^\delta(z) = \exp\left\{\frac{1}{2\pi}\int\frac{1+z\lambda}{\lambda-z}\ln|s(\lambda)|\frac{1-\eta[\delta^{-1}\operatorname{dist}(\lambda,\sigma_0)]}{1+\lambda^2}\,d\lambda\right\}.$$

Clearly, the conditions 1 and 2 are satisfied.

Theorem 12 (Pavlov (1979b)). *The operator $s_e^\delta(B)$ is bounded and $s_e^\delta(B)N_e \subset \tilde{N}_e$. For any element $\mathcal{F}$ in N_e the spectral resolution converges in the norm*

$$s_e^\delta(B)\mathcal{F} = -\int_{-\infty}^{\infty}\sum_{\nu\perp\operatorname{Ker}\Delta}\Delta_\nu(\lambda)s_\nu^{-1}\varphi_\nu(*,\lambda)\langle s_e^\delta(B)\mathcal{F},\psi_\pi\rangle_H\,d\lambda$$

and its sum, as $\delta \to 0$, is $\mathcal{F}$, that is,

$$\mathcal{F} = s\text{-}\lim_{\delta\to 0}\left\{-\int_{-\infty}^{\infty}\sum_{\nu\perp\operatorname{Ker}\Delta}\Delta_\nu(\lambda)s_\nu^{-1}(\lambda)s_e^\delta(\lambda)\varphi_\nu(*,\lambda)\langle\mathcal{F},\psi_\pi\rangle_H\,d\lambda\right\}.$$

3.9. Joint Completeness and the Basis Property. Let B be a model dissipative operator with inner characteristic function S whose both factors, the Blascke product Π, $\tilde{\Pi}$ and the singular one Θ, $\tilde{\Theta}$, are non-trivial in the canonical factorisation $S = \Pi\tilde{\Theta} = \Theta\tilde{\Pi}$. Here the family of root vectors of B is not complete in K, but the joint system of root vectors of B and B^+ could be. We say that this is a case of *joint completeness of the families of root vectors of B and B^+*. If the joint family is a Riesz basis for K, then we say that this is a case of *joint basis property* (see Pavlov (1971b) and Ivanov and Pavlov (1978)).

Considering the operators with the inner characteristic function $S = \Theta\tilde{\Pi}$, we can always work in terms of a one-component model, where $K = H_+^2 \ominus SH_+^2$ and the subspaces of the discrete spectra N_d and N_d^+ of the model operators B and B^+ have the form $N_d = \Theta(H_+^2 \ominus \tilde{\Pi}H_+^2)$ and $N_d^+ = H^2 \ominus \Pi H_+^2$, and their orthogonal complements are $N_s^+ = H_+^2\ominus\Theta H_+^2$ and $N_s = \Pi(H_+^2\ominus\tilde{\Theta}H_+^2)$, respectively. Obviously, the completeness of the joint system of root vectors of B and B^+, that is, the joint completeness, holds automatically when the angle between N_s^+ and N_s is positive. Adjoining the subspace SH_+^2 to N_d and N_s, and H_-^2 to N_d^+ and N_s^+, we can reduce the question of positiveness of the angles between the above subspaces to the simpler one of positiveness of the angles between ΘH_+^2 and ΠH_-^2 in the former case, and between ΘH_-^2 and ΠH_+^2 in the latter.

Usually in problems of spectral theory for differential operators where the question of joint completeness arises (for example, in the *Regge problem* (see Ivanov and Pavlov (1978) and Regge (1958)), there is an analytic, or even an entire, function such that multiplication by it in $L_2(\mathbb{R})$ maps the Hardy classes $H_\pm^2$ into subspaces forming a non-trivial angle. We formulate conditions for the joint completeness and basis property in terms of such a function, which is called a *generating function*.

Let $F : E \to E$ be an entire matrix-function whose determinant is not identically equal to zero and which admits the factorisations

$$F = \Pi \tilde{F}_e^{(+)} \text{ relative to the upper half-plane,}$$
$$F = \Theta \tilde{F}_e^{(-)} \text{ relative to the lower half-plane.}$$

$$(16)$$

Here Π and $(\Pi^+)^{-1}$ are Blaschke products and Θ and $(\Theta^+)^{-1}$ singular entire functions, and $F_e^{(\pm)}$ two-sided outer functions in the upper ans lower half-planes, respectively. The inner factors Π and Θ are assumed to be contracting in the upper half-plane.

Lemma 1 (Ivanov and Pavlov (1980)). *If D and D' are the domains of the operators of multiplication by F and $(F^{-1})^+$, respectively, in $L_2(\mathbb{R}, E)$, $D_\pm = D \cap H_\pm^2$, and $D'_\pm = D' \cap H_\pm^2$, then $\bar{D}_\pm = H_\pm^2$, $\bar{D}'_\pm = H_\pm^2$, $\overline{FD_+} = \Pi H_+^2$, $\overline{FD_-} = \theta H_-^2$, $\overline{(\bar{F}^+)^{-1} D'_+} = \theta H_+^2$, and $\overline{(F^+)^{-1} D'_-} = \Pi H_-^2$. If, in addition, the Hilbert operator bordered by multiplication by F and F^{-1} is bounded in L_2, that is,*

$$\left| \text{p.v.} \int_{-\infty}^{\infty} \frac{F(x)F^{-1}(y)}{x - y} u(y)dy \right|_{L_2} \leq C_F |\varphi|_{L_2},$$

$$(17)$$

then the angles β and β' between the pairs of subspaces $(\Pi H_+^2, \theta H_-^2)$ and $(\theta H_+^2, \Pi H_-^2)$ are positive ($\sin \beta, \sin \beta' \geq C_F^{-1}$) and the corresponding oblique projections are generated by the closure of the complete sets in $L_2(\mathbb{R}, E)$ of the constructions obtained by bordering the orthogonal projection P_- on the Hardy class H_-^2 by F and F^{-1} and by $(F^+)^{-1}$ and F^+, respectively, that is,

$$\mathcal{P}_{\theta H_-^2}^{\| \Pi H_+^2 \|} = F P_- F^1, \quad \mathcal{P}_{\Pi H_-^2}^{\| \theta H_+^2 \|} = (F^+)^{-1} P_- F^+.$$

In the scalar case, the condition (17) is equivalent to the well-known *Muckenhoupt condition* with weight function $w = |F|^2$

$$\sup_{\Delta \in \mathbb{R}} \left(\frac{1}{|\Delta|} \int_\Delta w \, dt \right) \left(\frac{1}{|\Delta|} \int_\Delta w^{-1} dt \right) < \infty.$$

In the matrix case, if the Muckenhoupt condition holds with weight matrix $w = F^+ F$, then the bordered Hilbert operator is bounded only under additional requirements (see, for example, Ivanov and Pavlov (1980)). In the problems of resonance scattering considered later, the generating function F arises from the very beginning as the value of the Jost solution at the origin. Then the boundedness of the Hilbert transform is no longer necessary in refined tests, since in the majority of interesting cases F and F^{-1} are uniformly bounded on the real axis.

Let $B : K \to K$ be a dissipative model operator with an inner characteristic function $S = \Pi\tilde{\Theta} = \Theta\tilde{\Pi}$.

Theorem 13 *((Pavlov (1979a), Khrushchev, Nikol'skij and Pavlov (1981)). If there is a generating function F admitting the factorisations (16) and satisfying the condition (17), then the joint system of root vectors of the operators B and B^+ is complete (jointly complete) in K and ω-linearly independent. If, in addition, each of the two systems of root vectors forms a Riesz basis for its linear hull, then the joint system is a Riesz basis for K.*

We point out that the geometric idea of applying a generating entire function to estimate the angle between subspaces in implicit form was mentioned in Regge (1958). The usefulness of this idea cannot be overestimated: it lies at the foundation of the tests for the basis property of families of exponentials and reproducing kernels established in Pavlov (1979a) and Khrushchev, Nikol'skij and Pavlov (1981).

A similar question can also be asked in the case of operators with an absolutely continuous spectrum: when is the joint system of eigenfunctions of the absolutely continuous spectrum of the model operators B and B^+ complete in the whole space? This question was answered for abstract operators in Pavlov and Strepetov (1986). It was shown there that joint completeness always occurs on the absolutely continuous spectrum; however, the subspaces of this spectrum always intersect, hence, the joint basis property does not hold. This fact has an obvious physical interpretation.

3.10. Partial Scattering, and Scattering for Dissipative Operators. The Lax-Phillips scheme in the scattering problem for a unitary group $U_t = e^{i\mathcal{L}t}$ with incoming and outgoing subspaces can be realized in the following standard manner.

Let $\overset{0}{U}_t = e^{i\mathcal{L}^0 t}$ be a unitary group in $D_- \oplus D_+$, coinciding with U_t on $D_\pm$ (the *"trivial coupling"*):

$$(U_t - \overset{0}{U}_t)|_{D_+} = 0, \quad t > 0; \qquad (U_t - \overset{0}{U}_t)|_{D_-} = 0, \quad t < 0.$$

Further, let $U^1_t = e^{i\mathcal{L}^1 t}$ be a unitary group acting on the shift-invariant subspace $K = H \ominus \{D_- \oplus D_+\} \equiv H_1$. If the operator $\mathcal{L}$ is *"congruent"* with $\mathcal{L}^0 \oplus \mathcal{L}^1$ in the sense that the wave operators $W_\pm(\mathcal{L}^0 \oplus \mathcal{L}^1, \mathcal{L})$ exist, then so does the corresponding scattering operator

$$s\text{-}\lim_{t \to \infty} e^{-i(\mathcal{L}^0 \oplus \mathcal{L}^1)t} e^{2i\mathcal{L}t} e^{-i(\mathcal{L}^0 \oplus \mathcal{L}^1)t};$$

this operator has the block form $S = \{S_{lk}\}$, l, $k = 0, 1$, where

$$S_{lk} = s\text{-}\lim_{t\to\infty} e^{-i\mathcal{L}^l t} P_{H_l} e^{2i\mathcal{L}t} P_{H_k} e^{-i\mathcal{L}^k t}, \quad l,\, k = 0, 1.$$

Of course, the corresponding S-matrix also has a block form. Its block S_{00} coincides with the *scattering matrix in the sense of Lax-Phillips*, that is, with the characteristic function of the generator of the semigroup of compressions $Z_t = \exp(iBt) = P_K e^{i\mathcal{L}t}|_K$ acting on $K = H_1$. When $\mathcal{L}^0 \oplus \mathcal{L}^1$ differs from $\mathcal{L}$ only by a finite-dimensional operator, the full S-matrix is, of course, unitary, but its blocks may be contracting operators. These blocks are called *partial S-matrices* corresponding to the given expansion of the non-perturbed operator in orthogonal terms.

When the dissipative generator of Z_t has an absolutely continuous spectrum, that is, $N_e \neq 0$, we may want to compare the unitary group U_t^1 with the semigroup Z_t using the tools of scattering theory. In this case it is convenient to use the definition of the wave operators and scattering matrix given by (see Pavlov (1982))

$$W_+(\mathcal{L}^1, B) = s\text{-}\lim_{t\to\infty} P_{H_1} U_{-t}^1 Z_t \mathcal{P}_{N_e},$$

$$W_-(B, \mathcal{L}^1) = s\text{-}\lim_{t\to\infty} \mathcal{P}_{N_e} Z_t U_{-t}^1 P_{H_1},$$

$$S(\mathcal{L}^1, B) = W_+(\mathcal{L}^1, B) W_-(B, \mathcal{L}^1).$$

Theorem 14. *If the pair $\mathcal{L}$, $\mathcal{L}^0 \oplus \mathcal{L}^1$ is congruent and the outer component $\mathcal{B}_e$ can be separated by means of a bounded projection $\mathcal{P}_e$, then the scattering operator $S(\mathcal{L}^1, \mathcal{B})$ exists and is given by the strong limit*

$$S(\mathcal{L}^1, \mathcal{B}) = s\text{-}\lim_{t\to\infty} e^{-i\mathcal{L}^1 t} P_{H_1} e^{2iBt} P_{H_1} e^{-i\mathcal{L}^1 t}.$$

This operator coincides with the element s_{11} of the full scattering operator, that is,

$$S(\mathcal{L}^1, \mathcal{B}) = s\text{-}\lim_{t\to\infty} e^{-i\mathcal{L}^1 t} P_{H_1} e^{2i\mathcal{L}t} P_{H_1} e^{-i\mathcal{L}^1 t}.$$

The proof is based on the fact that H_1 is a shift-invariant subspace for the unitary group $e^{i\mathcal{L}t}$ and that the inner component of $\mathcal{B}$ makes no contribution in the limit:

$$s\text{-}\lim_{t\to\infty} e^{iBt} \mathcal{P}_i = 0.$$

The fact that the S-matrix of the pair $(\mathcal{L}^1, \mathcal{B})$ is an object of selfadjoint theory serves as a basis for the computation of important characteristics of a dissipative operator such as the spectral shift function, and for the derivation of trace formulae (see Adamyan and Pavlov (1979)).

§4. Spectral Analysis of the Operators Arising in Resonance Scattering Problems

All such problems are connected with the analysis of hyperbolic equations of the form

$$u_{tt} + Lu = 0, \tag{18}$$

where L is a second-order elliptic selfadjoint operator acting on a Hilbert space H. In the majority of the cases considered below L is positive. Then in the space of the Cauchy data $U = (u, u_t) = (u_0, u_1)$ we can introduce the *energy metric*

$$|U|_{\mathcal{E}}^2 = \tfrac{1}{2}\{|\sqrt{L}\,u|_H^2 + |u_t|^2\},$$

which turns it into a Hilbert space $\mathcal{E}$. The *resolving operator U_t* that transfers the Cauchy data at the initial moment $t = 0$ to an arbitrary moment t is unitary in the energy metric and defines a "finite energy" solution (in the sense of distributions) of the equation (18), which for the Cauchy data can be rewritten in the form

$$\frac{1}{i}\frac{\partial}{\partial t}U = i\begin{pmatrix} 0 & -1 \\ L & 0 \end{pmatrix}U \equiv \mathcal{L}u. \tag{19}$$

The operator $\mathcal{L}$, conventionally called in what follows the *Lax-Phillips square root* of L, is selfadjoint in the space of the Cauchy data with finite energy and its spectral characteristics are directly connected with those of L.

The Lax-Phillips scheme described in the preseding section for the analysis of the scattering problem is applicable to the hyperbolic equation (18) when the resolving group $U_t = \exp(i\mathcal{L}t)$ has orthogonal incoming and outgoing subspaces. In the classical problem for the acoustic equation $u_{tt} - \Delta u = 0$ outside a compact obstacle in $\mathbb{R}^3$, Lax and Phillips studied the exponential decay of the energy in a neighbouring zone (see Lax and Phillips (1967)). The role of the incoming and outgoing subspaces $D_\pm$ is played here by the class of Cauchy data generated outside a fixed sphere $K_\alpha = \{x : |x| < a\}$ that includes the obstacle, and the solutions are of the incoming and outgoing wave type:

$$D_- = \{U = (u_0, u_1) : u(x, t) = 0, \ |x| + t \geq a\},$$
$$D_+ = \{U = (u_0, u_1) : u(x, t) = 0, \ |x| - t \geq a\}.$$

Lax and Phillips proved the exponential decay of the energy of the solutions of (19) in K_a for star-shaped obstacles; their method is based on the compactness of the compression $Z_t = P_K U_t|K$, $t \geq 0$, of the evolution operator on the shift-invariant subspace $K = \mathcal{E} \ominus \{D_- \oplus D_+\}$ consisting of the Cauchy data with finite energy concentrated entirely in K_a, that is, such that the first component is harmonic and the second one equal to zero outside K_a. Lax and Phillips proved that the matrix S corresponding to the scattering problem for the group U_t, in other words, the characteristic function of the generator B

of the corresponding contraction semigroup $Z_t = \exp(iBt)$, $t > 0$, coincides up to an exponential factor with the *quantum-mechanical s-matrix* $s(\omega, \nu, k)$, which describes the asymptotics at infinity of the scattered waves $\psi(x, \nu, k)$:

$$\psi(x, \nu, k) \simeq e^{-ik\langle x, \nu \rangle} + f(\omega, \nu, k)\frac{e^{ik|x|}}{4\pi|x|} \quad \text{as} \quad x \to \omega \cdot \infty,$$

$$s(\omega, \nu, k) = \delta(\omega + \nu) + \frac{ik}{2\pi}f(\omega, \nu, k),$$

$$S_a = \exp(2ika)s.$$

At the same time, the study of the spectral properties of the generator reduces to that of the analytic properties of the corresponding scattering matrix, which in itself is a very difficult problem. For the classical Lax-Phillips problem in $\mathbb{R}^3$, the spectral properties of the operator B have not been fully investigated so far. In the case of an ordinary differential operator in $L_2(\mathbb{R}, E)$ with a finite vector dimension d in $L_2(\mathbb{R}, E_d)$, the spectral analysis of the resonances is developed sufficiently far. We mention that the first step in the study of the properties of the eigenfunctions of the generator B in the simplest case was completed by Regge (1958), who, in fact, proved the joint completeness property for the operators B and B^+ (see Sect. 4.1 below).

All the resonance scattering problems discussed below include the spectral analysis of the corresponding dissipative operators. From the point of view of the theory of dissipative operators, these problems have the distinguishing characteristic that in them we are working from the very beginning with the dilation of the given dissipative operator B, which is the selfadjoint generator of the unitary evolution group whose compression yields the semigroup $\exp(iBt)$, $t \geq 0$. In accordance with the theorem of Adamyan and Arov (1966), the characteristic function of the dissipative operator arising in resonance scattering problems is an object of selfadjoint theory and coincides with the scatering matrix for a pair of selfadjoint operators. The corresponding eigenfunctions of the absolutely continuous spectrum (Sect. 4.2) are also obtained by a simple compression of the eigenfunctions of the original selfadjoint operator on the shift-invariant subspace.

4.1. Resonance Scattering for a Polar Matrix-Operator. Let $A(x)$ be a strictly positive, finite $d \times d$-matrix-function on the semi-axis $0 \leq x < \infty$, which is equal to one for $x > a$ and smooth on $0 < x < a$.

In the space $L_2(\mathbb{R}^+, A)$ of all vector-functions square-integrable with weight A over the semi-axis we consider the selfadjoint operator L given by the differential expression $-A^{-1}(x)\dfrac{d^2}{dx^2}$ and the boundary condition $u(0) = 0$. The resolving group of the corresponding wave equation $A(x)u_{tt} = u_{xx}$, $u(0, t) = 0$, is unitary in the space $\mathcal{E}$ of finite-energy Cauchy data with the metric $|U|_{\mathcal{E}}^2 = \frac{1}{2}\int\limits_0^\infty \{|u_0'|^2 + \langle A(x)u_1, u_1 \rangle\}dx$ and has orthogonal incoming and outgoing

B. S. Pavlov

subspaces $D_\pm$ that consist of the Cauchy data generating the incoming and outgoing d'Alembert waves for $x \geq a$, namely

$$D_-^a = \{(u_0, u_1) : u(x,t) = u(x+t),\ x \geq a,\ t > 0\},$$
$$D_+^a = \{(u_0, u_1) : u(x,t) = u(x-t),\ x \geq a,\ t > 0\}.$$

The shift-invariant subspace $K = \mathcal{E} \ominus \{D_- \oplus D_+\}$ consists of all the finite-energy data with a constant first component and a second one equal to zero for $x > a$.

The spectral analysis of the generator B of the compression semigroup

$$Z_t = \exp(iBt) = P_K U_t | K, \quad t \geq 0,$$

was introduced in Sect. 3 for the incoming spectral representation of the dilation, that is, the operator $\mathcal{L}$, which will now be constructed by means of a special system of eigenfunctions of the absolutely continuous spectrum (the scattered waves).

Let $E_a(x, k)$ be the *Jost solution* (see Agranovich and Marchenko (1960)), in other words, the solution of the homogeneous equation $E'' = k^2 A E$, which is equal to $\exp[-ik(x-a)]I$ for $x \geq a$. The value $E_a(0, k) \equiv M_a(k)$ of $E(x, k)$ at zero is called the *Jost function*. By the *scattered wave* we understand the solution of the homogeneous equation

$$\psi_a^-(x, k) = E_a(x, -k) - E_a(x, k) M_a^{-1}(k) M_a(-k).$$

The matrix coefficient

$$S_a(-k) = M_a^{-1}(k) M_a(-k)$$

is called the *reflection coefficient*. In Ivanov and Pavlov (1978) it is shown that the incoming spectral representation $\mathcal{T}_-$ of the group is given by

$$U = \begin{pmatrix} u_0 \\ u_1 \end{pmatrix} \xrightarrow{\mathcal{T}_-} \frac{1}{\sqrt{2\pi}} \lim_{R \to \infty} \int_0^R [iku_0 + u_1]\psi^-(x, k) A(x)\,dx \equiv \tilde{U}(k),$$

with inverse mapping

$$\tilde{U} \xrightarrow{\mathcal{T}_-^{-1}} U(x) = \frac{1}{\sqrt{2\pi}} \lim_{R \to \infty} \int_{-R}^R \begin{pmatrix} 1/(ik) \\ 1 \end{pmatrix} \psi^-(x, k) \tilde{U}(k)\,dk,$$

where the integral extends over a contour going round the point $k = 0$ and lying above it. The mapping $\mathcal{T}_-$ establishes the isometric correspondences (epimorphisms)

$$L_2(\mathbb{R}^+, A) \xrightarrow{\mathcal{T}_-} L_2(\mathbb{R}),$$
$$D_-^a \to H_-^2(E),$$
$$D_+^a \to S_a H_+^2(E).$$

The last assertion shows that the reflection coefficient is the characteristic function of the dissipative generator B of the compression group, which, according to the result of Adamyan and Arov (1966), is exactly what we expected. From the existence and uniqueness theorem for differential equations it follows that the Jost solution and function are entire exponential-type functions. Just as in the case of the Schrödinger matrix-operator (see Pavlov (1973a)), it can be proved that the roots of the former in the lower half-plane correspond to the exponentially decaying Jost solutions satisfying the boundary conditions at zero. Since in our case there are no such solutions, all the roots lie in the upper half-plane. Given that the coefficients of the equation are positive, the reflection coefficient satisfies $S(k) = S^+(-k)$; consequently, the roots of the s-matrix are distributed symmetrically with respect to the imaginary axis and the corresponding root subspaces coincide, that is, $\operatorname{Ker} S(k_n) = \operatorname{Ker} S^+(-\bar{k}_n)$. Finally, the s-matrix is unitary on the real axis and analytic in the upper half-plane, in other words, it is an inner function and can be factorised as the product of a singular inner factor and a Blaschke factor. The measure corresponding to the singular factor has a unique loading at infinity.

The factorisation of the scattering matrix is a difficult analytic problem. For the Schrödinger matrix equation, this problem was solved under the condition that the potential is a smooth Hermitian matrix-function with compact support, which vanishes at the rightmost point of its minimal support in a non-smooth manner (see Pavlov (1973a)). Here we formulate an analogous theorem for a polar operator.

With the matrix-function $A(x)$ we associate a monotonically increasing family of orthogonal projections in E that is maximal with respect to the property

$$\{A(x') - I\}P(x) = 0, \quad x' > x.$$

This family is normalised with respect to left continuity and, since E is finite-dimensional, has finitely many jumps P_S at growth points a_S distributed between zero and a, to the right of which $P(x) = I$. This family is called the *indicator family* of the weight matrix A. Without loss of generality, we may assume that $P(0) = 0$.

The indicator family shows how the dimension of the space increases when the weight matrix is non-trivial, as the motion proceeds from right to left along the x-axis. We assume that, in addition to the conditions of positiveness and compact support $(A(x) = I, x > a)$, the weight matrix has the following properties.

I. The function $\{A(x) - I\}P_S$ belongs to $C_{l'}$ on $(0, a_S)$.

II. There is an integer l, $l < l'$, such that near every point of growth a_S, $0 \le a_S - x \le 1$, of the indicator family the non-degeneracy condition

$$\left| \sum_{r=0}^{l} (a_S - x)^r \frac{d^r (A(x) - I)}{dx^r}(a_S) P_S e \right| \ge \delta_S |a_S - x|^l |P_S e|, \quad \delta_S > 0,$$

holds for any $e \in E$.

The conditions I and II mean that in every subspace $P_S E$ the weight function is equal to one at the rightmost point a_S of the support, exhibiting a discontinuity of a fairly low order. Weight functions with the above properties are obtained, for example, by the compression of smooth Hermitian matrix-functions $Q(x)$,

$$A(x) - I = P(x)Q(x)P(x),$$

by means of indicator families of projections such that at the points of jump of $P(x)$ the order of degeneracy of $P(x)Q(x)P(x)$ is less than that of the smoothness.

Theorem 15 (cf. Pavlov (1973a)). *If the weight function $A(x)$ satisfies the conditions* I *and* II, *then the s-matrix can be factorised in the form*

$$S^a(k) = \exp\left(ik \sum_S (a - a_S) P_S \right) \Pi_0(k) \exp\left(ik \sum_S (a - a_S) P_S \right),$$

where $\Pi_0(k)$ is a Blaschke factor and the exponentials are singular factors.

This theorem shows that the systems of eigenvectors of the operators B and B' are complete if and only if the weight function terminates at the right end-point of the interval $(0, a)$, that is, $P_a = I$.

Theorem 16 (Ivanov and Pavlov (1978), Pavlov (1973a)). *The operators B and $-B^+$ have a discrete spectrum, which coincides with the set of points k_n in the upper half-plane such that $\operatorname{Ker} S(k_n) \neq 0$. This set is distributed symmetrically with respect to the imaginary axis, and under the conditions* I *and* II *all its points except, possibly, finitely many are simple poles of the corresponding resolvent. If $e_n \in \operatorname{Ker} S(k_n) = \operatorname{Ker} S^+(-k_n)$ and $f_n \in \operatorname{Ker} S^+(k_n)$, then the normalised eigenfunctions of the operators B and B^+ are*

$$\psi_{k_n}(x) = \begin{cases} i\sqrt{\dfrac{\operatorname{Im} k_n}{\pi}} \begin{pmatrix} 1/(ik_n) \\ 1 \end{pmatrix} E(x, k_n) e_n, & x \leq a, \\[2.5em] i\sqrt{\dfrac{\operatorname{Im} k_n}{\pi}} \begin{pmatrix} 1/(ik_n) \\ 0 \end{pmatrix} e_n, & x > a, \end{cases}$$

$$\psi_{-\bar{k}_n}(x) = \begin{cases} i\sqrt{\dfrac{\operatorname{Im} k_n}{\pi}} \begin{pmatrix} 1/(i\bar{k}_n) \\ 1 \end{pmatrix} E(x, -\bar{k}_n) f_n, & x \leq a, \\[2.5em] i\sqrt{\dfrac{\operatorname{Im} k_n}{\pi}} \begin{pmatrix} 1/(i\bar{k}_n) \\ 0 \end{pmatrix} f_n, & x > a. \end{cases}$$

These functions satisfy the biorthogonality conditions

$$\langle \psi_{k_n}, \varphi_{\bar{k}_n} \rangle \varepsilon = \begin{cases} 0, & n \neq m, \\ \langle S'(k_n) e_n, f_n \rangle \cdot 2 \operatorname{Im} k_n, & n = m. \end{cases}$$

The spectral projections on the eigenspaces of the operator B, which correspond to the simple roots of the characteristic function, are

$$\mathcal{P}_{k_n}\begin{pmatrix} u_0 \\ u_1 \end{pmatrix} = \frac{1}{2\pi}\begin{pmatrix} 1/(ik_n) \\ 1 \end{pmatrix}\sum_{e_n \in \mathrm{Ker}\, S(k_n)} \langle S'(k_n)e_n, f_n\rangle^{-1}$$

$$\times\, E(x, k_n)e_n\langle ik_n u_0 + u_1, E(\cdot, -\bar{k}_n)f_n\rangle_{L_2(A)}.$$

Since the generator $\mathcal{L}$ of the unitary group is a dilation of the dissipative operator B, we have

$$(B - \lambda I)^{-1} = P_K(\mathcal{L} - \lambda I)^{-1}|K, \quad \mathrm{Im}\,\lambda < 0.$$

It is remarkable that an analogous but more complicated equality also holds for the operator L, whose resolvent at the point $\lambda = k^2$, $\mathrm{Im}\,k > 0$, is connected with the resolvent of $\mathcal{L}$ by the formula (see Ivanov and Pavlov (1980))

$$P_K\begin{pmatrix} (L - k^2 I)^{-1} & 0 \\ 0 & (L - k^2 I)^{-1} \end{pmatrix}P_K$$
$$= -\frac{(B + kI)^{-1} + (-B^+ + kI)^{-1}}{2k}. \tag{20}$$

From this formula it follows that the resolvent of L can be extended analytically to the non-physical sheet $\mathrm{Im}\,k < 0$ of the spectral variable; on this sheet, however, it is no longer the resolvent of any operator, but merely a linear combination of the resolvents of $-B$ and B^+. Nevertheless, the equality (20) enables us to compute the principal parts of the resolvent in terms of the root vectors of $-B$ and B^+.

Theorem 17 (Pavlov (1976), Ivanov and Pavlov (1978)). *The principal part of the resolvent of the operator L at the pole $\bar{k}_n$ for a smooth function u_0 such that* $\mathrm{supp}\, u_0 \subset [0, a]$ *has the form*

$$\frac{1}{2\pi}\cdot\frac{1}{2k(\bar{k}_n - k)}\sum_{e_n}\frac{E(x, -\bar{k}_n)f_n\langle u, E(\cdot, k_n)e_n\rangle_{L_2}}{\langle S'(k_n)e_n, f_n\rangle}.$$

Here e_n is an orthonormal basis for $\mathrm{Ker}\, S(k_n)$ *and* $f_n = S(k_n, e_n|S(k_n)e_n|^{-1})$.

A more refined analysis of the spectral properties of B is based on stronger assumptions on the coefficient A.

Theorem 18 (Ivanov and Pavlov (1978)). *If the matrix $\sqrt{A}$ is real-analytic on the interval $[0, a]$ and has no multiple eigenvalues, all the eigenvalues $\lambda_j(a)$ of the operator $A(a - 0)$ are different from unity, and all the numbers*

$$\beta_j = \frac{1}{2\int\limits_0^a \lambda_j(\xi)d\xi}\ln\left|\frac{\lambda_j(a) + 1}{\lambda_j(a) - 1}\right| \equiv \frac{1}{2\alpha_j}\ln\left|\frac{\lambda_j(a) + 1}{\lambda_j(a) - 1}\right|$$

are distinct, then

1) all sufficiently remote roots k_n of the S-matrix are simple and their set can be expanded in finitely many series

124   B. S. Pavlov

$$q_l^j = \frac{\pi}{\alpha_j}l + \gamma_j + i\beta_j + o(1), \quad \gamma_j = \begin{cases} 0, & \lambda_j(a) > 1, \\ \dfrac{\pi}{2\alpha_j}, & \lambda_j(a) < 1, \end{cases}$$

$$j = 1, 2, \ldots, \dim E, \quad l = 0, \pm 1, \ldots;$$

2) *the root vectors $e_l^j \in \operatorname{Ker} S(q_l^j)$ corresponding to these roots are asymptotically orthogonal, that is, $e_l^j \to e^j$ as $l \to \infty$ and $\langle e^j, e^{j'} \rangle = 0$ for $j \neq j'$, and the subspaces $\operatorname{Ker} S(q_l^j)$ are one-dimensional;*

3) *the system of root vectors of B is complete and forms a Riesz basis for the shift-invariant subspace.*

In the one-dimensional case $\dim E = 1$, the polar equation was discussed in Khrushchev, Nikol'skij and Pavlov (1981) and Pekker (1976). The latter studied the resonance scattering of waves for a spherically symmetric non-homogeneous density in $\mathbb{R}^3$. The variables can be separated in this problem, which reduces to an infinite sequence of problems corresponding to a fixed angular momentum for the coefficients u_l in the expansion of the solution in spherical functions Y_l, that is, $u = \sum\limits_{l=0}^{\infty} u_l(r) Y_l$. Here the functions $v_l = r u_l$ satisfy the wave equations

$$\frac{\partial^2 v_l}{\partial t^2} = \frac{1}{\rho^2(r)}\left[\frac{\partial^2 v_l}{\partial r^2} - \frac{l(l+1)}{r^2} v_l\right], \quad l = 0, 1, \ldots,$$

which, in the spaces $\mathcal{H}_l$ of Cauchy data $V \equiv (v, v_t) \equiv (v^0, v^1)$ with the metric

$$|V|_{\mathcal{H}_l}^2 = \frac{1}{2} \int\limits_0^{\infty} \left\{|v^0|^2 + \frac{l(l+1)}{r^2}|v^0|^2 + \rho^2(r)|v^1|^2\right\} dr,$$

generate a unitary dynamics with orthogonal incoming and outgoing subspaces $D_{\pm}^l$ consisting of the Cauchy data for the incoming and outgoing solutions with support in $(0, \infty)$. The density $\rho^2(r)$ is assumed to be positive, twice continuously differentiable on $[0, \infty)$, identically equal to unity for $x > a$, and to have one of the following three types of behaviour at a:

1) ρ is twice continuously differentiable on the closed interval $[0, a]$ and bounded, that is, $0 < c_1 \leq \rho(x) \leq c_2 < \infty$, $\rho(a - 0) \neq 1$;

2) $\rho(x) = (a - x)^{g-1} \rho_1(x)$, $1/2 < g < 1$, and $\rho_1(x)$ is twice continuously differentiable and bounded on $[0, a]$;

3) $\rho(x) = (a - x)^g \rho_1(x)$, $1 < g < \infty$, with the same conditions on ρ_1.

In the case 1), the distribution of the resonances, including those for higher moments $l > 0$, remains the same as for $l = 0$ and is given by the preceding theorem. In the cases 2 and 3 the following assertion holds.

Theorem 19 (Pekker (1976)). *For any moment number l, the resonances $k_n^{(2)}$ and $k_n^{(3)}$ tend asymptotically from the upper half-plane to the real axis along the lattice of integers, that is,*

$$k_n^{(2)} = -\left[\pi n + \frac{\pi}{2}(1 + 1/g)\right]\left(\int_0^a \rho(s)ds\right)^{-1} + O(1/n),$$

$$k_n^{(3)} = -\left[\pi n + \frac{\pi}{2}(1 - 1/g)\right]\left(\int_0^a \rho(s)ds\right)^{-1} + O(1/n^\alpha),$$

where $\alpha = 1$ for $g < 2$ and $\alpha = 2/g$ for $g > 2$.

The corresponding systems of resonance states are Riesz bases for their linear hulls and complete in the shift-invariant subspaces $K_l = \mathcal{H}_l \ominus \{D_+^l \oplus D_-^l\}$, and their corresponding orthogonalisers Φ^l can be represented in the form $I + V^l$, where V^l are compact operators in the symetrically normalised ideals $\mathfrak{S}_\alpha$, $\alpha = \alpha(g)$.

In Ivanov and Pavlov (1978), Pavlov (1971b) and Pekker (1976) the *Regge problem* was also discussed in its original formulation (see Regge (1958)). Regge sought conditions for the completeness of the family of solutions of a Sturm-Liouville problem with the spectral parameter in the boundary conditions, namely

$$-u'' = k^2 A^2(x)u, \quad 0 < x < b, \quad b > a,$$
$$u(0) = 0, \quad u'(b) + iku(b) = 0. \tag{21}$$

From Theorem 16 we see that the solutions of the problem (21) exist when k coincides with the resonances k_n, regardless of the choice of b, and are simply the second components of the eigenfunctions of the corresponding operator B^b, which is the generator of the compression of the unitary evolution group for the wave equation on the shift-invariant subspace that is the orthogonal complement $K_b = \mathcal{E} \ominus \{D_-^b \oplus D_+^b\}$ of the incoming and outgoing subspaces $D_\pm^b$ consisting of the Cauchy data for the incoming and outgoing solutions with support in (b, ∞). Theorem 18 also shows that, in view of the coincidence of the kernels at resonances symmetric with respect to the imaginary axis, that is, $\operatorname{Ker} S(k_n) = \operatorname{Ker} S^+(-\bar{k}_n)$, $k_n = -\bar{k}_n$, the eigenfunctions of the operator $-B^+$ at k_n are

$$\varphi_{k_n}(x) = \begin{cases} -i\sqrt{\dfrac{\operatorname{Im} k_n}{\pi}}\begin{pmatrix} 1 - /(ik_n) \\ 1 \end{pmatrix}E(x, k_n)e_n, & x \le b, \\ -i\sqrt{\dfrac{\operatorname{Im} k_n}{\pi}}\begin{pmatrix} -1/(ik_n) \\ 0 \end{pmatrix}e_n, & x > b, \end{cases}$$

which means that

$$\tfrac{1}{2}\{\psi_{k_n}(x) + \varphi_{k_n}(x)\} = \begin{cases} -i\sqrt{\dfrac{\operatorname{Im} k_n}{\pi}}\begin{pmatrix} 0 \\ 1 \end{pmatrix}E(x, k_n)e_n, & x \le b, \\ 0, & x > b, \end{cases}$$

$$\tfrac{1}{2}\{\psi_{k_n}(x) - \varphi_{k_n}(x)\} = \begin{cases} -i\sqrt{\dfrac{\operatorname{Im} k_n}{\pi}}\begin{pmatrix} 1/(ik_n) \\ 0 \end{pmatrix}E(x, k_n)e_n, & x \le b, \\ -i\sqrt{\dfrac{\operatorname{Im} k_n}{\pi}}\begin{pmatrix} 1/(ik_n) \\ 0 \end{pmatrix}e_n, & x > b, \end{cases}$$

From this we deduce the following assertion.

Theorem 20. *The one-component system $\{E(x, k_n)e_n\}$ is complete in $L_2(0, b)$ and $W_2^1(0, b)$ if and only if the joint system of eigenvectors of the operators B_b and B_b^+ is complete in the shift-invariant subspace, that is, if joint completeness holds.*

This assertion enables us to study the completeness (and basis property) of the system of solutions of the equation (21). Solving the corresponding problem for the Schrödinger equation with a potential with compact support which vanishes outside $(0, a]$, Regge (1958) showed that the family of solutions of a similar problem forms a complete system in $L_2(0, b)$ if $b \geq 2a$. The following assertion regarding the one-dimensional polar problem (21) ($\dim A = 1$) was proved in Khrushchev, Nikol'skij and Pavlov (1981).

If $a \leq b \leq a + \alpha$, then the system of solutions of problem (21) is complete in $L_2(0, b)$; if, in addition, $b = a + \alpha$, where $\alpha = \int\limits_0^a A(x)dx$, then this system is a Riesz basis.

The proof of this assertion in Khrushchev, Nikol'skij and Pavlov (1981) is based on a theorem on the equivalence of the Regge problem to that of joint completeness mentioned in the preceding section. More refined details can be found in Khrushchev (1985).

4.2. The Resonance Scattering Problem for the One-Dimensional Schrödinger Operator with a Matrix Potential.

This topic was studied in Pavlov (1973a), where the factorisation of the S-matrix and the series structure of its roots were described.

Theorem 21 (Pavlov (1973a)). *If the matrix potential V of the one-dimensional Schrödinger operator $Ly = -y'' + V(x)y$, $y(0) = 0$, in $L_2(0, \infty)$ is Hermitian and has compact support, that is, $V(x) = 0$ for $x > a$, and its associated matrix-function $\sigma_0(x) = \int\limits_x^\infty V(s)\,ds$ satisfies the conditions I and II in Sect. 4.1 with l replaced by $l + 1$, then the Jost function $M_a(k)$ normalised by means of the condition $E_a(k) = 1$, $E_a'(a, k) = -ik$, has as $|x| \to \infty$ the asymptotics*

$$M_a(k) = e^{ika}\left[1 + o(1) + \sum_j \exp(-2ika_j)\right]$$

$$\times \frac{1}{(2ik)^2}\left[V(a_j) + \frac{V'(a_j)}{2ik} + \cdots + \frac{V^{(e)}(a_j)}{(2ik)^e}\right]p_j.$$

Here the a_j are the growth points of the indicator family $P(x)$ of the potential and $p_j = P(a_j + 0) - P(a_j)$ their corresponding jumps.

If, in addition, $N_j^r = \bigcap\limits_{0 \le j \le r} \mathrm{Ker}\, V^{(s)}(a_j)$, $\Pi_j^r \equiv \Pi_{N_j^r}$ is the corresponding chain of orthogonal projections, $\Pi_j^{r-1} - \Pi_j^r = \pi_j^r$, $r \ge 1$, $p_j = \sum\limits_{r=1}^{l+1} \pi_j^r$, and the condition II is satisfied, that is,

$$\text{the operators } \pi_j^r V^{(s)} \pi_j^s \text{ are invertible in } \pi_j^s E,$$

then the roots k_n of the Jost function $M_a(k)$ are close to those of the matrix-function

$$f(k) = I + \sum_j \exp(-2ika_j) \frac{1}{(2ik)^2} \sum_{s=0}^{l} \frac{\pi_j^s V^{(s)} \pi_j^s}{(2ik)^s}$$

$$= \sum_{a_j > 0} \sum_{s=0}^{l} \left\{ \pi_j^s + e^{2ika_j} \frac{\pi_j^s V^{(s)}(a_j) \pi_j^s}{(2ik)^{s+2}} \right\};$$

consequently, they can be expanded in a series that is asymptotically close to the series $k_n \sim \dfrac{n\pi}{a_j} + \ln \dfrac{n\pi}{a_j} \cdot \dfrac{s+2}{2} + O(1)$ for the roots of the scalar equations

$$1 + \exp(-2ika_j)(2ik)^{-s-2}\lambda_j^s = 0,$$

where λ_j^s are the eigenvalues of the operator $\pi_j^s V^{(s)}(a_j)\pi_j^s$. The corresponding root vectors e_{k_n}, $|k_n| \gg 1$, of the Jost function are close to the eigenvectors of the $\pi_j^s V^{(s)}(a_j)\pi_j^s$.

From Theorem 21 we see that all the eigenvalues of the operators B and $-B^+$ except, perhaps, finitely many, are simple and can be expanded in asymptotically orthogonal series. We also see that these are not Carleson series, therefore, in the Schrödinger case the families of root vectors of B and $-B^+$ do not form a Riesz basis and do not even have the uniform minimality property. However, the expansions in these systems of elements in the shift-invariant subspace which are sufficiently smooth (and vanish fairly rapidly near the right-hand boundary of the support of the potential) converge in the energy norm.

Theorem 22. *If l is the number occurring in the condition of non-degeneracy of the potential jumps and u an element in $\mathcal{D}(\mathcal{L}^s) \cap N_d$, $s \ge 2(l+2)+1$, then the Fourier series of u with respect to the biorthogonal system of root vectors of the operators B and $-B^+$ converges in the energy norm.*

From Sect. 3.7 it follows that this series is summable in the sense of Abel for every element in K.

The important question of completeness of the system of root vectors of B and $-B^+$ is solved on the basis of Helson's test; completeness is guaranteed by the absence of a singular cofactor in the chracteristic function (see Helson (1964)).

Theorem 23 (Pavlov (1972)). *For the one-dimensional Schrödinger operator with a Hermitian matrix potential with compact support satisfying conditions* I *and* II, *the characteristic function of the operator* B^a *can be factorised in the form*

$$S_a = \Pi \exp(2ik) \sum_j (a - a_j) p_j = \exp(2ik) \sum_j (a - a_j) p_j \cdot \tilde{\Pi},$$

where Π *and* $\tilde{\Pi}$ *are Blaschke-Potapov products and* p_j *the indicator family of the potential. In particular, the systems of eigenfunctions of the operators* B *and* $-B^+$ *are complete in* K *if and only if the indicator family is trivial, that is,* $a_j = a$.

4.3. Resonance Scattering by an Arbitrary Potential.

In $L_2(-\infty, \infty)$ we consider the one-dimensional scalar Schrödinger operator with a nonnegative, locally bounded, measurable potential that vanishes on the left semi-axis, that is,

$$Ly = -y'' + q(x)y, \qquad q(x) = 0 \quad \text{for} \quad x < 0.$$

In the energy space $\mathcal{E}$ of the Cauchy data $U = (u, u_t)$ with norm

$$|U|_{\mathcal{E}}^2 = \frac{1}{2} \int_{-\infty}^{\infty} (|u_0'|^2 + q|u_0|^2 + |u_1|^2)dx,$$

the corresponding wave equation $u_{tt} + Lu = 0$ generates a unitary dynamics $U_t : U(0) \to U(t)$ with incoming and outgoing subspaces $D_{\mp}$ that consist of the Cauchy data for the incoming and outgoing d'Alembert waves $U(x \mp t)$ with supports on the left semi-axis. The corresponding shift-invariant subspace K consists of the finite energy data with support on the right semi-axis; in this context, u_0 is extended to the left semi-axis by a constant a and u_1 by zero, so that the entire energy of the data in K is located on the right semi-axis. The outgoing spectral representation of the generator $\mathcal{L}$ of the dynamics $U_t = \exp(i\mathcal{L}t)$ is constructed on the basis of the spectral representation of the operator L connected with the scattered waves; the latter are the solutions of the homogeneous equation $Lu = k^2 u$ whose form on the left semi-axis $(\text{Im } k = 0)$ is

$$\psi_+ = \chi_+(k, x) + S(k)\chi_-(k, x) = \exp(ikx) + S(k)\exp(-ikx).$$

The reflection coefficient S is determined from the condition that this solution should be proportional on the right semi-axis to the Weyl solution $\psi(x, \lambda) = \theta(x, \lambda) + m(\lambda)\varphi(x, \lambda)$, where $\theta(0, \lambda) = 1$, $\theta'(0, \lambda) = 0$, $\varphi(0, \lambda) = 0$, and $\varphi'(0, \lambda) = 1$. It turns out that $S(k)$ is connected with the Weyl function $m(\lambda)$ by the formula

$$S(k) = \frac{m(k^2) - i/k}{m(k^2) + i/k}, \quad k^2 = \lambda, \tag{22}$$

which means that $S(k)$ is an analytic contracting function in the upper half-plane.

The solutions ψ_+ constructed in this way are analytic functions of k in the upper half-plane for every fixed $x > 0$. Along with them we may consider the complex conjugate solutions $\psi_- = \bar{S}_+ = \bar{S}\chi_+ + \chi_-$ for k real. The latter are limit values of analytic functions in the lower half-plane $\operatorname{Im} k < 0$.

The function $S(k)$ arising here, called the *reflection coefficient*, is extended to the lower half-plane by means of the symmetry principle across the intervals of the real axis k where it is unitary, that is, across the complement of the absolutely continuous spectrum of the operator L_1 given in $L_2(0, \infty)$ by the same differential expression as L and a zero boundary condition at the origin. Together with L and L_1 we also consider the operator $L_0 = -\dfrac{d^2}{dx^2}$ with the condition $y' = 0$ at the origin. With L_0 and L_1 we also associate wave equations, which generate unitary dynamics $U_t^{0,1}$ in the corresponding energy spaces $\mathcal{E}_0$ and $\mathcal{E}_1$. The operators

$$\mathcal{L}_r = i \begin{pmatrix} 0 & -1 \\ L_r & 0 \end{pmatrix}, \quad L_r = L, L_0, L_1,$$

are the generators of these unitary groups.

Theorem 24 (Pavlov (1973b)). *The inversion formulae for the operators L_r, which are selfadjoint in the corresponding energy spaces of Cauchy data, are*

$$\begin{pmatrix} u_0 \\ u_1 \end{pmatrix} \overset{\mathcal{T}}{\longmapsto} \frac{1}{2} \begin{pmatrix} (iku_0 + u_1, \chi_+(k, *)) \\ (iku_0 + u_1, \chi_-(k, *)) \end{pmatrix} = \begin{pmatrix} \tilde{u}_0 \\ \tilde{u}_1 \end{pmatrix},$$

$$\begin{pmatrix} \tilde{u}_0 \\ \tilde{u}_1 \end{pmatrix} \overset{\mathcal{T}^{-1}}{\longmapsto} \frac{1}{2\pi} \int\limits_{-\infty}^{\infty} \begin{pmatrix} \dfrac{1}{ik}\chi_+(k, x) & \dfrac{1}{ik}\chi_-(k, x) \\ \chi_+(k, x) & \chi_-(k, x) \end{pmatrix} \begin{pmatrix} 1 & \bar{S}(k) \\ S(k) & 1 \end{pmatrix} \begin{pmatrix} \tilde{u}_0 \\ \tilde{u}_1 \end{pmatrix} dk,$$

$$E \overset{\mathcal{T}}{\longmapsto} L_2\left(\frac{1}{2\pi} \begin{pmatrix} 1 & \bar{S} \\ S & 1 \end{pmatrix} \right),$$

$$U_t \longmapsto e^{ikt}$$

for any Cauchy data with finite energy (all integrals are understood as strong limits of the corresponding integrals over finite intervals),

$$\begin{pmatrix} u_0 \\ u_1 \end{pmatrix} \overset{\mathcal{T}_1}{\longmapsto} (iku_0 + u_1, (1 - s)^{-1}\varphi) = \tilde{u},$$

$$\tilde{u} \overset{\mathcal{T}_1^{-1}}{\longmapsto} \frac{2}{\pi} \int\limits_{-\infty}^{\infty} (1 - |S(k)|^2) \begin{pmatrix} \dfrac{1}{ik} \cdot \dfrac{\varphi}{1 - S} \\ \dfrac{\varphi}{1 - S} \end{pmatrix} \tilde{u}(k)dk,$$

$$A_1 \overset{\mathcal{T}_1}{\mapsto} L_2\left(\frac{2}{\pi}(1 - |S(k)|^2)\right),$$

$$U_t^1 \overset{\mathcal{T}_1}{\mapsto} e^{ikt}$$

for arbitrary data in the absolutely continuous subspace A_1 of the operator L_1, $A_1 \subset H_1$, and

$$\begin{pmatrix} u_0 \\ u_1 \end{pmatrix} \overset{\mathcal{T}_0}{\mapsto} \left(iku_0 + u_1, \frac{\chi_+ + \chi_-}{2}\right) = \tilde{u},$$

$$\tilde{u} \overset{\mathcal{T}_0^{-1}}{\mapsto} \frac{2}{\pi}\int_{;-\infty}^{;\infty} \begin{pmatrix} \cos kx \\ \dfrac{ik}{\cos kx} \end{pmatrix} \tilde{u}(k)dk,$$

$$H_0 \overset{\mathcal{T}_0}{\mapsto} L_2,$$

$$U_t^0 \overset{\mathcal{T}_0}{\mapsto} e^{ikt}$$

for arbitrary data in H_0.

We see that the unitary groups U_t and U_t^0 have general incoming and outgoing subspaces consisting of the Cauchy data for the incoming and outgoing waves on the left semi-axis. This enables us to define wave operators for the pair of dynamics U_t and U_t^0, and to construct the scattering matrix $S(U_0, U)$. We can also compare the pair of dynamics $\{U_t^0 \oplus U_t^1\}$ and U_t, and, constructing the wave operators, we can compute the "full" scattering matrix $S(U^0 \oplus U^1, U; k)$.

Theorem 25 (Pavlov (1976)). *The full matrix is*

$$S(U^0 \oplus U^1, U; k) = \begin{pmatrix} S(k) & 1 \\ \dfrac{1 - S(k)}{1 - \bar{S}(k)}(1 - |S|^2) & -\bar{S}(k)\dfrac{1 - S(k)}{1 - \bar{S}(k)} \end{pmatrix},$$

$$S(U^0, U) = S(k).$$

The full S-matrix is unitary with respect to the metric $\begin{pmatrix} 1 & 0 \\ 0 & 1 - |S|^2 \end{pmatrix}$.

Going over to the study of the dissipative generator of the contraction semigroup of the dynamics U_t on the shift-invariant subspace $K = \mathcal{E} \ominus \{D_- \oplus D_+\}$, we first mention that the inversion operator $\mathcal{T}$ in the theorem, which is connected with the Jost solutions, yields a symmetric spectral representation of the unitary group. Using the theorems on the spectral analysis of the model operator given in Sect. 3 and the inversion formulae, we arrive at the following assertion.

Theorem 26 (Pavlov (1972,1976)). *The reflection coefficient S is the characteristic function of the operator B. The absolutely continuous spectrum of B consists of the set of intervals on the real axis obtained as the closure of* *the intervals where $1 - |S(k)|^2 > 0$. The role of eigenfunctions of the continuous spectrum of B is played by the functions*

$$\psi_k(x) = \begin{cases} \begin{pmatrix} 1/(ik) \\ 1 \end{pmatrix} \chi_+(x,k), & x > 0, \ k = k + i0, \\ \begin{pmatrix} 1/(ik) \\ 0 \end{pmatrix}, & x < 0, \end{cases}$$

and the eigenfunctions of the operator $-B^+$ are

$$\varphi_{-k}(x) = \begin{cases} \begin{pmatrix} 1/(ik) \\ 1 \end{pmatrix} \chi_-(x,k), & x > 0, \ k = k + i0, \\ \begin{pmatrix} 1/(ik) \\ 0 \end{pmatrix}, & x < 0. \end{cases}$$

The spectral projections corresponding to the intervals of the absolutely continuous spectrum that do not contain spectral singularities are written in the form $(x > 0)$

$$\mathcal{E}^B_\Delta \begin{pmatrix} u_0 \\ u_1 \end{pmatrix} = -\frac{1}{2\pi} \int_\Delta \begin{pmatrix} 1/(ik) \\ 1 \end{pmatrix} \chi_+(x,k) \langle iku_0 + u_1, \tfrac{1}{2}\chi_-(*,k)\rangle \frac{1 - |S(k)|^2}{S(k)} \, dk.$$

The eigenvalues of the operator B coincide with the roots k_n of the s-matrix, that is, $S(k_n) = 0$. They are distributed in the upper half-space $\operatorname{Im} k_n > 0$ and their corresponding eigenfunctions are

$$\psi_{k_n} = \begin{cases} i \begin{pmatrix} 1/(ik_n) \\ 1 \end{pmatrix} \sqrt{\frac{\operatorname{Im} k_n}{\pi}} \, \chi_+(k_m, x), & x > 0, \\ i \begin{pmatrix} 1/(ik_n) \\ 0 \end{pmatrix} \sqrt{\frac{\operatorname{Im} k_n}{\pi}}, & x > 0. \end{cases}$$

The eigenfunctions of the operator $-B^+$ corresponding to the eigenvalues $-\bar{k}_n$ are

$$\varphi_{-\bar{k}_n} = \begin{cases} i \begin{pmatrix} 1/(i\bar{k}_n) \\ 1 \end{pmatrix} \sqrt{\frac{\operatorname{Im} k_n}{\pi}} \, \chi_-(\bar{k}_m, x), & x > 0, \\ i \begin{pmatrix} 1/(i\bar{k}_n) \\ 0 \end{pmatrix} \sqrt{\frac{\operatorname{Im} k_n}{\pi}}, & x > 0. \end{cases}$$

They are biorthogonal if the roots are simple, that is, if $S'(k_n) \neq 0$, then

$$\langle \psi_{k_n}, \varphi_{-\bar{k}_m} \rangle_\mathcal{E} = -2i \operatorname{Im} k_n S'(k_n) \delta_{mn},$$

and in this case the eigenprojections of the operator B are written in the form

$$\mathcal{P}_{k_n} \begin{pmatrix} u_0 \\ u_1 \end{pmatrix} = \begin{pmatrix} 1/(ik_n) \\ 1 \end{pmatrix} \chi_+(k_n, x) \langle iku_0 + u_1, \tfrac{1}{2}\chi_-(\bar{k}_n, x)\rangle \frac{1}{\pi i S'(k_n)}, \quad x < 0.$$

From Theorem 26 it follows that the eigenfunctions of the absolutely continuous spectrum of the given dissipative operator are obtained by a simple compression on the shift-invariant subspace of the eigenfunctions $\begin{pmatrix} 1/(ik) \\ 1 \end{pmatrix} \chi_+$ of the generator $\mathcal{L}$ of the original dynamics. This is a reflection of the general result (Theorem 10), according to which a canonical system of eigenfunctions of a dissipative operator is obtained by a compression on the shift-invariant subspace of the radiating eigenfunctions of the dilation.

Theorem 26 enables us to apply to B the general theorems on the spectral properties of dissipative operators mentioned in Sect. 3. It is clear that the detailed study of the analytic properties of the reflection coefficient and the investigation of the location of its roots (the eigenvalues of B) in the complex plane and of the spectral singularities on the real axis are central questions of the spectral analysis of this operator. On the basis of this analysis we can develop a scattering theory for the pair B, $\mathcal{L}_1$ in accordance with Theorem 14 (in Sect. 3.10). The S-matrix occurring here differs only by a unitary factor from the s-matrix in the "Lax-Phillips channel" $\partial_+ \oplus \partial_-$:

$$S(\mathcal{L}, B) = s\text{-}\lim_{t \to \infty} e^{-i\mathcal{L}_1 t} P_K e^{2i\mathcal{L}t} P_K e^{-i\mathcal{L}_1 t} = T_1^{-1}\left\{ -\bar{S}(k)\frac{1 - S(k)}{1 - \bar{S}(k)} \right\} T_1.$$

This is a general property, which can be proved in terms of a model for abstract operators of the form $\mathcal{L}_0 \oplus \mathcal{L}_1$ comparable to $\mathcal{L}$ (Pavlov (1975a)).

A detailed analysis of the complex roots of the reflection coefficient for a smooth, real and periodic potential on the right semi-axis with discontinuities at the integral values of x, that is, $q(x) = q(x+1)$, $q(1-0) \neq q(1+0) = q(0)$, can be found in Pavlov and Smirnov (1977). In this case, the *Weyl function* on the right semi-axis can be computed in terms of standard solutions by means of the formula

$$m(k) = -\frac{\varphi' + \theta}{\varphi} \pm \sqrt{\left(\frac{\varphi' - \theta}{2\varphi}\right)^2 - \frac{1}{\varphi^2}}, \quad \varphi = \varphi(1-0), \quad \theta = \theta(1-0).$$

Here the square root is determined from the condition

$$\mathrm{Im}\left\{ \sqrt{\left(\frac{\varphi' - \theta}{2}\right)^2 - 1} + \frac{i}{k}\theta' \right\} > 0.$$

The characteristic function of B in this case is

$$S(k) = \frac{k[\varphi' + \theta + \sqrt{(\varphi' - \theta)^2 - 4} + i\theta']}{k[\varphi' + \theta + \sqrt{(\varphi' - \theta)^2 - 4} - i\theta']}.$$

Theorem 27 (Pavlov and Smirnov (1977)). *Let $q(0) = q_0$, $q(1) = q_1$, $q'(0) = q_0'$, $q'(1) = q_1'$, and $\int_0^1 q\, dx = Q$. If $q_0/q_1 > 1$, then, as $n \to \infty$, in the upper half-plane there is an infinite sequence of roots*

$$k_n = n\pi + \frac{i}{2}\ln\frac{q_0}{q_1} - \frac{1}{2\pi n}\left\{\frac{q_0'}{2q_0} - \frac{q_1'}{\alpha q_1} + \frac{Q}{q_0}(q_0 + q_1)\right\} + O\left(\frac{1}{n^2}\right).$$

In addition, there may be roots k_n' close to the points $k_n^0 = n\pi - Q/(2n\pi)$. In this case only finitely many roots k_n' may be real.

We remark that the first series of roots lies asymptotically over the left end-points of the gaps at a finite height of $\dfrac{i}{2}\ln\dfrac{q_0}{q_1}$, while the second one lies near these points k_n^0.

If we take the potential on the right semi-axis to be the interval $[0, N]$ of a periodic potential, then the corresponding coefficient has the form

$$S_N = \frac{\rho^{-N}(k) - \rho^N(k)}{\rho^{-N}(k)S_1^{-1}(k) - \rho^N S_2^{-1}(k)},$$

where $S_l = \dfrac{m_l + ik}{m_l + ik}$, m_l, $l = 1, 2$ are the Weyl functions for the corresponding periodic problems on $(0, \infty)$ and $(-\infty, 0)$, respectively, and $-i\ln\rho = p(k)$ is the quasimomentum corresponding to the given periodic potential. We see that the reflection coefficient vanishes at the points where $\sin Np(k) = 0$. In every band of the periodic potential where p changes on an interval of length π there are exactly N such points, and they divide the band into equal parts in the scale of the quasimomentum (see Pavlov and Smirnov (1977)).

In Pavlov (1974) the spectral properties are discussed of a dissipative operator whose characteristic function is, conversely, the coefficient of transition to the problem of one-dimensional scattering with an arbitrary real potential satisfying the condition $\int |q(x)|(1 + |x|)dx < \infty$. The spectrum of this operator always has an absolutely continuous component and may also have a discrete one, corresponding to the roots in the complex plane of the transition coefficient. The eigenfunctions of this operator are obtained from the standard Jost solutions by means of an orthogonal projection in the Hardy classes of functions depending on the spectral parameter k.

4.4. The Automorphic Wave Equation. It is well known that the Lobachevskij plane realised as the upper half-plane $\operatorname{Im} z > 0$ with the measure $y^{-2}dx\,dy = d\mu$ is the homogeneous space of the group $G = \mathrm{SL}(2\mathbb{R})$ of all 2×2-matrices with determinant equal to unity. G acts on this space as the group of Moebius transformations $z \mapsto gz = (az + c)(bz + d)^{-1}$ with real coefficients and determinant $ad - bc = 1$. The measure μ and the operator $-q^2\Delta - \frac{1}{4}I = L$ are invariant with respect to G. Let Γ be a subgroup of G such that its fundamental domain F is non-compact: it has a unique point on the "absolute" $y = 0$ and the volume of F is finite, that is, $\int_F d\mu < \infty$. An example of such subgroup is the modular group with generators $z \mapsto -1/z$ and $z \mapsto z + 1$. The corresponding fundamental domain is the strip $F = \{z : |\operatorname{Re} z| < 1, |z| > 1\}$.

In F we consider functions that are restrictions to F of smooth automorphic functions u, $u(z) = u(\gamma z)$, $\gamma \in \Gamma$, defined on the plane. The operator L is essentially selfadjoint on such functions (see Faddeev (1967)); we denote by L its closure and consider the *automorphic wave equation* $u_{tt} - Lu = 0$. In the space of all the Cauchy data $U = (u, u_t) = (u_0, u_1)$ with finite energy

$$|U|_{\mathcal{E}}^2 = \frac{1}{2} \int_F \{|\nabla u_0|^2 - (4y^2)^{-1}|u_0| + y^{-2}|u_1|^2\} dx\, dy, \qquad (23)$$

the wave equation generates a formally unitary dynamics $U_t : (u, u_t)(0) \mapsto (u, u_t)(t)$ with incoming and outgoing subspaces $D_{\mp}$ consisting of the Cauchy data of the incoming and outgoing waves

$$u_{\mp}(y, t) = \exp(\mp t/2)\Phi(y e^{\pm t}$$

with support on the half-strip $F \cap \{y > 1\}$. Unfortunately, the energy metric (23) is indefinite since in this case the operator L_Γ has a unique negative eigenvalue $\lambda = -1/4$ whose eigenfunction is identically constant. Imposing the condition of orthogonality to unity on each component of the Cauchy data with finite energy (23), we obtain an energy space $\hat{\mathcal{E}}$ with a positive metric, in which our wave equation also generates a unitary dynamics U_t with incoming and outgoing subspaces $\hat{D}_{\pm} = \{(u_0, u_1) \in D_{\pm}, u_{0,1} \perp 1\}$ (see Pavlov and Faddeev (1972)). Along with the above dynamics, we also consider the non-perturbed one U_t^0, defined by the one-dimensional wave equation $u_{tt} - y^2 u_{yy} + (1/4)u = 0$ in the space of all the Cauchy data $U = (u, u_t)$ component-wise orthogonal to unity, equipped with the energy norm

$$|U|_{\hat{\mathcal{E}}}^2 = \frac{1}{2} \int_0^\infty \left\{|u_y|^2 - \frac{1}{4y^2}|u|^2 + \frac{1}{y^2}|u_t|^2\right\} dy.$$

The incoming and outgoing subspaces $D_{\pm}$ of the groups U_t and U_t^0 coincide, which enables us to use the Lax-Phillips approach. The corresponding scattering matrix is computed explicitly in the form of the reflection coefficient (see Faddeev (1967)), and is expressed in terms of the Riemann ζ-function or, better, in terms of the entire function

$$\xi(s) = s(s - 1)\pi^{-s/2}\Gamma(1/(2s))\zeta(s),$$

also introduced by Riemann, by means of the relation

$$S(k) = \frac{\xi(-2ik)}{\xi(2ik)} \cdot \frac{k - i/2}{k + i/2}. \qquad (24)$$

We mention that all the zeros of the ξ-function lie in the strip $0 < \mathrm{Re}\, s < 1$ and coincide with the "critical" zeros of the ζ-function. The set of roots has two symmetry axes, namely $\mathrm{Re}\, s = 1/2$ and $\mathrm{Im}\, s = 0$.

From the general theory, developed in Gel'fand (1952), it follows that the S-matrix (24) is analytic in the upper half-plane and that all its roots k_n except $i/2$ lie in the strip $0 < \mathrm{Im}\, k < 1/2$ and are distributed symmetrically with respect to the lines $\mathrm{Im}\, k = 1/4$ and $\mathrm{Re}\, k = 0$. They are uniquely connected with the roots s_n of the ζ-function in the "critical" strip by the formula $k_n = is_n/2$. At the imaginary infinity, the S-matrix has the asymptotics

$$S(i\varkappa) \approx (\varkappa/\pi)^{1/2};$$

consequently, it does not contain an exponential factor (a singular inner factor), which means that it is a purely Blaschke product. In other words, the system of root vectors of the dissipative contracting generator $Z_t = e^{iBt} = P_K U_t|_{\hat{K}}$ is complete.

Theorem 28 (Pavlov and Faddeev (1972)). *The Riemann hypothesis about the roots of the ζ-function is true if and only if the energy of the semigroup Z_t decreases exponentially with exponent $1/4$ as $t \to \infty$, that is,*

$$\overline{\lim_{t \to \infty}} \frac{1}{t} \ln |Z_t (B + iI)^{-1}| = -\frac{1}{4}.$$

The Riemann hypothesis in the weaker version $2\gamma < \mathrm{Re}\, s_n < 1 - 2\gamma$ is equivalent to the exponential decay of the semigroup Z_t with index γ, $\gamma < 1/4$.

4.5. The Partial s-Matrix for the Acoustic Equation and the Schrödinger Equation. In Sect. 3.10 we introduced the concept of partial s-matrix as a block of the full s-matrix, which was assumed to exist. In reality, the Lax-Phillips approach enables us to define the concept of partial s-matrix and prove its existence independently, without reference to the existence of the full s-matrix.

We assume that the two unitary groups $U_t = e^{i\mathcal{L}t}$ and $U_t^0 = e^{i\mathcal{L}^0 t}$ have incoming and outgoing subspaces $D_\pm$ and $D_\pm^0$ in the pair of Hilbert spaces H and H^0, which can be identified by means of the asymptotically isometric operator $\mathcal{J} : |\mathcal{J}U_{\pm t}^0 f_\pm|_H \to |f|_{H^0}$, $f_\pm \in D_\pm^0$:

$$D_+ = \lim_{t \to \infty} \overline{U(-t)JU^0(t)D_t^0},$$

$$D_- = \lim_{t \to \infty} \overline{U(t)JU^0(-t)D_-^0}.$$

We denote by $\mathcal{H}_\pm$ and $\mathcal{H}_\pm^0$ the minimal reducing subspaces of U_t and U_t^0 that contain $D_\pm$ and $D_\pm^0$, respectively.

Theorem 29 (Pavlov (1982)). *If*

$$\int_0^\infty |(\mathcal{L}\mathcal{J} - \mathcal{J}\mathcal{L}^0)(\mathcal{L}^0 \mp \mu I)^{-1} \exp(\pm i\mathcal{L}^0 t)P_\pm^0| dt < \infty, \qquad (25)$$

then there exist, and are isometric and complete, the "partial" wave operators

$$W_+(\mathcal{L}, \mathcal{L}^0) = s\text{-}\lim_{t\to\infty} U(-t)\mathcal{J}P_+^0 U^0(t),$$

$$W_-(\mathcal{L}, \mathcal{L}^0) = s\text{-}\lim_{t\to\infty} U(t)\mathcal{J}P_-^0 U^0(-t),$$

$$|W_\pm f|_\mathcal{H} = |P_{\mathcal{H}_\pm^0} f|_{\mathcal{H}_0}, \quad W_\pm \mathcal{H}_\pm^0 = \mathcal{H}_\pm,$$

and the corresponding partial scattering operator

$$S = W_-^+ W_+ : \mathcal{H}_+^0 \to \mathcal{H}_-^0$$

commutes with $\mathcal{L}^0$, consequently, it acts as multiplication by an operator-valued contracting function $s(k)$, called the partial scattering matrix.

If the channels $D_\pm^0$ are orthogonal, $\mathcal{J}D_\pm^0 = D_\pm$, and

$$U(\pm t)\mathcal{J} - \mathcal{J}U^0(\pm t))|_{D_\pm^0} = 0, \quad t > 0, \tag{26}$$

then the partial s-matrix is analytic and bounded in the upper half-plane.

If the conditions

$$|P_-^0 U^0(t)I^+ \mathcal{J}U^0(T)P_+^0|_{\mathcal{H}^0} < Ce^{-\gamma(t-T)}, \quad t > T, \tag{27}$$

of exponential asymptotic orthogonality of $D_\pm^0$ and the condition

$$|(\mathcal{L}\mathcal{J} - \mathcal{J}\mathcal{L}^0)(\mathcal{L}^0 - \mu J \pm i\delta])^{-1}U^0(\pm t)P_\pm^0|_\mathcal{H} \leq C\delta^{-1}e^{-\gamma t}, \quad t > T, \tag{28}$$

of closeness of U_t and U_t^0 on $D_\pm^0$ are satisfied, then the partial s-matrix can be extended analytically to the strip $0 < \operatorname{Im} k < \gamma/2$ and is uniformly bounded in every interior strip $0 < \operatorname{Im} k \leq \gamma'/2, 0 < \gamma' < \gamma$.

We mention that the replacement of the exponential on the right-hand side in the last two conditions by power functions $t^{-l}, l > 2$, leads to the smoothness of the corresponding s-matrix; more precisely, we find that $S \in C^{l-2}$.

The above theorem can be applied directly to the wave equations $u_{tt} - \Delta u + qu = 0, q \geq 0$, or $\rho^2 u_{tt} = \Delta u$ in $\mathbb{R}^3$ if their coefficients—the potential q or the density ρ—have the correct behaviour on cones whose bases $\omega_\pm$ are domains on the unit sphere in $\mathbb{R}^3$, that is,

$$\Psi_\pm = \{x : \langle x, \nu \rangle \geq R^\pm(\nu), \nu \in \omega_\pm\}.$$

Thus, the conditions (27) and (28) are satisfied for the equation $-\Delta u + qu + u_{tt} = 0, q \geq 0$, if in the cones $\Omega_\pm$ the potential satisfies

$$0 \leq q(x) \leq C_\pm \exp\{-\gamma \sup_{\nu \in \omega_\pm} \langle x, \nu \rangle\} \cdot |x|^{-2}$$

and $\mathcal{J}$ is taken to be the natural identity operator.

The "partial" point of view in respect of the s-matrix is useful when solving factorisation problems.

Thus, for example, if the density ρ in the equation $\rho^2 u_{tt} = \Delta u$ is equal to unity outside the convex domain Ω with boundary $|x| = R(\nu)$, which is the convex hull of the support $\text{supp}\{\rho - 1\}$, then the Schrödinger scattering matrix exists (see Pavlov (1982)) and is given by a distribution kernel $s(\nu, \nu', k)$ expressed in the usual way in terms of the scattering amplitude by

$$s(\nu, \nu', k) = \delta(\nu + \nu') + \frac{ik}{2\pi} f(\nu, \nu', k).$$

We introduce the infinitesimal incoming and outgoing subspaces $D_\pm^{\nu\pm}$ (in the directions $\nu_\pm$) consisting of the Cauchy data of the incoming and outgoing waves in these directions.

Theorem 30 (Pavlov (1982)). *The partial s-matrix corresponding to the subspaces $D_\pm^{\nu\pm}$ is connected with the full Schrödinger S-matrix by the equality*

$$\hat{S}(\nu, \nu', k) = \exp(ik)\{R(\nu) + R(\nu')\}S(\nu, \nu', k), \quad \nu, \nu' \in \omega_\pm,$$

and is an inner function in the upper half-plane $\text{Im}\, k > 0$.

The formal integral operator with distribution kernel $\hat{S}(\nu, \nu', k)$ is the full S-matrix in the Lax-Phillips scattering scheme, associated with the incoming and outgoing subspaces outside the domain Ω with boundary $|x| = R(\nu)$.

§5. A Dissipative Schrödinger Operator

In problems of resonance scattering we deal from the very beginning with a selfadjoint operator, which is the dilation of the given dissipative one. The situation is different for the *dissipative Schrödinger operator*

$$Lu = -\Delta u + qu.$$

The standard method for the spectral analysis of such an operator used to consist in the successive construction of the resolvent and the study of its denominator. For complex (not necessarily dissipative) potentials decaying fairly rapidly at infinity, this method has been used to find conditions under which the discrete spectrum is finite (see Pavlov (1961,1962)). These conditions are exact in a certain sense (Pavlov (1966,1967,1968)). However, the next step, which involves the proof of some analogue of the spectral theorem, is extremely complicated in this technique. This step can be completed only in the case of finitely many eigenvalues and spectral singularities.

For dissipative Schrödinger operators there is another way of obtaining a spectral resolution, namely, by constructing a dilation and the corresponding functional model (see Sect. 3). The central point in this method is the construction of a minimal selfadjoint dilation and the study of the corresponding scattering matrix, which is the characteristic function of the given operator.

The realisation stages of this technique are described in Sect. 5.3 and Sect. 5.4. In Sect. 5.1 we investigate the structure of the spectrum of the Schrödinger operator with a rapidly decaying complex potential.

5.1. The Spectrum and Uniqueness Theorems. For the ordinary differential operator $L_h y = -y'' + qy$ in $L_2(0, \infty)$ with a bounded complex decaying potential and the complex boundary condition $y'(0) - hy(0) = 0$ at the origin, an important concept is the so-called *Weyl solution*, which is a solution of the homogeneous equation $Ly = \lambda y$, square-integrable over $(0, \infty)$ and depending analytically on the spectral parameter λ. This solution is uniquely defined for an appropriate normalisation (see Titchmarsh (1946)). For example, for summable potentials it is convenient to take such a solution to be the Jost solution $f(k, x)$, $k = \sqrt{\lambda}$, which has exponential asymptotics at infinity, that is, $f(x, k)e^{-ikx} \to 1$ as $x \to \infty$. This solution satisfies the Volterra integral equation

$$f(x, k) = \exp(ikx) + \int\limits_x^\infty k^{-1} \sin k(t - x) q(t) f(t, k) dt, \quad \operatorname{Im} k \geq 0.$$

The eigenvalues $\lambda_s = k_s^2$ of the operator l_k correspond to the complex roots of the Jost function $D_h(k) = f'(0, k) - hf(0, k)$, and the spectral singularities to its real roots. The smoothness of the Jost function is determined by the rate of decay of the potential. Thus, for potentials with several finite moments the following assertion holds.

Lemma 2 (Pavlov (1966)). *We have*

$$\sup_{\operatorname{Im} k \geq 0} |D_h(k) - ik| < \infty, \qquad \sup_{\operatorname{Im} k \geq 0} \left| \frac{dD_h(k)}{dk} \right| < \infty,$$

$$\sup_{\operatorname{Im} k \geq 0} \left| \frac{d^r}{dk^r} D_h(k) \right| \leq 2^r \left[\frac{1}{2} \int\limits_0^\infty |q| x^r dx + \frac{b}{r+1} \int\limits_0^\infty |q| x^{r+1} dx \right],$$

where

$$b = 2 \int\limits_0^\infty |q| dx + |h| \exp\left(\int\limits_0^\infty |q| x \, dx \right).$$

In particular, if $|q(x)| \leq \text{const} \cdot \exp(-\delta x^\alpha)$, $0 < \alpha < 1$, then the Jost function is of Gevrey class G_β, $\beta = \alpha(1 - \alpha)^{-1}$, that is,

$$\sup_{\operatorname{Im} k \geq 0} |D_h^{(k)}(k)| \leq CQ^r r! r^{2/\beta}.$$

By analogy with the arguments of Beurling and Carleson (see Carleson (1952)), using Jensen's inequality we arrive at the following assertion.

Lemma 3 (Pavlov (1966)). *If f is a regular function in the unit disk, continuous up to the boundary circle with all its derivatives up to order N, and $\sup\limits_{|z| \leq 1} |g^{(r)}(z)| \leq A_r$, $r = 0, 1, \ldots, N \leq \infty$, then $g(z) \equiv 0$ provided that $g(z)$*

vanishes together with all its derivatives up to order N on a set F of measure zero on the circle, satisfying the condition

$$\int\limits_{0}^{1} \ln T(s)\, d\varphi_F(s) = -\infty,$$

where $T(s) = \inf\limits_{0 \le r \le N}(r!)^{-1} A_r s^r$ and $\varphi_F(s)$ is the measure of an s-neighbourhood of the set F.

An analogous assertion holds for the half-plane, which is obtained from the above lemma by means of a conformal mapping of the argument and guarantees what is now termed "quasi-analyticity in the half-plane".

The last two lemmas yield a number of assertions concerning the structure of the set σ_∞^h of spectral singularities of "infinite order" of the operator l_h, where the Jost function D_h vanishes together with all its derivatives. We mention that this set also contains all the accumulation points of the eigenvalues. We give an example.

Theorem 31 (Pavlov (1966)). *If the potential q satisfies the inequality $|q(x)| \le C \exp(-\delta x^\alpha)$, $0 < \alpha < 1/2$, then the image of the set σ_∞^h in the plane k lies on the real axis, is bounded, closed, has measure zero, and the family of all its finite contiguity intervals l_ν satisfies the condition $\sum |l_\nu|^{(1-2\alpha)/(1-\alpha)} < \infty$.*

If the decay condition holds with exponent $\alpha = 1/2$, then all the spectral singularities of the operator L_h are of a finite order a and the total number of spectral singularities and eigenvalues of L_h is finite.

In the three-dimensional case there holds an analogue of Theorem 31 (see Pavlov (1966)), whose proof is also based on the uniqueness theorem, but where the role of the Jost function is played by the *Fredholm determinant* $D(\sqrt{\lambda})$ of the Lippman-Schwinger equation iterated once, that is,

$$v(x,y,\lambda) = v_0(x,y,\lambda) + \int\limits_{\mathbb{R}^3} I(x,z,\lambda)q(z)v(z,y,\lambda)\,dz,$$

with

$$I(x,z,\lambda) = \frac{1}{16\pi^2} \int \frac{e^{i\sqrt{\lambda}(|x-s|+|s-z|)}}{|x-s||s-z|} q(s)\,ds,$$

and where Lemma 1 is replaced by the following assertion.

Lemma 4 (Pavlov (1966)). *If the potential q satisfies the estimates*

$$\sup_x |q(x)|(1 + |x|^{3+\varepsilon})|x|^r = C_r < \infty, \quad r = 0, 1, \ldots,$$

then the Fredholm ratio $D(k)$ is a regular function of the variable k in the upper half-plane, satisfying

140   B.S. Pavlov

$$\sup_{\operatorname{Im} k \geq 0} |D^{(r)}(k)| \leq F_0 6^{2r} e^r r! Q(r),$$

where F_0 is a constant depending only on C_0 and

$$Q(r) = \max_{\substack{r_i \geq 1 \\ \sum_i r_i = r}} \prod_i (r_i!)^{-1}[C_{r_i} + \sqrt{C_{2r_i}}].$$

The above result (the theorem in Pavlov (1974)) can in fact be sharpened if we use more exact uniqueness theorems (see Khrushchev (1977)). In this way we can obtain an exact description (see Khrushchev (1984)) of the sets of spectral singularities of operators with potentials that decay faster than $\exp(-\delta x^\alpha)$, $0 < \alpha < 1/2$.

The scheme described here for the application of uniqueness theorems to the study of the accumulation of the discrete spectrum and of the spectral singularities of infinite multiplicity can also be applied to other non-selfadjoint operators, in particular, to operators of higher order. More interesting results have been obtained for the periodic Schrödinger operator with a rapidly decaying complex perturbation of the potential (see Zheludev (1967)) and to the Friedrichs model with a complex kernel (Naboko (1974)).

It is interesting that the results mentioned above are already exact in the class of one-dimensional Schrödinger operators with a real, rapidly decaying potential and a complex boundary condition. The construction of corresponding examples makes use of the technique of the inverse scattering problem and is based on the fact that the Weyl functions corresponding to various boundary conditions are connected in a linear-fractional manner (see Titchmarsh (1946)). Consequently, the spectrum of the operator L_h is computed as the set of the singularities of the *Weyl function $m_h(\lambda)$* given by

$$m_h(\lambda) = -h D_h^{-1}(k) D_{-1/h}(k) = -\frac{1 + h m_\infty(\lambda)}{m_\infty(\lambda) - h}.$$

The Weyl function m_∞ of a selfadjoint operator with a first homogeneous boundary condition $y(0) = 0$ is constructed so that its imaginary part produces the spectral function

$$\rho_\infty(\lambda) = -\lim_{\delta \to 0} \int_0^\infty \operatorname{Im} m_\infty(u + i\delta)\, du$$

and generates a potential of the required class.

It turns out that it suffices to construct a function $m(k)$ such that

1) m is regular in the upper half-plane and in some neighbourhoods of the points $k = 0$ and $k = \infty$;

2) $m(0) \neq 0$, $m'(0) \neq 0$;

3) $km(k) \to -i$ as $|k| \to \infty$;

4) $\operatorname{Im} m(i\varkappa) = 0$, $\varkappa > 0$;

5) $\operatorname{Im} m(-k) = -\operatorname{Im} m(k) > 0$, $k > 0$.

Lemma 5 (Pavlov (1967)). *If the function m satisfies the conditions 1)–5) and is smooth, that is, $m \in \mathbb{C}^N$, $N \geq 3$, then there are a Schrödinger differential operator with a smooth real potential q that has $N - 2$ finite moments, and a real number a_0 such that the function*

$$-[m(\sqrt{\lambda})]^{-1} + a_0$$

is the Weyl function of this operator with the boundary condition $y(0) = 0$.

If the function $m(k)$ belongs to the Gevrey class G_β, that is,

$$\sup_{\operatorname{Im} k \geq 0} |m^{(r)}(k)| \leq Ar!r^{2/\beta}B^r, \quad 0 < \beta < 1,$$

then the generated potential satisfies the condition

$$|q(x)| \leq c\exp\{-\delta x^\alpha\}, \quad \alpha = \beta/(\beta + 1).$$

The last assertion can be used to prove the existence of one-dimensional Schrödinger operators with a real, rapidly decaying potential and a complex boundary condition at the origin, whose spectral structure looks exotic compared with that of selfadjoint operators of the same type. A simple example (see Pavlov (1967)) is obained if m is taken to be

$$iA\left[U\left(\frac{\varkappa k - i}{ik\varkappa - 1}\right) - U(-i)\right], \quad \varkappa > 0,$$

where U is a regular function in the disk $|z| < 1$ given by

$$U(z) = \int_0^{z} e^{-a/(1-\zeta^2)^\beta} \cos\frac{b}{(1-\zeta^2)^\beta}\, d\zeta, \quad 0 < \beta < 1;$$

here $\beta = \alpha(1-\alpha)^{-1}$, $a = 1/32$, $b = 1 - \beta/32$, and the branch of $(1 - \zeta^2)^\beta$ is chosen from the condition of positiveness on the diameter $1 - \zeta^2 > 0$. The functions U and m are of Gevrey class G_β. For the special choice $A = \varkappa[2U'(-i)]^{-1}$, $\varkappa > 0$, the function m satisfies the conditions 1)–5) and takes the value

$$\tfrac{1}{2}i\varkappa[U(1) - U(-i)][U'(-1)]^{-1}$$

infinitely many times. The Schrödinger operator with a smooth potential q, $|q(x)| \leq C\exp[-\delta x^\alpha]$, $0 < \alpha < 1/2$, corresponding to this m has a countable sequence of simple eigenvalues accumulating from the lower half-plane to the point $\lambda = \varkappa^{-2}$.

More refined examples of operators with infinite (complete) sets of eigenvalue accumulation points are given in Pavlov (1968). In Khrushchev (1984) it is shown that practically any set of "non-uniqueness" of Gevrey class G_β,

$\beta = \alpha/(1-\alpha)$, lying on the positive semi-axis of the spectral parameter λ, can be the set of eigenvalue accumulation points of an operator with a potential q, $|q(x)| \leq C \exp[-\delta x^{\alpha}]$.

5.2. The Dilation and the Characteristic Function. From Sect. 5.1 we see that the structure of the discrete spectrum of the simplest Schrödinger operator with a rapidly decaying real potential and a complex boundary condition can be non-trivial. Here we develop the spectral theory of such an operator in terms of a functional model. We do this by following the presentation in Pavlov (1976,1977).

In $L_2(0,\infty)$ we consider the differential expression $-y'' + q(x)y$ with a real, locally bounded potential, for which the case of the limit point holds. Prescribing at zero the additional complex boundary condition

$$(y' - hy)|_0 = 0, \quad 2\operatorname{Im} h = g^2, \quad g > 0,$$

we obtain a fully defined operator L_h, which is a simple dissipative operator that differs from a selfadjoint one L_0 only in the one-dimensional subspace. Restricting L_0 to a symmetric operator L_{00} defined on the lineal D_0 of all functions in $D(L_0)$ vanishing together with their derivatives at $x = 0$, we form the adjoint L^+. We then extend the construction to a selfadjoint operator $\mathcal{L}$ on a space $\mathcal{H}$ as follows. Let $v_+ \in W_2^1(0,\infty) \subset L_2(0,\infty) = D_+$ and $v_- \in W_2^1(-\infty,0) \subset L_2(-\infty,0) = D_-$. We define the action of $\mathcal{L}$ on the lineal $D(\mathcal{L})$ of all vectors $u = (v_-, v, v_+)$ in $\mathcal{H} = D_- \oplus L_2(0,\infty) \oplus D_+$ satisfying the boundary conditions

$$(v' - hv)|_0 = gv_-(0), \quad (v' - \bar{h}v)|_0 = gv_+(0)$$

by writing

$$\mathcal{L} \begin{pmatrix} v_- \\ v \\ v_+ \end{pmatrix} = \begin{pmatrix} -dv_-/id\xi \\ L^+v \\ -dv_+/id\xi \end{pmatrix},$$

thus defining an extension from $L_2(0,\infty)$ to $\mathcal{H}$ of the symmetric operator L_{00}.

Theorem 32 (Pavlov (1976)). *The operator $\mathcal{L}$ is the minimal selfadjoint dilation of L_h.*

To construct a model for L_h and compute its characteristic function, we need to construct the scattered waves $U_\lambda^{\mp}$ generated by the incidence of a monochromatic wave from D_- (D_+).

Let m_∞ be the Weyl function of the selfadjoint operator L_0, and let $\chi = \theta + m\varphi$, where $\theta(0) = \varphi'(0) = 1$ and $\theta'(0) = \varphi(0) = 0$, be the corresponding Weyl solution. It is easy to verify that for any real λ the vector-functions

$$u_\lambda^-(x,\xi) = \begin{pmatrix} e^{-\lambda\xi} \\[4pt] \dfrac{g\chi(x,\lambda-i0)}{m_\infty(\lambda-i0)-h} \\[10pt] \dfrac{m_\infty(\lambda-i0)-\bar h}{m_\infty(\lambda-i0)-h}\,e^{-i\lambda\xi} \end{pmatrix},$$

$$u_\lambda^+(x,\xi) = \begin{pmatrix} \dfrac{m_\infty(\lambda+i0)-h}{m_\infty(\lambda+i0)-\bar h}\,e^{-i\lambda\xi} \\[10pt] \dfrac{g\chi(x,\lambda+i0)}{m_\infty(\lambda+i0)-\bar h} \\[6pt] e^{-i\lambda\xi} \end{pmatrix}$$

satisfy the equation $\mathcal{L}u_\lambda^\pm = \lambda u_\lambda^\pm$ and the boundary conditions. The same can similarly be verified for the vector-functions

$$u_\lambda^<(x,\xi) = \begin{pmatrix} 0 \\ \varphi_h \\ e^{-i\lambda\xi} \end{pmatrix}, \quad u_\lambda^>(x,\xi) = \begin{pmatrix} e^{-i\lambda\xi} \\ \varphi_{\bar h} \\ 0 \end{pmatrix},$$

where φ_h and $\varphi_{\bar h}$ satisfy the equation $L^+\varphi_h = \lambda\varphi_h$, the boundary conditions

$$(\chi'_h - h\varphi_h)|_0 = 0, \quad (\varphi_{\bar h} - \bar h\varphi_{\bar h})|_0 = 0,$$

and the normalisation conditions

$$(\varphi'_h - \bar h\varphi_h)|_0 = g, \quad (\varphi'_{\bar h} - h\varphi_{\bar h})|_0 = g.$$

We write $S = \dfrac{m_\infty - h}{m_\infty - \bar h}$. Then the following assertion holds.

Theorem 33 (Pavlov (1976)). *Each of the systems*

$$\{u_\lambda, u_\lambda^<\}, \quad \{u_\lambda^|, u_\lambda^>\}, \quad \{u_\lambda^<, u_\supset^>\}$$

is complete in the Hilbert space $\mathcal{H}$ of dilations. In addition, all three are orthonormal, the first two with weight matrix $\mathrm{diag}\{1, 1-|S|^2\}$ *and the last one with weight matrix* $\begin{pmatrix} 1 & \bar S \\ S & 1 \end{pmatrix}$. *The spectral representations are also expressed in terms of these systems, as follows:*
1) the incoming representation $\mathcal{T}_-$

$$\mathcal{T}_- : U \mapsto \tilde U_-(\lambda) = \begin{pmatrix} \langle u, u_\lambda^-\rangle_\mathcal{H} \\ \langle u, u_\lambda^<\rangle_\mathcal{H} \end{pmatrix}, \quad D_- \mapsto \begin{pmatrix} H_-^2 \\ 0 \end{pmatrix},$$

$$e^{i\mathcal{L}t} \mapsto e^{i\lambda t}, \quad \langle f,g\rangle = \frac{1}{2\pi}\int\limits_{-\infty}^{\infty} \left\langle \begin{pmatrix} 1 & 0 \\ 0 & 1-|S|^2 \end{pmatrix}\tilde f_-(\lambda), \tilde g_-(\lambda)\right\rangle d\lambda,$$

$$f = \frac{1}{2\pi}\int\limits_{-\infty}^{\infty} u_\lambda^-\langle f, u_\lambda^-\rangle_\mathcal{H}\,d\lambda + \frac{1}{2\pi}\int\limits_{1-|S|^2>0} (1-|S|^2)u_\lambda^<\langle f, u_\lambda^<\rangle_\mathcal{H}\,d\lambda;$$

2) *the outgoing representation $\mathcal{T}_+$*

$$\mathcal{T}_+ : U \mapsto \tilde{U}_+(\lambda) = \begin{pmatrix} \langle u, u_\lambda^+ \rangle_{\mathcal{H}} \\ \langle u, u_\lambda^> \rangle_{\mathcal{H}} \end{pmatrix}, \quad D_+ \mapsto \begin{pmatrix} H_+^2 \\ 0 \end{pmatrix},$$

$$e^{i\mathcal{L}t} \mapsto e^{i\lambda t}, \quad \langle f, g \rangle = \frac{1}{2\pi} \int\limits_{-\infty}^{\infty} \left\langle \begin{pmatrix} 1 & 0 \\ 0 & 1 - |S|^2 \end{pmatrix} \tilde{f}_-(\lambda), \tilde{g}_-(\lambda) \right\rangle d\lambda,$$

$$f = \frac{1}{2\pi} \int\limits_{-\infty}^{\infty} u_\lambda^+ \langle f, u_\lambda^+ \rangle_{\mathcal{H}} d\lambda + \frac{1}{2\pi} \int\limits_{1-|S|^2>0} (1 - |S|^2) u_\lambda^> \langle f, u_\lambda^> \rangle_{\mathcal{H}} d\lambda;$$

3) *the symmetric representation $\mathcal{T}$*

$$\mathcal{T} : U \mapsto \tilde{U}(\lambda) = \begin{pmatrix} \langle u, u_\lambda^< \rangle \\ \langle u, u_\lambda^> \rangle \end{pmatrix},$$

$$D_+ \mapsto \begin{pmatrix} H_+^2 \\ 0 \end{pmatrix}, \quad D_- \mapsto \begin{pmatrix} 0 \\ H_-^2 \end{pmatrix}, \quad e^{i\mathcal{L}t} \mapsto e^{i\lambda t},$$

$$\langle f, g \rangle = \frac{1}{2\pi} \int\limits_{-\infty}^{\infty} \left\langle \begin{pmatrix} 1 & \bar{S} \\ S & 1 \end{pmatrix} \tilde{f}_\lambda, \tilde{g}_\lambda \right\rangle d\lambda,$$

$$f = \frac{1}{2\pi} \int\limits_{-\infty}^{\infty} \{U_\lambda^<, U_\lambda^>\} \begin{pmatrix} 1 & \bar{S} \\ S & 1 \end{pmatrix} \tilde{U}_\lambda d\lambda.$$

It can be seen that $S = (m_\infty - h)(m_\infty - \bar{h})$, which is the transition coefficient from the incoming channel to the outgoing one, is the principal parameter of the above spectral representations. The next assertion follows directly from the results in Sect. 3.

Theorem 34 (Pavlov (1976)). *The transition coefficient $S(\lambda)$ is the characteristic function of the operator L_h, and the family φ_h of compressions on $L_2(0, \infty)$ of its radiating eigenfunctions is a canonical family of eigenfunctions of the absolutely continuous spectrum.*

All the facts of spectral analysis mentioned in Sect. 3 now carry over to the case of the operator L_h. It is clear that full information on the spectral properties of L_h is contained in its characteristic function. Nikol'skij and Khrushchev (1987) give an exact description of the class of all characteristic functions of the Schrödinger operators L_h with a complex boundary condition at zero and a real, rapidly decaying potential.

Theorem 34 (Nikol'skij and Khrushchev (1987)). *If the potential of the operator L_h is a real function and satisfies the condition*

$$|q(x)| < C \exp(-\delta x^\alpha), \quad 0 < \alpha < 1,$$

then its characteristic function S, as a function of the spectral parameter λ, satisfies

1) $|S^{(r)}(\lambda)| \leq CQ^r r! r^{r/\beta}$, $\beta = \alpha/(1-\alpha)$,
2) $|S(\lambda)| = 1$, $\lambda > 0$,
3) $|S(\lambda)| < 1$, $\operatorname{Im}\lambda > 0$,
4) $|S'(0)| \neq 0$,
5) $S(\lambda) = 1 + \dfrac{a}{i\lambda} + o\left(\dfrac{1}{\lambda}\right)$, $\lambda \to \infty$, $a > 0$.

Conversely, any function with the properties 1)–5) *and analytic at infinity is the characteristic function of some dissipative Schrödinger operator with a potential of the above class, and is the restriction to the real axis of a meromorphic function.*

Based on the last assertion, we can fully describe the sets of eigenvalues of operators like L_h with rapidly decaying potentials as sets of the roots of the corresponding characteristic functions (see Nikol'skij and Khrushchev (1987)). For example, the following assertion holds.

Theorem 35 (Nikol'skij and Khrushchev (1987)). *If E is an arbitrary compact subset in the upper half-plane such that all its accumulation points lie on the positive semi-axis and*

$$\sum_{\lambda \in E} \operatorname{Im}\lambda + \int\limits_{\mathbb{R}} \log \frac{1}{\operatorname{dist}(x, E)} \cdot \frac{dx}{1 + x^2} < \infty,$$

then there is a dissipative Schrödinger operator with a decaying real potential whose set of eigenvalues coincides with E.

The condition of potential decay has a strong influence on the spectral properties of L_h.

Theorem 36 (Pavlov (1975b)). *If L_h is a differential operator with a real potential q that has a finite moment $\int\limits_{0}^{\infty} x|q(x)|dx < \infty$, then the inner component can be separated by means of a bounded projection if and only if that component is finite-dimensional.*

This assertion means that if we try to obtain the separability of a component of the operator L_h only by using conditions on the decay of the potential, then we are forced to impose the restrictions in Theorem 31, which guarantee the quasi-analiticity of the characteristic function in the half-plane. We mention that this property (ensured, for example, by the rapid decay of the potential, that is, $|q(x)| \leq C \exp\{-\delta x^{1/2}\}$) implies that this function does not have singular divisors with singularities at a finite distance. The singularities at infinity are excluded by the asymptotics of the Weyl function m_∞. Thus, the following assertion holds.

146   B. S. Pavlov

Theorem 37 (Pavlov (1975b)). *If the real potential q of the operator L_h satisfies the condition $|q(x)| \leq c\exp\{-\delta x^{1/2}\}$, then the system of its eigenfunctions and root functions of the discrete spectrum together with the eigenfunctions of the absolutely continuous spectrum is complete in $L_2(0,\infty)$, and the inner subspace can be separated by means of a bounded projection.*

On the other hand, for any $\alpha \in (0,1/2)$ and any number $\lambda_0 \in (0,\infty)$ there is an operator of the form L_h with a real potential satisfying $|q(x)| \leq C\exp\{-\delta x^\alpha\}$ and such that for some complex boundary condition it has infinitely many simple eigenvalues with accumulation point λ_0, while the joint family of eigenfunctions of the continuous and discrete spectra of L_h is not complete in $L_2(0,\infty)$.

We mention that the second assertion of the theorem is simply proved, by making a minor modification in the construction in Lemma 5. When the function $U(z)$ used there is replaced by the new one

$$U_\varepsilon(z) = \int\limits_0^z \exp\{-\varepsilon(1 - \zeta^2)^{-1}\}U'(\zeta)d\zeta$$

in the formula for $m(k)$, which now becomes $m_\varepsilon(k)$, we obtain the characteristic function

$$S_{\varepsilon,h}(\lambda) = \frac{m_\infty^\varepsilon(\lambda) - h}{m_\infty^\varepsilon(\lambda) - \bar{h}} = \frac{1 + m^\varepsilon(\sqrt{\lambda})(a_0 - h)}{1 + m^\varepsilon(\sqrt{\lambda})(a_0 - \bar{h})},$$

which has the singular inner divisor $S^0(\lambda) = \exp\{i\varepsilon(1 - \lambda\varkappa^2)^{-1}\}$. Furthermore, it turns out that after division by this factor, $S_{\varepsilon,h}$ can be factorised as the product of an outer factor and a Blaschke factor. The close connection noted above between the behaviour of the various factors of the characteristic function is typical only in the case when the latter is smooth, that is, for decaying potentials. In general, even for real potentials the situation is different: the set of accumulation points of the eigenvalues may fill an interval, and this is consistent with separability and completeness.

Theorem 38 (Pavlov (1975b)). *There is a real, infinitely differentiable potential q such that for the operator L_0 there holds the case of the limiting point, and for some complex boundary condition $y'(0) = hy(0)$ the eigenvalues of L_h fill an interval, the joint system of eigenfunctions of the discrete and continuous spectra is complete in $L_2(0,\infty)$, and L_h itself is similar to a normal operator.*

We conclude by giving formulae for the spectral projections of L_h.

Theorem 39 (Pavlov (1976)). *If L_h is a dissipative operator with a real potential q such that the case of the Weyl limiting point holds for the operator L_0, then a canonical system of eigenfunctions of its absolutely continuous spectrum consists of the solutions v_λ of the homogeneous equation*

$-y'' + q(x)y = \lambda y$ *satisfying the boundary condition* $v'_\lambda(0) - hv_\lambda(0) = 0$ *and the normalisation condition* $v'_\lambda(0) - \bar{h}v_\lambda(0) = g = \sqrt{2\,\mathrm{Im}\,h}$. *The spectral projections on a closed interval* ω *of the absolutely continuous spectrum which does not contain spectral singularities are bounded and their kernels are given by*

$$\mathcal{E}_\omega(x, x') = \frac{1}{2\pi} \int\limits_\omega \varphi_h(x, \lambda)\varphi_h(x', \lambda)\frac{1 - |S(\lambda)|^2}{S(\lambda)}\, d\lambda$$

(here we have used the fact that, since the potential is real, $\bar{\varphi}_h(x', \lambda) = \varphi_{\bar{h}}(x', \lambda)$*).*

The eigenfunctions of the discrete spectrum of L_h *coincide with the square-integrable solutions of the type* φ_h, *and the kernels of their corresponding projections are computed (for simple eigenvalues* λ, *that is,* $s'(\lambda) \neq 0$*) by means of the formula*

$$\mathcal{P}_\lambda(x, x') = i\frac{\varphi_h(x, \lambda)\varphi_h(x', \lambda)}{S'(\lambda)}.$$

5.3. The Three-Dimensional Schrödinger Operator with a Complex Potential.
In this case the non-selfadjointness defect is infinite; however, assuming that the potential $q(x) + i\rho(x)$ is bounded and that its imaginary part $\rho(x) = \frac{1}{2}g^2(x) \geq 0$ tends to zero at infinity, we can show that the lower half-plane $\mathrm{Im}\,\lambda < 0$ is free of the spectrum, while the upper one contains only countably many eigenvalues. From Holmgren's uniqueness theorem it follows that the Schrödinger operator

$$L = -\Delta + [q(x) + \frac{i}{2}g^2(x)]$$

with a potential of the class mentioned above is simple in $L_2(\mathbb{R}^3)$. Completing this space with incoming and outgoing channels $D_- = L_2(-\infty, 0; L_2(\mathrm{supp}\,g))$ and $D_+ = L_2(0, \infty; L_2(\mathrm{supp}\,g))$, we consider the operator $\mathcal{L}$ that acts on a sufficiently smooth vector-function $U = (v_-, v, v_+)$ in $\mathcal{H} = D_- \oplus L_2(\mathbb{R}^3) \oplus D_+$ such that $v_+(0, x) - v_-(0, x) = ig(x)u(x)$ according to the formula

$$\mathcal{L}\begin{pmatrix} v_- \\ v \\ v_+ \end{pmatrix} = \begin{cases} -\dfrac{1}{i}\dfrac{dv}{d\xi}(\xi, x), \\[2mm] \mathrm{Re}\,Lv + \dfrac{g(x)}{2}[v_-(0, x) + v_+(0, x)], \\[2mm] -\dfrac{1}{i}\dfrac{dv_+}{d\xi}(\xi, x). \end{cases}$$

Theorem 40 (Pavlov (1977)). *The operator* $\mathcal{L}$ *is the minimal selfadjoint dilation of* L.

The eigenfunctions of the dilation $\mathcal{L}$ can be written in terms of the limit values of the kernel of the resolvent of the operators L and L^* as

$$\lim_{\lambda \to \lambda + i0}\{(L - \lambda I)^{-1}\}(x, s) = R^L_{\lambda + i0}(x, s).$$

These limit values exist if they exist for the kernel of the resolvent of the operator $-\Delta + q$. In what follows we assume that this condition is satisfied.

Theorem 41 (Pavlov (1977)). *The vector-functions*

$$U^-(x,\xi;\lambda,s) = \frac{1}{\sqrt{2\pi}} \begin{pmatrix} \delta(x-s)\exp(-i\lambda\xi) \\ -R^L_{\lambda-i0}(x,s)g(s) \\ [\delta(x-s) - ig(x)R^L_{\lambda-i0}(x,s)g(s)]\exp(-i\lambda\xi) \end{pmatrix}$$

are incoming eigenfunctions of the dilation. The system $\{U^-(\cdot,\cdot;\lambda,s)\}$, $\lambda \in \mathbb{R}$, $s \in \operatorname{supp} G$, *is orthonormal, that is,*

$$\langle Y^-(\cdot,\cdot;\lambda,s), U^-(\cdot,\cdot;\lambda',s')\rangle = \delta(\lambda-\lambda')\delta(s-s'),$$

and complete in $\mathcal{H}_- = V_t \exp(i\mathcal{L}t)D_-$. *The vector-functions*

$$U^+(x,\xi;\lambda,s) = \frac{1}{\sqrt{2\pi}} \begin{pmatrix} [\delta(x-s) + ig(x)R^{L^*}_{\lambda+i0}(x,s)g(s)]\exp(-i\lambda\xi) \\ -R^{L^*}_{\lambda+i0}(x,s)g(s) \\ \delta(x-s)\exp(-i\lambda\xi) \end{pmatrix}$$

are outgoing eigenfunctions of the dilation. The system $\{U^+\}$ *is orthonormal and complete in* $\mathcal{H}_+ = V_t \exp(i\mathcal{L}t)D_+$.

The operator-function S with kernel

$$s(x,s,z) = \delta(x-s) + ig(x)R^{L^*}_z(x,s)g(s)$$

is analytic, is a contraction in the upper half-plane, and has scalar multiples if some power of the operator $T_z = gR^{L^*}_z g$ has a finite trace. S is the characteristic function of the operator L and establishes a correspondence between the incoming and outgoing eigenfunctions on the domain $\sigma_1 \subset \mathbb{R}$ where S is unitary, given by

$$U^+(\cdot,\cdot;\lambda,t) = \int_{\operatorname{supp} g} U^-(\cdot\cdot;\lambda,h)s(h,t,\lambda)dh$$

(this equality is understood in the sense of distributions).

The radiating eigenfunctions of the dilation have the form

$$U^<_\pi(x,\xi,\lambda) = \frac{1}{\sqrt{2\pi}} \begin{pmatrix} 0 \\ u^<_\pi(x,\lambda) \\ \pi\exp(-i\lambda\xi) \end{pmatrix},$$

where $u^<_\pi$ is the unique solution of the equation $Lu^< = \lambda u^<$, normalised by means of the condition $igu^< = \pi$. The absorbing eigenfunctions of the dilation are constructed analogously from the solutions of the adjoint equation $L^*u^> = \lambda u^>$ satisfying the normalisation condition $igu^> = -\nu$; they are

$$U_\nu^>(x,\xi,\lambda) = \frac{1}{\sqrt{2\pi}} \begin{pmatrix} \nu \exp(-i\lambda\xi) \\ u_\lambda^>(x,\nu) \\ 0 \end{pmatrix}.$$

Here $\{\pi\}$ and $\{\nu\}$ are orthonormal families that yield the polar representation $S(\lambda) = \sum s_\pi \nu \langle\,\cdot\,;\pi\rangle,\ s_\pi > 0$.

Theorem 42 (Pavlov (1977)). *The compressions $U_\pi^<$ and $U_\nu^>$ of the incoming and outgoing eigenfunctions of the dilation on $L_2(\mathbb{R}^3)$ form canonical systems of eigenfunctions of the absolutely continuous spectrum of the operators L and L^*, respectively. Furthermore, the following properties hold.*

1) The absolutely continuous spectrum σ_e of L and L^* coincides with the subset of the real axis where the characteristic function is not unitary, that is, $\Delta(\lambda) > 0$ and $\tilde{\Delta}(\lambda) > 0$, where $\Delta(\lambda) = I - S^+(\lambda)S(\lambda)$ and $\tilde{\Delta}(\lambda) = I - S(\lambda)S^+(\lambda)$.

2) The functions $u_\pi^<$ and $u_\nu^>$ satisfy the biorthogonality conditions, that is, the distributional equality

$$\langle u_\pi^>(\cdot,\lambda'), u_{\nu'}^<(\cdot,\lambda')\rangle = -\delta(\lambda - \lambda')\frac{\langle S\pi, \nu'\rangle}{1 - s_\pi^2}, \quad s_\pi < 1.$$

3) The spectral projections of the operator L corresponding to the closed intervals ω of the absolutely continuous spectrum not containing spectral singularities can be written in the form of integral operators with kernels

$$\mathcal{E}_\omega(x,x') = -\int_\omega \sum_{s_\pi < 1} u_\pi^<(x,\lambda)\overline{u_\nu^>(x',\lambda)}\,\frac{1 - s_\pi^2}{s_\pi}\,d\lambda.$$

4) The eigenfunctions $u_\pi^<$ of the discrete spectrum of L corresponding to the simple eigenvalues λ_k (the simple roots of the characteristic function) and the eigenfunctions $u_\nu^>$ of the discrete spectrum of L^* corresponding to the complex-conjugate eigenvalues are biorthogonal, and the corresponding spectral projections are given by the kernels

$$\mathcal{P}_{\lambda_k}(x,x') = 2\pi i \sum_\pi \frac{u_\pi^<(x,\lambda_k)\overline{u_\nu^>(x',\lambda_k)}}{\langle S'(\lambda_k)\pi, \nu\rangle}.$$

Here the summation extends over the vectors π that form an orthonormal basis for the polar representation $S(\lambda_k) = \sum s_\pi \nu\langle\,\cdot\,,\pi\rangle$ of the characteristic function at the point λ_k, to which there correspond zero singular numbers. The vectors ν can be completed by continuity, use being made of the polar representation near λ_k.

The study of the spectral properties of L and L^* continues according to the scheme set out in Sect. 2 and Sect. 3. We do not formulate the corresponding theorems here since they are obtained as a direct application to this case of the abstract assertions discussed in Sect. 2 and Sect. 3. We mention only that the detailed investigation of the spectral properties of L and L^* is based on the preliminary study of the corresponding characteristic function, which can be expressed in terms of the resolvent of L^* by means of the formula

$$S_\lambda = I + igR^{L^*}_{\lambda+i0}g, \quad S : L_2(\operatorname{supp} g) \to L_2(\operatorname{supp} g).$$

It can be seen that this characteristic function arises as an object of the selfadjoint theory and is the scattering matrix of the dilation, in other words, the coefficient of transition from the incoming channel to the outgoing one. Using the fact that $L_2(\operatorname{supp} g)$ is embedded as a subspace in $L_2(\mathbb{R}^3)$, we can replace $R^{L^*}_{\lambda+i0}$ by the resolvent $R^{\mathcal{L}}_{\lambda+i0}$ of the selfadjoint dilation, that is,

$$S_\lambda = I + igR^{\mathcal{L}}_{\lambda+i0}g.$$

References*

Adamyan, V.M., Arov, D.Z. (1966): On unitary couplings of semi-unitary operators. Mat. Issled. *1*, no. 2, 3–64. English transl.: Transl., II. Ser., Am. Math. Soc. *95*, 75–129 (1970). Zbl. 258.47012

Adamyan, V.M., Pavlov, B.S. (1979): A trace formula for dissipative operators. Vestn. Leningr. Univ. 1979, No. 7, Mat. Mekh. Astron. No. 2, 5–9. English transl.: Vestn. Leningr. Univ., Math. *12*, 85–91 (1980). Zbl. 419.47012

Agranovich, Z.S., Marchenko, V.A. (1960): The Inverse Problem of Scattering Theory. Izd. Khar'kov. Univ., Khar'kov. Zbl. 98,60

Beurling, A. (1949): On two problems concerning linear transformation in Hilbert space. Acta Math. *81*, 239–255. Zbl. 33,377

Carleson, L. (1952): Sets of uniqueness for functions regular in the unit circle. Acta Math. *87*, 325–345. Zbl. 46,400

Carleson, L. (1962): Interpolation by bounded analytic functions and the corona problem. Ann. Math., II. Ser. *76*, 547–559. Zbl. 192,168

Faddeev, L.D. (1967): Expansion in the eigenfunctions of the Laplace operator in the fundamental domain of a discrete group in the Lobachevskij plane. Tr. Mosk. Mat. O.-va *17*, 323–350. English transl.: Trans. Mosc. Math. Soc. *17*, 357–386 (1969). Zbl. 201,416

Gel'fand, I.M. (1952): On the spectrum of non-selfadjoint differential operators. Usp. Mat. Nauk *7*, No. 6, 183–184. Zbl. 48,96

Helson, H. (1964): Lectures on Invariant Subspaces. Academic Press, New York London. Zbl 119,113

Ivanov, S.A., Pavlov, B.S. (1978): Carleson resonance series in the Regge problem. Izv. Akad. Nauk SSSR, Ser. Mat. *42*, 26–55. English transl.: Math. USSR, Izv. *12*, 21–51 (1978). Zbl. 375.47021

Ivanov, S.A., Pavlov, B.S. (1980): Vector systems of exponentials and the zeros of entire matrix-functions. Vestn. Leningr. Univ., Ser. I 1980, No. 1, 25–31. English transl.: Vestn. Leningr. Univ., Math. *13*, 31–38 (1981). Zbl. 446.46045

---

* For the convenience of the reader, references to reviews in Zentralblatt für Mathematik (Zbl.), compiled by means of the MATH database, have, as far as possible, been included in this bibliography.

Katsnel'son, V.E. (1967): On conditions for the basis property of the system of root vectors of some classes of operators. Funkts. Anal. Prilozh. *1*, No. 2, 39–51. English transl.: Funct. Anal. Appl. *1*, 122–132 (1967). Zbl. 172,174

Khrushchev, S.V. (=Hruščev, S.V.) (1977): Sets of uniqueness for Gevrey classes. Ark. Mat. *15*, 253–304. Zbl. 387.30021

Khrushchev, S.V. (1984): Spectral singularities of dissipative Schrödinger operator with rapidly decreasing potential. Indiana Univ. Math. J. *33*, 613–638. Zbl. 548.34022

Khrushchev, S.V. (1985): The Regge problem for strings, unconditionally convergent eigenfunction expansions and unconditional bases of exponentials in $L_2(-T,T)$. J. Oper. Theory *14*, 67–85. Zbl. 577.34020

Khrushchev, S.V., Nikol'skij, N.K., Pavlov, B.S. (1981): Unconditional bases of exponentials and reproducing kernels. In: Lect. Notes Math. *864*, 214–335. Zbl. 466.46018

Lax, P.D., Phillips, R. (1967): Scattering Theory. Academic Press, New York London. Zbl. 186,163

Levin, B.Ya. (1956): Distribution of the Roots of Entire Functions. GITTL, Moscow. English transl.: Akademie-Verlag, Berlin 1962. Zbl. 111,73

Lidskij, V.B. (1962): On the summability of series of principal vectors of non-selfadjoint operators. Tr. Mosk. Mat. O.-va *11*, 3–35. English transl.: Am. Math. Soc., Transl., II. Ser. *40*, 193–228 (1964). Zbl. 117,331

Livšic, M.S. (1966): Operators, Oscillations, Waves (Open Systems). Nauka, Moscow. English transl.: Transl. Math. Monogr. *34*. Zbl. 143,367

Lyantse, V.E. (1964a): On a differential operator with spectral singularities. I. Mat. Sb., Nov. Ser. *64*, 521–561. English transl.: Transl., II. Ser., Am. Math. Soc. *60*, 185–225 (1967). Zbl. 127,39

Lyantse, V.E. (1964b): On a differential operator with spectral singularities. II. Mat. Sb., Nov. Ser. *65*, 47–103. English transl.: Transl., II. Ser., Am. Math. Soc. *60*, 227–283 (1967). Zbl. 127,39

Marchenko, V.A. (1960): Expansion in the eigenfunctions of non-selfadjoint singular second-order differential operators. Mat. Sb., Nov. Ser. *52*, 739–788. English transl.: Transl., II. Ser., Am. Math. Soc. *25*, 77–130 (1963). Zbl. 113,72

Martirosyan, R.M. (1957): On the spectra of the non-selfadjoint differential operator $-\Delta + q$ in three-dimensional space. Izv. Akad. Nauk Arm. SSR, Fiz.-Mat. *10*, 85–111. Zbl. 78,279

Naboko, S.N. (1974): On the Friedrichs non-selfadjoint model. Zap. Nauchn. Semin. Leningr. Otd. Mat. Inst. Steklova *39*, 40–58. English transl.: J. Sov. Math. *8*, 27–41 (1977). Zbl. 346.47042

Naboko, S.N. (1980): A functional model in perturbation theory and its applications in scattering theory. Tr. Mat. Inst. Steklova *147*, 86–114. English transl.: Proc. Steklov Inst. Math. *147*, 85–116 (1981). Zbl. 445.47010

Naimark, M.A. (1954): Study of the spectrum and expansion in the eigenfunctions of a non-selfadjoint differential operator on the semi-axis. Tr. Mosk. Mat. O.-va *3*, 187–270. Zbl. 56,311

Nikol'skij, N.K., Khrushchev, S.V. (1987): A functional model and some problems of spectral theory of functions. Tr. Mat. Inst. Steklova *176*, 97–210. English transl.: Proc. Steklov Inst. Math. *176*, 101–214 (1988). Zbl. 649.47010

Nikol'skij, N.K., Pavlov, B.S. (1970): Bases of eigenvectors of completely non-unitary contractions and the characteristic function. Izv. Akad. Nauk SSSR, Ser. Mat. *34*, 90–133. English transl.: Math. USSR, Izv. *4*, 91–134 (1971). Zbl. 232.47026

Pavlov, B.S. (1961): On the non-selfadjointness of the operator $-y'' + q(x)y$ on the semi-axis. Dokl. Akad. Nauk SSSR *141*, 807–810. English transl.: Sov. Math., Dokl. *2*, 1565–1568 (1962). Zbl. 209,451

Pavlov, B.S. (1962): On the spectral theory of non-selfadjoint differential operators. Dokl. Akad. Nauk SSSR *146*, 1267–1270. English transl.: Sov. Math., Dokl. *3*, 1483–1487 (1963). Zbl. 128,81

Pavlov, B.S. (1966): On a non-selfadjoint Schrödinger operator. Probl. Mat. Fiz. *1*, 102–132. Zbl. 171,86

Pavlov, B.S. (1967): On a non-selfadjoint Schrödinger operator. II. Probl. Mat. Fiz. *2*, 133–157. English transl.: Spectral Theory and Problems in Diffraction, Topics in Math. Phys. *2* (1968). Zbl. 189,381

Pavlov, B.S. (1968): On a non-selfadjoint Schrödinger operator. III. Probl. Mat. Fiz. *3*, 59–80. English transl.: Spectral Theory and Problems in Diffraction, Topics in Math. Phys. *3*, 53–71 (1969). Zbl. 191,101

Pavlov, B.S. (1971a): On the completeness of the set of resonance states for a system of differential equations. Dokl. Akad. Nauk SSSR *196*, 1272–1275. English transl.: Sov. Math., Dokl. *12*, 352–356 (1971). Zbl. 232.47016

Pavlov, B.S. (1971b): On the joint completeness of the system of eigenfunctions of a contraction and its adjoint. Probl. Mat. Fiz. *5*, 101–112. Zbl. 302.47010

Pavlov, B.S. (1972): The continuous spectrum of resonances on a non-physical sheet. Dokl. Akad. Nauk SSSR *206*, 1301–1304. English transl.: Sov. Math., Dokl. *13*, 1417–1421 (1972). Zbl. 321.47002

Pavlov, B.S. (1973a): The factorisation of the scattering matrix and the series structure of its roots. Izv. Akad. Nauk SSSR, Ser. Mat. *37*, 217–246. English transl.: Math. USSR, Izv. *7*, 215–245 (1974). Zbl. 313.34014

Pavlov, B.S. (1973b): On the one-dimensional scattering of plane waves by an arbitrary potential. Teor. Mat. Fiz. *16*, 105–115. English transl.: Theor. Math. Phys. *16*, 706–713 (1974). Zbl. 289.47006

Pavlov, B.S. (1973c): Spectral analysis of a differential operator with a "spread" boundary condition. Probl. Mat. Fiz. *6*, 101–119. Zbl. 284.47029

Pavlov, B.S. (1974): On the operator-theoretic meaning of the transition coefficient. Probl. Mat. Fiz. *7*, 102–126. R. Zh. Mat. 1975, 8 B719

Pavlov, B.S. (1975a): The calculation of loss in scattering problems. Mat. Sb., Nov. Ser. *97*, 77–93. English transl.: Math. USSR, Sb. *26*, 71–87 (1976). Zbl. 325.47007

Pavlov, B.S. (1975b): On conditions for the separability of the spectral components of a dissipative operator. Izv. Akad. Nauk SSSR, Ser. Mat. *39*, 123–148. English transl.: Math. USSR, Izv. *9*, 113–137 (1976). Zbl. 317.47006

Pavlov, B.S. (1975c): Expansion in the eigenfunctions of the completely continuous spectrum of a dissipative operator. Vestn. Leningr. Univ., Ser. I 1975, No. 1, 130–137. English transl.: Vestn. Leningr. Univ., Math. *8*, 135–143 (1980). Zbl. 308.47034

Pavlov, B.S. (1976): Dilation theory and the spectral analysis of non-selfadjoint differential operators. Proc. 7th Winter School, Drogobych, 1974. TsEMI, Moscow, 3–69. English transl.: Transl., II. Ser., Am. Math. Soc. *115*, 103–142 (1981). Zbl. 516.47007

Pavlov, B.S. (1977): A selfadjoint dilation of a dissipative Schrödinger operator and the expansion in its eigenfunctions. Mat. Sb., Nov. Ser. *102*, 511–536. English transl.: Math. USSR, Sb. *31*, 457–478 (1977). Zbl. 356.47007

Pavlov, B.S. (1979a): The basic property of a system of exponentials and Muckenhoupt's condition. Dokl. Akad. Nauk SSSR *247*, 37–40. English transl.: Sov. Math., Dokl. *20*, 655-659 (1979). Zbl. 429.30004

Pavlov, B.S. (1979b): A functional model and spectral singularities. Probl. Mat. Fiz. *9*, 113–121. English transl.: Sel. Math. Sov. *6*, 37–44 (1987). Zbl. 494.47013

Pavlov, B.S. (1982) An analyticity condition for the partial scattering matrix. Probl. Mat. Fiz. *10*, 183–208. English transl.: Sel. Math. Sov. *5*, 279–296 (1986). Zbl. 514.47008

Pavlov, B.S., Faddeev, L.D. (1972): Scattering theory and automorphic functions. Zap. Nauchn. Semin. Leningr. Otd. Mat. Inst. Steklova *27*, 161–193. English transl.: J. Sov. Math. *3*, 522–548 (1975). Zbl. 335.35004

Pavlov, B.S., Smirnov, N.V. (1977): Resonance scattering by a one-dimensional crystal and a thin film. Vestn. Leningr. Univ., Ser. I 1977, No. 3, 71–80. English transl.: Vestn. Leningr. Univ., Math. *10*, 307–318 (1982). Zbl. 372.34014

Pavlov, B.S., Strepetov, A.V. (1986): Joint completeness in the case of the continuous spectrum. Funkts. Anal. Prilozh. *20*, 33–36. English transl.: Funct. Anal. Appl. *20*, 27–30 (1986). Zbl. 606.47006

Pekker, M.A. (1976): Resonances in the scattering of acoustic waves by a spherical non-homogeneity of the density. Proc. 7th Winter School, Drogobych, 1974. TsEMI, Moscow, 70–100. English transl.: Transl., II. Ser., Am. Math. Soc. *115*, 143–164 (1980). Zbl. 463.35065

Potapov, V.P. (1955): The multiplicative structure of analytic non-stretching matrix-functions. Tr. Mosk. Mat. O.-va *4*, 125–236. English transl.: Transl., II. Ser., Am. Math. Soc. *15*, 131–243 (1960). Zbl. 66,60

Regge, T. (1958): Analytic properties of the scattering matrix. Nuovo Cimento *8*, 671–679. Zbl. 80,419

Sadovnichij, V.A. (1973): Analytic Methods in the Spectral Theory of Differential Operators. Izd. Moskov. Gos. Univ., Moscow

Sakhnovich, L.A. (1968): Dissipative operators with an absolutely continuous spectrum. Tr. Mosk. Mat. O.-va *19*, 213–270. English transl.: Trans. Mosc. Math. Soc. *19*, 233–297 (1968). Zbl. 179,468

Smirnov, V.I. (1932): Sur les formules de Cauchy et de Green et quelques problèmes qui s'y rattachent. Izv. Akad. Nauk SSSR, Ser. Mat. *7*, 337–372. Zbl. 5,107

Szökefalvi-Nagy, B., Foiaş, C. (1970): Analyse Harmonique des Opérateurs de l'Espace de Hilbert. Academiai Kiado, Budapest. Zbl. 201,450

Titchmarsh, E.C. (1946): Eigenfunction Expansions Associated with Second-Order Differential Equations. Clarendon Press, Oxford. Zbl. 61,135

Trejl', S.R. (1986): A spatial-compact system of eigenvectors forms a Riesz basis if it is uniformly minimal. Dokl. Akad. Nauk SSSR *288*, 308–312. English transl.: Sov. Math., Dokl. *33*, 675–679 (1986). Zbl. 628.46008

Zheludev, V.A. (1967): On the eigenvalues of the perturbed Schrödinger operator with a periodic potential. Probl. Mat. Fiz. *2*, 108–123. English transl.: Spectral Theory and Problems in Diffraction, Topics in Math. Phys. *2* (1968). Zbl. 167,443

III. Index Theorems

B.V. Fedosov

Translated from the Russian
by C. Constanda

Contents

Introduction

The theory of the index of elliptic operators has for a long time been developed in parallel within the framework of two branches of mathematics that, traditionally, are regarded as quite far apart. One of them is the the theory of elliptic equations and boundary value problems—in particular, the theory of singular integral equations. The other is topology and algebraic geometry, where very specific elliptic operators have been considered. A significant role in bringing these two domains together was played by Gel'fand (1960), who posed the problem of topological classification of elliptic operators, in particular, the computation of the index in topological terms. The latter was fully solved by Atiyah and Singer in 1963. The Atiyah-Singer theorem has generated a tremendous amount of interest, which has continued to this day and has exercised an immense influence on the subsequent development and convergence of the theory of differential equations and topology. Thus, for example, the necessity to extend the class of deformations of elliptic operators has led to new algebras of pseudodifferential operators (PDOs). In topology, the Atiyah-Singer theorem has stimulated the further development of K-theory.

A detailed proof of the Atiyah-Singer theorem is given in Atiyah and Singer (1968a,b), as well as in the book Palais (1965), where one can find all the necessary prerequisites from topology and the theory of partial differential equations. These surveys also contain an analysis of examples of elliptic operators that have already become classical types, such as de Rham and Dolbeault complexes, the Hirzebruch signature operator and the Dirac operator. A more accessible (but also less complete) presentation for the beginner can be found in the monograph Booss (1977).

Much effort has been expended to search for alternative proofs of the index theorem. First of all, one should mention the important paper Atiyah, Bott and Patody (1973), where the so-called heat equation method was developed. This method has been applied repeatedly by other authors in various versions. A probabilistic approach, also based on the heat equation, was proposed in the recent papers Bismut (1984). A direct proof that does not make use of a topological technique is given in Fedosov (1974) and Hörmander (1979). Another interesting direct proof of the index theorem for the Dirac operator, which is based on quantisation on supermanifolds, has been proposed recently in Getzler (1983).

The Atiyah-Singer theorem has given rise to various generalisations. We indicate the most important ones.

1. The equivariant index theorem (Atiyah and Segal (1968)), in particular, the Atiyah-Bott fixed point theorem (Atiyah and Bott (1967)).

2. The index of a family of elliptic operators (Atiyah and Singer (1971a)).

3. The index of real elliptic operators (Atiyah and Singer (1971b)).

4. The L^2-index of elliptic operators on noncompact manifolds, and applications to group representation theory (Atiyah (1976), Atiyah and Schmid (1977), Connes and Moscovici (1982)).

5. The index of almost periodic and stochastic operators, and index theory in von Neumann II-type factors (Fedosov and Shubin (1978), Coburn, Moyer and Singer (1973)).

6. The index theorem on manifolds with boundary, announced by Atiyah and Bott in 1964. Its proof was given much later by Boutet de Monvel (1971). A detailed discussion of this problem can be found in the recent monograph Rempel and Schulze (1982). We mention that the remarkable Boutet de Monvel algebra of boundary value problems was introduced chiefly for the needs of index theory.

7. The index of Toeplitz operators (Boutet de Monvel (1979), Boutet de Monvel and Guillemin (1989)).

8. Interesting connections have been discovered recently between the theory of the index of elliptic operators with geometric quantisation (Kirillov (1985)), the method of orbits in representation theory (Kirillov (1972)), and the spectral theory of operators with periodic bicharacteristics (Colin de Verdier (1979)). Thus, in Berline and Vergne (1985) Kirillov's character formula is interpreted as a formula for the equivariant index of the Dirac operator. In Boutet de Monvel and Guillemin (1989) the multiplicity of the eigenvalues of Hamiltonians with periodic trajectories is expressed as the index of some elliptic operator on the manifold of the orbits.

9. Recently (Fedosov (1986,1989)) the author has proposed a new approach to deformation quantisation (Bayen, Flato, Fronsdal, Lichnerowicz and Sternheimer (1978). In particular, an index theory is constructed in the algebra of quantum observables. The conditions that ensure that the index is an integer yield quantisation conditions, which were absent in the deformation approach.

10. The index theory for elliptic operators has found various interesting applications in contemporary physics. The reader may become acquainted with them in Schwarz (1981), Alvarez-Gaumé (1984) and Getzler (1983).

We comment briefly on the nature of the applications of the index theorem. The majority of them consist in proving existence theorems for solutions, which follow from the positiveness of the index. A classical example of such an argument is the Riemann-Roch theorem. More modern examples are the proof by Atiyah and Schmid of the existence of discrete series of irreducible representations and Schwarz's proof of the existence of instantons and the computation of the dimension of the instanton manifold (Schwarz (1981)).

There are also applications of an entirely different type in the case when the index is an integer. These can be clarified by using the Dirac operator as an example. According to the Atiyah-Singer formula, the index of an operator is equal to the so-called A-genus of a manifold, which, generally speaking, is not an integer. Consequently, if the A-genus is not an integer, then there is no spinor structure on this manifold. A more meaningful example is connected with deformation quantisation, mentioned in 9 above: if the index is not an integer, then the algebra of quantum observables does not admit an operator represenation.

The brevity of this paper does not permit us to consider all the above questions. Thus, 3, 4, 8 and 10 are completely left out. The first and second chapters, where we discuss the Atiyah-Singer theorem and its generalisations mentioned in 1, 2 and 5, aim to acquaint the reader with these topics and also to prepare him for Chapter 3, in which we conduct a fairly detailed analysis of results related to 9. This affects the nature of the presentation. In particular, we make systematic use of the language of connections and characteristic classes.

Chapter 1
The Atiyah-Singer Theorem

§1. The Index of Fredholm Operators

1.1. Fredholm Operators. Let H^0 and H^1 be Hilbert spaces. We consider closed linear operators $A : H^0 \to H^1$ with a dense domain $D(A)$. The subspace $\operatorname{Ker} A = \{u \in H^0 : Au = 0\}$ is called the *kernel* of A, and the quotient space $\operatorname{Coker} A = H^1/\operatorname{Im} A$, where $\operatorname{Im} A$ is the image of the operator, is called the *cokernel*. A is called a *Fredholm operator* if its kernel and cokernel are finite-dimensional. The *index* of a Fredholm operator is defined by

$$\operatorname{ind} A = \dim \operatorname{Ker} A - \dim \operatorname{Coker} A .$$

We indicate some simple properties of Fredholm operators.

1. The adjoint $A^* : H^1 \to H^0$ of a Fredholm operator is also a Fredholm operator.

2. There hold the direct orthogonal decompositions $H^0 = \operatorname{Ker} A \oplus \operatorname{Im} A^*$ and $H^1 = \operatorname{Ker} A^* \oplus \operatorname{Im} A$. In particular, the images $\operatorname{Im} A$ and $\operatorname{Im} A^*$ are closed subspaces.

3. There is a bounded operator $R_0 : H^1 \to H^0$ such that $1 - R_0 A$ and $1 - A R_0$ are orthogonal projections on $\operatorname{Ker} A$ and $\operatorname{Ker} A^*$, respectively.

From the property 2 it follows that the spaces $\operatorname{Coker} A$ and $\operatorname{Ker} A^*$ are isomorphic, so that the index can also be given by the alternative formula

$$\begin{aligned}
\operatorname{ind} A &= \dim \operatorname{Ker} A - \dim \operatorname{Ker} A^* \\
&= \dim \operatorname{Ker} A^* A - \dim \operatorname{Ker} A A^* = -\operatorname{ind} A^* .
\end{aligned} \tag{1.1}$$

We sketch the proof of these properties. The equality $\dim \operatorname{Coker} A = d$ means that there are elements $v_1, v_2, \ldots, v_d \in H^1$ such that any $v \in H^1$ can be represented uniquely in the form $v = Au + c_1 v_1 + \cdots + c_d v_d$, where u is orthogonal to $\operatorname{Ker} A$. The operator $A : (\operatorname{Ker} A)^\perp \oplus \mathbb{C}^d \to H^1$ acting according to the rule $(u, c_1, c_2, \ldots, c_d) \mapsto Au + c_1 v_1 + \cdots + c_d v_d$ is densely defined, closed, and has an everywhere defined inverse. By Banach's theorem, the inverse A^{-1} is bounded. Hence, there is a constant $k > 0$ such that the a priori estimate $\|u\|_0 \le k\|Au\|_1$ holds for any $u \in D(A) \cap (\operatorname{Ker} A)^\perp$ (the subscripts 0 and 1 refer to the norms in H^0 and H^1, respectively). From this estimate it follows that $\operatorname{Im} A$ is closed and that $H^1 = \operatorname{Ker} A^* \oplus \operatorname{Im} A$. The operator R_0 coincides with A^{-1} on $\operatorname{Im} A$ and is extended by zero to $\operatorname{Ker} A^*$. From the same a priori estimate, rewritten in the form $(A^* Au, u) \ge (u, u)/k^2$, we also deduce that the image $\operatorname{Im} A^*$ is closed and that $H^0 = \operatorname{Ker} A \oplus \operatorname{Im} A^*$, which shows that A^* is a Fredholm operator.

We mention that the concepts of Fredholm operator and index can also be defined for operators on Banach spaces, but we confine ourselves to the case of Hilbert spaces. We also mention that often the definition of a Fredholm operator asks A to be bounded. Although this requirement implies no loss of generality, since a closed, densely defined operator is bounded in the graph norm, sometimes it is an inconvenience. Other names used for a Fredholm operator are Φ-operator and Noether operator.

We need some information on trace class operators. An extensive discussion can be found in Gohberg and Krein (1965) (see also Shubin (1978), Appendix 3). Let $T : H^0 \to H^1$ be a compact operator in Hilbert spaces. We denote by $s_i > 0$ the non-zero eigenvalues of T^*T and introduce the *trace norm* $\|T\|_{\mathrm{tr}}$ by the equality $\|T\|_{\mathrm{tr}} = \sum_{i=1}^{\infty} s_i^{1/2}$. T is called a *trace class operator* if its trace norm is finite. Trace class operators form a two-sided ideal in the algebra of bounded operators: if in the sequence

$$H^0 \xrightarrow{A} H^1 \xrightarrow{T} H^2 \xrightarrow{B} H^3$$

both A and B are bounded and T is a trace class operator, then $\|BTA\|_{\mathrm{tr}} \leq \|A\|\,\|B\|\,\|T\|_{\mathrm{tr}}$, where $\|\cdot\|$ is the operator norm. The adjoint of a trace class operator is also a trace class operator and has the same trace norm.

The *trace* of a trace class operator $T : H \to H$ is defined by

$$\mathrm{tr}\,T = \sum_{i=1}^{\infty}(Te_i, e_i) = \sum_{k=1}^{\infty} \lambda_k \ ,$$

where e_i is an arbitrary orthonormal basis for H and λ_k are the eigenvalues of T counted with regard for their multiplicity. The first equality is the definition of the trace (it does not depend on the choice of basis), while the second one represents Lidskij's theorem on trace. The trace is a linear functional on the space of trace class operators, it is bounded in the trace norm, and has the property that $\mathrm{tr}\,AT = \mathrm{tr}\,TA$ for any bounded operator A and any trace class operator T. This property can be sharpened: if AB and BA are trace class operators (A and B may be unbounded), then $\mathrm{tr}\,AB = \mathrm{tr}\,BA$. The equality follows from Lidskij's theorem, since the non-zero eigenvalues of AB and BA and their orders of multiplicity coincide. It is also obvious that $\mathrm{tr}\,T^* = \overline{\mathrm{tr}\,T}$.

For a trace class operator T with a continuous kernel $T(x,y)$ in $L^2(\mathbb{R}^n)$, the trace can be expressed in the integral form

$$\mathrm{tr}\,T = \int T(x,x)\,dx \ .$$

The following two assertions, which express the index as a difference of traces, form the starting point of many investigations regarding the index. The first assertion lies at the basis of the "heat equation" method (see Atiyah, Bott and Patody (1973)), while the second one is encountered so frequently that it could be attributed to mathematical folklore.

Theorem 1.1. *If A is a Fredholm operator, A^*A and AA^* have a discrete spectrum, and e^{-A^*At} and e^{-AA^*t} are trace class operators for any $t > 0$, then*

162 B.V. Fedosov

$$\operatorname{ind} A = \operatorname{tr} e^{-A^*At} - \operatorname{tr} e^{-AA^*t} \, . \tag{1.2}$$

The equality (1.2) follows from the fact that the non-zero eigenvalues of A^*A and AA^* coincide, and the multiplicities of the zero eigenvalues are equal to $\dim \operatorname{Ker} A^*A$ and $\dim \operatorname{Ker} AA^*$, respectively.

Theorem 1.2. *A closed, densely defined operator $A : H^0 \to H^1$ is a Fredholm operator if and only if there is an operator $R : H^1 \to H^0$ such that $1 - RA$ and $1 - AR$ are trace class operators. In this case,*

$$\operatorname{ind} A = \operatorname{tr}(1 - RA) - \operatorname{tr}(1 - AR) \, . \tag{1.3}$$

Necessity is proved by the property 3. Sufficiency follows from the inclusions $\operatorname{Ker} A \subset \operatorname{Ker} RA$ and $\operatorname{Im} A \supset \operatorname{Im} AR$, and the fact that RA and AR satisfy Fredholm's theorems (since they differ from the identity by a compact operator). The equality (1.3) holds for $R = R_0$ in the property 3 and does not depend on the choice of R, since $(R - R_0)A$ and $A(R - R_0)$ are trace class operators with the same trace.

The operator R is called a *parametrix (regulariser)*. Sometimes the term "parametrix" is used for an operator with a weaker property, namely, that $1 - RA$ and $1 - AR$ are compact.

1.2. Properties of the Index. From (1.1) it follows that the index of a self-adjoint Fredholm operator is zero. As (1.3) shows, the index of operators in finite-dimensional spaces is equal to $\dim H^0 - \dim H^1$ and does not depend on the operator. It is also obvious that $\operatorname{ind} A_1 \oplus A_2 = \operatorname{ind} A_1 + \operatorname{ind} A_2$. Here $A_1 : H_1^0 \to H_1^1$, $A_2 : H_2^0 \to H_2^1$, and $A_1 \oplus A_2 : H_1^0 \oplus H_2^0 \to H_1^1 \oplus H_2^1$ is the direct sum of the operators.

The stability property. Let A and R be a Fredholm operator and its parametrix, and suppose that $B : H^0 \to H^1$ satisfies $\|RB\| < 1$ and $\|BR\| < 1$. Then $A - B$ is also a Fredholm operator and $\operatorname{ind}(A - B) = \operatorname{ind} A$.

To prove this, we remark that a parametrix R_1 of $A - B$ is given by the Neumann series

$$R_1 = R \sum_{n=0}^{\infty} (BR)^n = \sum_{n=0}^{\infty} (RB)^n R \, .$$

Then

$$1 - R_1(A - B) = \sum_{n=0}^{\infty} (RB)^n (1 - RA) \, ,$$

$$1 - (A - B)R_1 = (1 - AR) \sum_{n=0}^{\infty} (BR)^n \, .$$

These are trace class operators, since they contain trace class factors. Furthermore, the traces of $(RB)^n(1 - RA)$ and $(1 - AR)(BR)^n$ coincide for $n \geq 1$, since the factors under the trace sign can be interchanged cyclically.

Consequently, we deduce that the index of a family $A(t)$ of bounded Fredholm operators which is continuous in the operator norm does not depend on the parameter t, although the dimensions of the kernel and cokernel may vary. To put it briefly, the index is a homotopic invariant.

The logarithmic property. Suppose that in the sequence

$$H^0 \xrightarrow{A_1} H^1 \xrightarrow{A_2} H^2$$

A_1 and A_2 are Fredholm operators. Then so is $A_2 A_1$, and ind $A_2 A_1 =$ ind $A_1 +$ ind A_2.

Indeed, let R_1 and R_2 be parametrices. Then $R = R_1 R_2$ is a parametrix for $A_2 A_1$. Next,

$$\begin{aligned} \operatorname{tr}(1 - R_1 R_2 A_2 A_1) &= \operatorname{tr}(1 - R_1 A_1) + \operatorname{tr} R_1 (1 - R_2 A_2) A_1 \\ &= \operatorname{tr}(1 - R_1 A_1) + \operatorname{tr}(1 - R_2 A_2) - \operatorname{tr}(1 - R_2 A_2)(1 - A_1 R_1) \,. \end{aligned}$$

Similarly,

$$\begin{aligned} \operatorname{tr}(1 &- A_2 A_1 R_1 R_2) \\ &= \operatorname{tr}(1 - A_2 R_2) + \operatorname{tr}(1 - A_1 R_1) - \operatorname{tr}(1 - R_2 A_2)(1 - A_1 R_1) \,, \end{aligned}$$

from which the assertion follows.

The multiplicative property. Let $H_1 \otimes H_2$ be the *tensor product of Hilbert spaces*, that is, the Hilbert space generated by the formal products $u_1 \otimes u_2$, $u_1 \in H_1$, $u_2 \in H_2$ (which satisfy the bilinearity properties), equipped with the inner product $(u_1 \otimes u_2, v_1 \otimes v_2) = (u_1, v_1)(u_2, v_2)$. If A_1 and A_2 are linear operators in H_1 and H_2, respectively, then $A_1 \otimes A_2$ is defined on elements of the form $u_1 \otimes u_2$, $u_1 \in D(A_1)$, $u_2 \in D(A_2)$, by the formula $(A_1 \otimes A_2)(u_1 \otimes u_2) = A_1 u_1 \otimes A_2 u_2$. If, in addition, A_1 and A_2 are trace class operators, then so is $A_1 \otimes A_2$, and $\operatorname{tr} A_1 \otimes A_2 = \operatorname{tr} A_1 \operatorname{tr} A_2$.

Let $A_1 : H_1^0 \to H_1^1$ and $A_2 : H_2^0 \to H_2^1$ be Fredholm operators. We define a product $\#$ by

$$A_1 \# A_2 = \begin{pmatrix} A_1 \otimes 1 & -1 \otimes A_2^* \\ 1 \otimes A_2 & A_1^* \otimes 1 \end{pmatrix}, \tag{1.4}$$

where 1 denotes the identity operator in the corresponding space.

This operator acts from the space $(H_1^0 \otimes H_2^0) \oplus (H_1^1 \otimes H_2^1)$ into the space $(H_1^1 \otimes H_2^0) \oplus (H_1^0 \otimes H_2^1)$. It is also a Fredholm operator, and its index is equal to ind A_1 ind A_2. Indeed, the operators $(A_1 \# A_2)^* (A_1 \# A_2)$ and $(A_1 \# A_2)(A_1 \# A_2)^*$ are given by the matrices

$$\begin{pmatrix} A_1^* A_1 \otimes 1 + 1 \otimes A_2^* A_2 & 0 \\ 0 & A_1 A_1^* \otimes 1 + 1 \otimes A_2 A_2^* \end{pmatrix},$$

$$\begin{pmatrix} A_1 A_1^* \otimes 1 + 1 \otimes A_2^* A_2 & 0 \\ 0 & A_1^* A_1 \otimes 1 + 1 \otimes A_2 A_2^* \end{pmatrix}.$$

It is easy to see that the kernel of $A_1^* A_1 \otimes 1 + 1 \otimes A_2^* A_2$ is generated by elements of the form $u_1 \otimes u_2$, where $u_1 \in \operatorname{Ker} A_1$ and $u_2 \in \operatorname{Ker} A_2$; consequently, its

164 B.V. Fedosov

dimension is equal to $\dim \operatorname{Ker} A_1 \dim \operatorname{Ker} A_2$. The discussion for the remaining diagonal elements of these matrices is similar. This shows immediately that $\operatorname{ind} A_1 \# A_2 = \operatorname{ind} A_1 \operatorname{ind} A_2$. This relation can also be obtained from (1.2).

We also mention some important properties of the product $A_1 \# A_2$ for bounded Fredholm operators. Up to homotopies, this product is commutative and distributive over direct sum with respect to either factor. In addition, if one of the factors is invertible, then so is the product.

§2. Elliptic Pseudodifferential Operators

2.1. Basic Results in the Theory of PDOs. Unless otherwise stipulated, in what follows we consider smooth compact manifolds without boundary, equipped with a Riemannian metric. We use the standard notation TM, T^*M, $T^*M \setminus 0$, $S(M)$ and $B(M)$, respectively, for the tangent and cotangent bundles (which are identified by means of the Riemannian metric), the bundle of nonzero covectors $\xi \neq 0$, and the bundles of unit spheres $|\xi| = 1$ and balls $|\xi| < 1$; also, we denote by π be the projection of these bundles on M.

Let E be a smooth complex vector bundle over M and $C^\infty = C^\infty(E) = C^\infty(M, E)$ the space of its smooth sections. We assume that the bundle is *Hermitian*, that is, the fibres are equipped with a Hermitian inner product $\langle \, , \, \rangle$, which enables us to define an inner product on the space of sections by setting $(u, v) = \int \langle u(x), v(x) \rangle \, dx$, where dx is the Riemannian element of volume on M, and to consider the space $L^2 = L^2(E) = L^2(M, E)$ of square-integrable sections. The lifting of E to T^*M or $T^*M \setminus 0$ is denoted by π^*E or, simply, by E if this does not create ambiguity.

There are excellent handbooks on the theory of pseudodifferential operators (PDOs), such as Shubin (1978), Hörmander (1984–1985) and Treves (1980), to which we refer the reader. We need only the definition of the principal symbol and some basic theorems on the action of PDOs in Sobolev spaces. We restrict our attention to so-called classical PDOs. Typical representatives of classical PDOs are differential operators and their inverses. By the *principal symbol* of a PDO $A : C^\infty(E^0) \to C^\infty(E^1)$ of order m we understand a function $\sigma(A) = a(x, \xi)$ on $T^*M \setminus 0$ which is positively homogeneous of degree m with respect to ξ, takes values in the homomorphism of the fibres $a(x, \xi) : E_x^0 \to E_x^1$, and is constructed as follows. Let $(x_0, \xi_0) \in T^*M \setminus 0$, and let u_0 be a vector in $E_{x_0}^0$, $u(x)$ a section of E^0 with support in the neighbourhood of x_0 and such that $u(x_0) = u_0$, and $f(x)$ a function whose differential is $df|_{x_0} = \xi_0$. Then

$$a(x_0, \xi_0)u_0 = \lim_{\lambda \to +\infty} \lambda^{-m} e^{-i\lambda f(x_0)} A(u(x)e^{i\lambda f(x)})|_{x=x_0} . \qquad (1.5)$$

In terms of local coordinates and local reference frames of E^0 and E^1, $a(x, \xi)$ is given by a matrix function. In particular, if A is a differential operator defined in terms of local coordinates by $A = \sum_{|\alpha| \le m} a_\alpha(x)D^\alpha$, where $D = -i(\partial/\partial x)$, then $a(x, \xi) = \sum_{|\alpha| = m} a_\alpha(x)\xi^\alpha$.

The correspondence $\sigma : A \mapsto a$ between operators and their principal symbols has the following properties:

1. σ is surjective, and two operators of order m with an identical principal symbol differ by an operator of order $m - 1$;

2. $\sigma(AB) = \sigma(A)\sigma(B)$;

3. if $A^* : C^\infty(E_1) \to C^\infty(E^0)$ is the formal adjoint of A defined by $(Au, v) = (u, A^*v)$ for any $u, v \in C^\infty$, then $\sigma(A^*) = a^*$, where a^* is the homomorphism adjoint to a with respect to the Hermitian metric $\langle \, , \, \rangle$ in the fibres.

A PDO $A : C^\infty(E^0) \to C^\infty(E^1)$ of order m can be extended to a closed operator from $L^2(E^0)$ to $L^2(E^1)$; in this case the formal adjoint A^* can be extended to an adjoint operator $A^* : L^2(E^1) \to L^2(E^0)$. The PDO is bounded in L^2 if $m \leq 0$, compact if $m < 0$, and a trace class operator if $m < -n$, where $n = \dim M$.

It is convenient to consider PDOs in the *Sobolev spaces* $H^s = H^s(E) = H^s(M, E)$. These spaces are defined as the completion of C^∞ with respect to the inner product $(u, v)_s = (\Lambda_s u, \Lambda_s u)$, $s \in \mathbb{R}$, where Λ_s is some selfadjoint positive definite PDO with principal symbol $|\xi|^s$. In particular, H^0 coincides with L^2, the space H^{s_2} is compactly embedded in H^{s_1} if $s_2 > s_1$, the intersection of all the H^s coincides with C^∞, and their union coincides with the space $\mathcal{D}'$ of generalized functions (distributions).

From the properties of PDOs in L^2 we deduce the following properties of PDOs in Sobolev spaces.

1. For any s_0, s_1 and m, an operator A of order m can be extended to a closed operator $A : H^{s_0}(E^0) \to H^{s_1}(E^1)$.

2. The operator is bounded for $m \leq s_0 - s_1$, compact for $m < s_0 - s_1$, and a trace class operator for $m \leq s_0 - s_1 - n$.

3. The adjoint $(A)^* : H^{s_1}(E^1) \to H^{s_0}(E^0)$ of $A : H^{s_0}(E^0) \to H^{s_1}(E^1)$ is the closure of the operator $(A)^* = \Lambda_{s_0}^{-2} A^* \Lambda_{s_1}^2$, where A^* is the formally adjoint PDO.

In particular, an operator A of order m is bounded from H^s to H^{s-m} for any s.

2.2. The Index of Elliptic PDOs. A PDO A is called *elliptic* if its principal symbol is invertible on $T^*M \setminus 0$, that is, the homomorphism $a : \pi^* E^0 \to \pi^* E^1$ is an isomorphism. A remarkable property of elliptic operators is their Fredholm character in Sobolev spaces. To prove this, we construct a parametrix for an elliptic operator $A : H^{s_0}(E^0) \to H^{s_1}(E^1)$. Let R_0 be a PDO with principal symbol a^{-1}. Then the operators $1 - R_0 A$ and $1 - A R_0$ are of order -1. We set

$$R = R_0 \sum_{k=0}^{N} (1 - A R_0)^k = \sum_{k=0}^{N} (1 - R_0 A)^k R_0 \, , \qquad (1.6)$$

where $N \geq \dim M$. Then the operators

$$1 - RA = (1 - R_0 A)^{N+1}, \qquad 1 - AR = (1 - AR_0)^{N+1}$$

are of order $-(N+1) < -\dim M$; consequently, they are trace class operators in H^{s_0} and H^{s_1}, respectively.

It turns out that the index of A does not depend on the Sobolev spaces where the operator is considered. Indeed, if $u \in H^s$ is a solution of the elliptic equation $Au = 0$, then $u = (1 - RA)u$, which implies that $u \in H^{s+N+1}$. Iterating, we deduce that u belongs to all the Sobolev spaces, that is, $u \in C^\infty$. This argument is called a *regularisation theorem*. Thus, the kernel of A in Sobolev spaces coincides with that of the operator $A : C^\infty(E^0) \to C^\infty(E^1)$. It can be proved analogously that the kernel of the adjoint operator is isomorphic to that of the formal adjoint $A^* : C^\infty(E^1) \to C^\infty(E^0)$, so that

$$\operatorname{ind} A = \dim \operatorname{Ker} A - \dim \operatorname{Ker} A^* ,$$

where A and the formally adjoint A^* are considered in C^∞.

Apart from the properties listed in 1.1, the index of an elliptic operator also has the specific *stability property*.

Theorem 1.3. *The index of an elliptic operator A depends only on the restriction of the principal symbol a to the manifold $S(M)$, and remains constant under any deformations of A that preserve ellipticity if, in addition, the function $a|_{S(M)}$ is deformed continuously.*

To prove this, first we remark that the operator may be assumed to be of order 0. Indeed, multiplying an elliptic operator A of order m by an elliptic selfadjoint operator Λ_{-m} with principal symbol $|\xi|^{-m}$, we obtain an operator of order 0. In view of the logarithmic property, this process does not alter the index, since Λ_{-m} is selfadjoint. The principal symbol of an operator of order zero does not depend on $|\xi|$ and can be regarded as a function on $S(M)$.

Now let A_1 and A_2 be elliptic operators of order 0 with principal symbols a_1 and a_2, respectively. We assume that the principal symbols are sufficiently close, that is, the function $a(t) = (1-t)a_1 + ta_2$ is invertible on $S(M) \times [0,1]$. Then $A(t) = (1-t)A_1 + tA_2$ is a family of Fredholm operators in L^2 continuous with respect to the operator norm, and, by the stability property stated in 1.1, it has a constant index. This yields the statement of the theorem.

The above assertion leads to a new, topological point of view regarding the index of elliptic operators. Suppose, for simplicity, that the bundles E^0 and E^1 are trivial. Then the principal symbol $a(x, \xi)$ of an elliptic operator A of order 0 is a non-degenerate matrix of order $N = \dim E^0 = \dim E^1$ which depends on the point $(x, \xi) \in S(M)$. In this way, we have a mapping $a : S(M) \to \mathrm{GL}(N, \mathbb{C})$, and the index depends only on the homotopic class of this mapping. The order N of a may be increased as desired. To this end, we need to change from A to $A \oplus 1$, where 1 is the identity operator in the trivial one-dimensional bundle, which, obviously, is elliptic and of index zero. Then $\operatorname{ind}(A \oplus 1) = \operatorname{ind} A$, and the principal symbol of $A \oplus 1$ is $a \oplus 1 \in \mathrm{GL}(N+1, \mathbb{C})$. This means that we can arbitrarily increase the orders of the matrices by deforming the principal symbol. Such homotopic classes, where deformations can increase the order of the matrices, are called *stable*. Thus, we arrive at the conclusion that the index is an integer-valued function on the set of stable homotopic classes of mappings of $S(M)$ in the group of non-degenerate matrices. There arises the problem of expressing this function explicitly in terms of topological invariants. Such an expression is usually caled the *topological index* $\mathrm{ind}_t A$ of the operator A, as opposed to the analytic index defined by (1.1).

In the general case when E^0 and E^1 are nontrivial bundles, the topological index is defined by means of the stable homotopic class of the mapping $a : S(M) \to \mathrm{Iso}(\pi^* E^0, \pi^* E^1)$ in the space of isomorphisms of the bundles $\pi^* E^0$ and $\pi^* E^1$.

We conclude this discussion with a remark on the multiplicative property of the index of elliptic operators. Let $A_i : C^\infty(E_i^0, M_i) \to C^\infty(E_i^1, M_i)$, $i = 1, 2$, be elliptic PDOs of the same order on manifolds M_1 and M_2, respectively. We lift E_i^0 and E_i^1 to the direct product $M_1 \times M_2$ and define the bundles $\mathcal{E}^0 = (E_1^0 \otimes E_2^0) \oplus (E_1^1 \otimes E_2^1)$ and $\mathcal{E}^1 = (E_1^1 \otimes E_2^0) \oplus (E_1^0 \otimes E_2^1)$ over $M_1 \times M_2$. As in 1.1, we consider the operator $A_1 \# A_2 : C^\infty(\mathcal{E}^0, M_1 \times M_2) \to C^\infty(\mathcal{E}^1, M_1 \times M_2)$ given by the matrix (1.4). The operator $A_1 \otimes 1$ acts on the section $u_1(x) \otimes u_2(y)$ only with respect to the variable x, that is, $(A_1 \otimes 1) u_1(x) \otimes u_2(y) = (A_1 u_1)(x) \otimes u_2(y)$, with similar expressions for the remaining elements of (1.4). Then, as was proved in 1.1, $\mathrm{ind}\, A_1 \# A_2 = \mathrm{ind}\, A_1 \,\mathrm{ind}\, A_2$. However, here we encounter a technical difficulty connected with the fact that $A_1 \# A_2$ is not a PDO. Indeed, the natural candidate for the role of principal symbol is

$$
a_1 \# a_2 = \begin{pmatrix} a_1(x,\xi) \otimes 1 & -1 \otimes a_2^*(y,\eta) \\ 1 \otimes a_2(y,\eta) & a_1^*(x,\xi) \otimes 1 \end{pmatrix} ,
$$

which is not a smooth function on $T^*(M_1 \times M_2) \setminus 0$; it is not even defined for $\xi = 0$, $\eta \neq 0$ or $\xi \neq 0$, $\eta = 0$ (except in the case of differential operators). Thus, we cannot associate with $A_1 \# A_2$ a homotopic class of mappings of $S(M_1 \times M_2)$ in the group of non-degenerate matrices, in terms of which the topological index should be defined.

This difficulty is overcome as follows. We consider operators A_1 and A_2 of order $m \geq 1$. Then, extending the principal symbols by zero to the zero sections of the bundles $T^* M_1$ and $T^* M_2$, we obtain continuous, even Lipschitz, functions a_1 and a_2 on $T^* M_1$ and $T^* M_2$, respectively, and, at the same time, a Lipschitz function $a_1 \# a_2$ on $T^*(M_1 \times M_2)$. This function is invertible on $S(M_1 \times M_2)$ and defines a homotopic class of mappings of $S(M_1 \times M_2) \to \mathrm{Iso}(\pi^* \mathcal{E}^0, \pi^* \mathcal{E}^1)$. Let A be any elliptic PDO on $M_1 \times M_2$ with principal symbol in this homotopic class. The question arises whether the indices of the operators A and $A_1 \# A_2$ are equal or not. This question is answered affirmatively by means of a smoothing technique: we construct a sequence of smooth functions $a_n(x, \xi, y, \eta)$ on $T^*(M_1 \times M_2) \setminus 0$, positively homogeneous of order m in ξ and η, which converges uniformly to $a_1 \# a_2$ on $S(M_1 \times M_2)$, and the corresponding sequence of PDOs A_n, which converges to $A_1 \# A_2$ in the norm of operators from $H^s(M_1 \times M_2, \mathcal{E}^0)$ to $H^{s-m}(M_1 \times M_2, \mathcal{E}^1)$. Then the stability property of the index in 1.1 and Theorem 1.3 guarantee that the indices of A and $A_1 \# A_2$ coincide. More details can be found in Palais (1965).

2.3. Elliptic Complexes. A sequence of PDOs of the same order m

$$0 \to C^\infty(E^0) \xrightarrow{A_0} C^\infty(E^1) \xrightarrow{A_1} \ldots \xrightarrow{A_{n-1}} C^\infty(E^n) \to 0$$

is called a *complex* if $A_{i+1}A_i = 0$. A complex is called *elliptic* if the corresponding sequence of symbols

$$0 \to \pi^*E^0 \xrightarrow{a_0} \pi^*E^1 \xrightarrow{a_1} \ldots \xrightarrow{a_{n-1}} \pi^*E^n \to 0$$

is exact for $(x, \xi) \in T^*M \setminus 0$. We recall that a sequence is exact when the equation $a_i u_i = u_{i+1}$ with respect to u_i is solvable if and only if $a_{i+1}u_{i+1} = 0$.

An elliptic operator may be regarded as a particular case of two-term elliptic complex

$$0 \to C^\infty(E^0) \xrightarrow{A} C^\infty(E^1) \to 0 .$$

Conversely, an elliptic complex can be associated with an elliptic operator which is equivalent in some sense to the complex. This association is realized by means of Hodge's theory.

We consider the bundles $\mathcal{E} = \bigoplus_{i=0}^{n} E^i$, $\mathcal{E}^+ = \bigoplus_{k \geq 0} E^{2k}$ and $\mathcal{E}^- = \bigoplus_{k \geq 0} E^{2k+1}$, and the operators A and A^* in $C^\infty(\mathcal{E})$ defined by the matrices

$$A = \begin{pmatrix} 0 & 0 & \ldots & 0 & 0 \\ A_0 & 0 & \ldots & 0 & 0 \\ 0 & A_1 & \ldots & 0 & 0 \\ \vdots & \vdots & \ddots & \vdots & \vdots \\ 0 & 0 & \ldots & A_{n-1} & 0 \end{pmatrix}, \quad A^* = \begin{pmatrix} 0 & A_0^* & 0 & \ldots & 0 \\ 0 & 0 & A_1^* & \ldots & 0 \\ \vdots & \vdots & \vdots & \ddots & \vdots \\ 0 & 0 & 0 & \ldots & A_{n-1}^* \\ 0 & 0 & 0 & \ldots & 0 \end{pmatrix},$$

where $A_i^* : C^\infty(E^{i+1}) \to C^\infty(E^i)$ is formally adjoint to A_i. From the definition of a complex it follows that $A^2 = (A^*)^2 = 0$. The operator $\Delta = (A + A^*)^2 = AA^* + A^*A$ is called the *Hodge-Laplace operator* and is given by a diagonal matrix with elements $\Delta_0 = A_0^*A_0$, $\Delta_n = A_{n-1}A_{n-1}^*$ and $\Delta_k = A_k^*A_k + A_{k-1}A_{k-1}^*$ for $k = 1, 2, \ldots, n-1$. It is clear that A and A^* commute with Δ.

We can convince ourselves that the condition of ellipticity of the complex is equivalent to the ellipticity of the operator Δ, as well as that of $A + A^*$. The sections $u \in \mathrm{Ker}\, \Delta$ are called *harmonic*. Since Δ is elliptic, the harmonic sections form a finite-dimensional space $\mathrm{Ker}\, \Delta \subset C^\infty(\mathcal{E})$. In view of the equality $(\Delta u, u) = (Au, Au) + (A^*u, A^*u)$, the condition $\Delta u = 0$ means that $Au = 0$ and $A^*u = 0$. Hence, $\mathrm{Ker}\, \Delta = \mathrm{Ker}\, A \cap \mathrm{Ker}\, A^*$.

For an arbitrary section $u \in C^\infty(\mathcal{E})$ we denote by u_0 the orthogonal projection of u on $\mathrm{Ker}\, \Delta$ in $L^2(\mathcal{E})$. Since Δ is selfadjoint and elliptic, the equation

$\Delta v = u - u_0$ has a smooth solution. Consequently, we obtain an expression for u as a sum of three orthogonal terms

$$u = u_0 + AA^*v + A^*Av \,, \tag{1.7}$$

which is called the *Hodge-de Rham decomposition*. This yields *Hodge's theorem*, which states that the spaces $\operatorname{Ker}\Delta_i$ are isomorphic to the factor-spaces $H^i = \operatorname{Ker}A_i/\operatorname{Im}A_{i-1}$, called *cohomologies of the complex*. Indeed, if $u \in \operatorname{Ker}A$, then, applying A to (1.7), we find that $A\Delta v = \Delta Av = 0$, from which it follows that $Av \in \operatorname{Ker}\Delta \subset \operatorname{Ker}A^*$. Then $A^*(Av) = 0$, and (1.7) reduces to two terms: $u = u_0 + A(A^*v)$. This equality means that u and u_0 belong to the same coset in $\operatorname{Ker}A/\operatorname{Im}A$, which proves the theorem. In particular, Hodge's theorem implies that the cohomologies H^i are finite-dimensional.

With an elliptic complex we associate the elliptic operator $\mathcal{A} : C^\infty(\mathcal{E}^+) \to C^\infty(\mathcal{E}^-)$ defined as the restriction of $A + A^*$ to $C^\infty(\mathcal{E}^+) \subset C^\infty(\mathcal{E})$. Then $\mathcal{A}^* : C^\infty(\mathcal{E}^-) \to C^\infty(\mathcal{E}^+)$ coincides with the restriction of $A + A^*$ to $C^\infty(\mathcal{E}^-)$. We have

$$\operatorname{Ker}\mathcal{A} = \bigoplus_{k\geq 0} \operatorname{Ker}(A + A^*)|_{C^\infty(E^{2k})} = \bigoplus_{k\geq 0} \operatorname{Ker}\Delta_{2k} \,,$$

$$\operatorname{Ker}\mathcal{A}^* = \bigoplus_{k\geq 0} \operatorname{Ker}(A + A^*)|_{C^\infty(E^{2k+1})} = \bigoplus_{k\geq 0} \operatorname{Ker}\Delta_{2k+1} \,.$$

From this we obtain

$$\operatorname{ind}\mathcal{A} = \sum_{i=0}^{n}(-1)^i \dim\operatorname{Ker}\Delta_i = \sum_{i=0}^{n}(-1)^i \dim H^i \,.$$

This number is also called the *Euler characteristic* of the complex.

§3. Characteristic Classes and Elements of K-Theory

In this section we introduce the topological prerequisites necessary in the construction of the topological index. A detailed presentation of connection theory can be found in Lichnerowicz (1955), of characteristic classes in Dupont (1978), and of K-theory in Atiyah (1967), Bott (1967a), as well as in the book Husemöller (1966).

3.1. Connections and Characteristic Classes. Let E be a complex m-dimensional bundle over an n-dimensional, not necessarily compact, manifold M, $\operatorname{Hom}(E,E)$ the bundle of the fibre homomorphisms of E, $\Lambda = \bigoplus_{p=0}^{n} \Lambda^p$ the bundle of exterior differential forms, and $\Lambda^\pm$ the direct sum of the Λ^p for even and odd values of p, respectively. We consider differential p-forms with values in E and in $\operatorname{Hom}(E,E)$, which are the sections of the bundles $E \otimes \Lambda^p$ and

$\text{Hom}(E, E) \otimes \Lambda^p$, respectively. In the local reference frame $e = \{e_1, e_1, \ldots, e_m\}$ over a neighbourhood $U \subset M$, a p-form $u \in C^\infty(E \otimes \Lambda^p)$ can be rewritten as $u = e_\alpha u^\alpha$ (summation on α from 1 to m), where u^α are scalar complex p-forms on U, or $u = e\{u\}$, for short, where e is the row consisting of the basis vectors and $\{u\}$ the column vector consisting of the forms u^α. Similarly, $A \in C^\infty(\text{Hom}(E, E) \otimes \Lambda^q)$ can be given in the local reference frame by a matrix $\{A\}$ whose elements A^α_β are scalar q-forms on U. If $u \in C^\infty(E \otimes \Lambda^p)$ and A and B are differential q- and r-forms, respectively, with values in $\text{Hom}(E, E)$, then the product $A \wedge u$ is defined as the product of a matrix by a column vector, that is, $(A \wedge u)^\alpha = A^\alpha_\beta \wedge u^\beta$, while the product $A \wedge B$ can be defined as a product of matrices: $(A \wedge B)^\alpha_\beta = A^\alpha_\gamma \wedge B^\gamma_\beta$; the form $[A, B] = A \wedge B - (-1)^{qr} B \wedge A$ is called the *commutator* of A and B. When we change to a different reference frame $e' = ef$, where f is a non-degenerate transition matrix function, we have $\{u\}' = f^{-1}\{u\}$ and $\{A\}' = f^{-1}\{A\}f$. Differential forms with values in $\text{Hom}(E^0, E^1)$ and actions on them are defined similarly.

By a *connection* ∂ on the bundle E we understand a linear differential first order operator $\partial : C^\infty(E) \to C^\infty(E \otimes \Lambda^1)$ satisfying *Leibniz's rule*

$$\partial(\varphi u) = d\varphi u + \varphi \partial u \, ,$$

where $u \in C^\infty(E)$, φ is a scalar function, and $d\varphi$ its differential. Here the 1-form ∂u is called the *covariant differential* of u. When the bundle is equipped with a Hermitian inner product $\langle \, , \, \rangle$, the connection is called *Hermitian* (that is, it preserves the Hermitian structure) if for any $u, v \in C^\infty(E)$

$$d\langle u, v \rangle = \langle \partial u, v \rangle + \langle u, \partial v \rangle \, .$$

In the local reference frame e, the covariant differential is uniquely determined by Leibniz's rule if the covariant differentials ∂e_α of the basis sections are given. Expanding ∂e_α in the basis e_α, we obtain $\partial e_\alpha = e_\beta \Gamma^\beta_\alpha$, or $\partial e = e\Gamma$, for short, where Γ is a matrix of 1-forms on U called the local *connection form*. We mention that the matrices Γ do not determine a global section of the bundle $\text{Hom}(E, E) \otimes \Lambda^1$, since if we go over to a different reference frame, then

$$\Gamma' = f^{-1}\Gamma f + f^{-1}df \, ,$$

that is, the transformation law is not the same as that for the sections $A \in C^\infty(\text{Hom}(E, E) \otimes \Lambda^1)$. If the connection ∂ is Hermitian, then in an orthonormal reference frame the matrix Γ is skew-Hermitian: $\Gamma^* = -\Gamma$.

If ∂_0 and ∂_1 are two connections on E with local connection forms Γ^0 and Γ^1, respectively, then from the transformation law it follows that $\Delta\Gamma = \Gamma^1 - \Gamma^0$ is a globally defined section of the bundle $\text{Hom}(E, E) \otimes \Lambda^1$. Conversely, if ∂_0 is any connection, then any other connection has the form $\partial_1 u = \partial_0 u + Au$, where $A \in C^\infty(\text{Hom}(E, E) \otimes \Lambda^1)$. As a corollary, we deduce that the set of connections is convex. We mention without proof that on any bundle there exists a connection, which is Hermitian if the bundle is Hermitian.

By means of a connection we can differentiate not only the sections of the bundle E, but also objects associated with it, for example, the sections of $E \otimes \Lambda$ or $\mathrm{Hom}(E, E) \otimes \Lambda$. The covariant differential can be extended to them so that Leibniz's rule is preserved, that is,

$$\partial(A \wedge B) = (\partial A) \wedge B + (-1)^p A \wedge \partial B ,$$

where p is the degree of A. In addition, if A and B are scalar forms, then the covariant differential becomes equal to the exterior differential d. Thus, for example, for $u = e_\alpha u^\alpha \in C^\infty(E \otimes \Lambda)$ we have

$$\partial u = \partial e_\alpha \wedge u^\alpha + e_\alpha du^\alpha = e_\alpha(du^\alpha + \Gamma^\alpha_\beta \wedge u^\beta) = e(d\{u\} + \Gamma \wedge \{u\}) .$$

For $A \in C^\infty(\mathrm{Hom}(E, E) \otimes \Lambda^p)$ the covariant differential is defined by means of the equality

$$(\partial A)u = \partial(Au) - (-1)^p A \partial u ,$$

where u is an arbitrary section of the bundle E. In a local frame this yields

$$\{\partial A\} = d\{A\} + \Gamma \wedge \{A\} - (-1)^p\{A\} \wedge \Gamma = d\{A\} + [\Gamma, \{A\}] .$$

Similarly, if $A \in C^\infty(\mathrm{Hom}(E^0, E^1) \otimes \Lambda^p)$, where E^0 and E^1 are bundles with connections ∂_0 and ∂_1, respectively, then ∂A is defined by

$$(\partial A)u = \partial_1(Au) - (-1)^p A \partial_0 u .$$

For the sections $A \in C^\infty(\mathrm{Hom}(E, E) \otimes \Lambda)$ we can define a scalar form $\mathrm{tr}\, A$, equal, by definition, to the sum $\mathrm{tr}\{A\}$ of the diagonal elements of $\{A\}$, which, in view of the transformation law, does not depend on the choice of local reference frame. Since the trace of the commutator of matrix forms is zero, it follows that

$$\mathrm{tr}\, \partial A = \mathrm{tr}\, d\{A\} = d\,\mathrm{tr}\, A .$$

The form $\det A$ for sections $A \in C^\infty(\mathrm{Hom}(E, E) \otimes \Lambda^+)$ is defined analogously. The evenness of the forms is essential here since, in view of the commutativity of the algebra Λ^+, the order of the terms plays no role in the computation of the determinant.

We define the *curvature of a connection* ∂. The operator $\partial^2 = \partial\partial : C^\infty(E \otimes \Lambda^p) \to C^\infty(E \otimes \Lambda^{p+2})$, which at a first glance appears to be a second order differential operator, is in fact an operator of order zero, that is, $\partial^2 u = \Omega u$, where Ω is a 2-form with values in $\mathrm{Hom}(E, E)$, called the curvature of the connection. Indeed, in a local reference frame we have

$$\{\partial^2 u\} = d\{\partial u\} + \Gamma \wedge \{\partial u\}$$
$$= d(d\{u\} + \Gamma u) + \Gamma \wedge (d\{u\} + \Gamma\{u\}) = (d\Gamma + \Gamma \wedge \Gamma)\{u\} .$$

Thus, in such a frame

$$\{\Omega\} = d\Gamma + \Gamma \wedge \Gamma = d\Gamma + \tfrac{1}{2}[\Gamma, \Gamma] .$$

Then $\partial^2 A = [\Omega, A]$ for $A \in C^\infty(\mathrm{Hom}(E, E) \otimes \Lambda^p)$ and $\partial^2 A = \Omega_1 \wedge A - A \wedge \Omega_0$ for $A \in C^\infty(\mathrm{Hom}(E^0, E^1) \otimes \Lambda^p)$, where Ω_0 and Ω_1 are the curvatures of ∂_0 and ∂_1, respectively.

A connection is called *flat* if its curvature is zero, and *Abelian* if its curvature is a multiple of a scalar form. In the case of an Abelian connection, $\partial^2 A = 0$ for any $A \in C^\infty(\mathrm{Hom}(E, E) \otimes \Lambda)$, since the commutator with a scalar form Ω is zero. In a trivial bundle with a globally defined reference frame e we can give a flat connection by setting $\Gamma = 0$. We also mention that any connection in a one-dimensional bundle is Abelian.

The basic properties of curvature are expressed by Bianchi's identity and the formula for the variation of the curvature. Let $u \in C^\infty(E)$ be an arbitrary section. Then for the form $\partial^3 u$ we can obtain the two expressions

$$\partial^3 u = \partial(\partial^2 u) = \partial \Omega u + \Omega \wedge \partial u , \qquad \partial^3 u = \partial^2(\partial u) = \Omega \wedge \partial u ,$$

from which, by identification, we arrive at *Bianchi's identity* $\partial \Omega = 0$.

Now let $\partial_t u = \partial_0 u + \Delta \Gamma(t) u$ be a smooth connection depending on a parameter $t \in [0, 1]$. Then $(\partial_t u)^\cdot = \Delta \dot\Gamma(t) u = \dot\Gamma(t) u$, where $\dot\Gamma(t)$ is a globally defined section of the bundle $\mathrm{Hom}(E, E) \otimes \Lambda^1$ and the dot denotes differentiation with respect to t. From this we obtain

$$(\Omega(t) u)^\cdot = (\partial_t \partial_t u)^\cdot = \dot\Gamma(t) \wedge \partial_t u + \partial_t(\dot\Gamma(t) u) = (\partial_t \dot\Gamma(t)) u .$$

Consequently, we deduce that the variation of the curvature is given by $\dot\Omega(t) = \partial_t \dot\Gamma(t)$.

We now define the *characteristic classes* of vector bundles. Let ∂ be a connection in a bundle E, Ω its curvature, and $\omega = -\Omega/(2\pi i)$. We set $\omega^k = \omega \wedge \omega \wedge \ldots \wedge \omega$ (k factors) and introduce the scalar 2-forms $\psi_k = \mathrm{tr}\, \omega^k$, called the *Adams forms*. From Bianchi's identity it follows that these forms are closed: $d\psi_k = \mathrm{tr}\, \partial \omega^k = k \, \mathrm{tr}\, \omega^{k-1} \partial \omega = 0$. Consequently, they define $2k$-dimensional cohomology classes of M. We claim that these classes are independent of the choice of connection ∂. Indeed, any two connections ∂_0 and ∂_1 can be joined by the linear homotopy $\partial_t = (1 - t)\partial_0 + t\partial_1$. Let $\Omega(t)$ be the curvature of ∂_t. Then, by the formula for the variation of the curvature,

$$\begin{aligned}
(\mathrm{tr}\, \Omega^k(t))^\cdot &= k \, \mathrm{tr}\, \Omega^{k-1}(t) \wedge \partial_t \dot\Gamma(t) \\
&= k \, \mathrm{tr}\, \partial_t(\Omega^{k-1}(t) \wedge \dot\Gamma(t)) = k d\, \mathrm{tr}\, \Omega^{k-1}(t) \wedge \dot\Gamma(t) ,
\end{aligned}$$

from which it follows that the variations of the Adams forms are exact forms.

Along with the Adams forms we also consider the *Chern forms*, defined as the coefficients of the characteristic polynomial $\det(\lambda - \omega) = \sum_{k=0}^{m} \lambda^{m-k} c_k$. It is clear that the ψ_k and c_k can be expressed in terms of each other polynomially by means of Newton's formulae (which connect the sums of the powers of $\lambda_1, \lambda_2, \ldots, \lambda_m$ with elementary symmetric functions). In this way, the Chern forms c_k also define cohomology classes independent of connection. These are called the *Chern characteristic classes* of the bundle E.

In general, for any analytic function $f(z)$ or formal power series we can define non-homogeneous even-dimensional cohomology classes $\operatorname{tr} f(\omega)$ and $\det f(\omega)$, called the *additive characteristic class* and the *multiplicative characteristic class* of the bundle E, respectively (the series are terminating ones because M is finite-dimensional). We denote them by $f(E)$ with the signs $+$ for additive classes and $\times$ for multiplicative classes, as necessary. All of them can be expressed polynomially in terms of the Adams or Chern classes; therefore, they do not depend on connection.

The following caracteristic classes are encontered most frequently:

$C(E) = 1 + c_1 + c_2 + \cdots + c_m = \det(1 - \omega)$—the *complete Chern class*;

$\operatorname{ch} E = \operatorname{tr} e^{\omega}$—the *Chern character*;

$T(E) = \det(\omega/(1 - e^{-\omega}))$—the *Todd class*;

$A(E) = \det((\omega/2)/(\operatorname{sh}(\omega/2)))^{1/2}$—the *A-class*.

We list the properties of characteristic classes.

1. *Natural behaviour under mappings.* Let $\varphi : M \to N$ be a smooth mapping of manifolds. For any bundle E over N the mapping φ induces a bundle $\varphi^* E$ over M. In addition, φ defines a cohomology homomorphism $\varphi^* : H(N) \to H(M)$. Then

$$f(\varphi^* E) = \varphi^* f(E)$$

for any characteristic class.

2. For a direct sum of bundles, the additive classes are summed while the multiplicative classes are multiplied together.

3. Let E^* be the dual bundle of E. The fibre E^*_x at a point x is the space of linear forms on E_x. Then the characteristic class of E^* for a function $f(z)$ coincides with the characteristic class of E for $f(-z)$.

In addition, the class $\operatorname{ch} E$ has some important specific properties.

4. *Multiplicativity with respect to tensor products:* $\operatorname{ch} E_1 \otimes E_2 = \operatorname{ch} E_1 \operatorname{ch} E_2$. Indeed, let ∂_1 and ∂_2 be connections on the bundles E_1 and E_2, respectively. Then on $E = E_1 \otimes E_2$ we have the connection $\partial = \partial_1 \otimes 1 + 1 \otimes \partial_2$ acting on sections of the form $u_1 \otimes u_2$ according to the formula

$$\partial(u_1 \otimes u_2) = \partial_1 u_1 \otimes u_2 + u_1 \otimes \partial_2 u_2 \, ,$$

which implies that

$$\partial^2(u_1 \otimes u_2) = \Omega_1 u_1 \otimes u_2 + u_1 \otimes \Omega_2 u_2 \, .$$

Let ω_1, ω_2 and ω be the corresponding forms for E_1, E_2 and $E = E_1 \otimes E_2$, respectively. Then $\omega = \omega_1 \otimes 1 + 1 \otimes \omega_2$, which yields

$$e^{\omega} = e^{\omega_1} \otimes e^{\omega_2} \, . \tag{1.8}$$

Taking traces, we arrive at the required equality.

5. Let $\Lambda^k(E)$ be the kth exterior power of E, that is, the bundle of anti-symmetric tensors in the kth tensor power $\overset{k}{\bigotimes} E = E \otimes E \otimes \cdots \otimes E$ (k factors);

also, let $\Lambda^\pm(E)$ be the direct sum of the $\Lambda^k(E)$ with k even (odd), and $\Lambda(E)$ the direct sum for all k. Then

$$\operatorname{ch}\Lambda^+(E^*) - \operatorname{ch}\Lambda^-(E^*) = c_m(E)\mathcal{T}^{-1}(E) . \tag{1.9}$$

To verify this, we compute $\operatorname{ch}\Lambda^k(E)$. Let ω_k be the curvature form (divided by $-2\pi i$) for the bundle $\overset{k}{\bigotimes} E$. Just as in the case of (1.8), we obtain

$$e^{\omega_k} = e^\omega \otimes e^\omega \otimes \ldots \otimes e^\omega \qquad (k \text{ factors}) .$$

Projecting this equality on the space of skew-symmetric tensors and taking traces, we find that $\operatorname{ch}\Lambda^k(E) = \operatorname{tr}\Lambda^k e^\omega$, which is the sum of the diagonal minors of order k in the matrix e^ω. Then

$$\operatorname{ch}\Lambda^+(E) - \operatorname{ch}\Lambda^-(E) = \sum_{k=0}^{m}(-1)^k\operatorname{tr}\Lambda^k e^\omega = \det(1 - e^\omega) .$$

Rewriting this equality for E^* and making use of the property 3, we arrive at (1.9).

For real bundles E, the characteristic classes are defined as characteristic classes of the complexification, that is, formally by means of the same equalities $f_+(E) = \operatorname{tr} f(-\Omega/(2\pi i))$ and $f_\times(E) = \det f(-\Omega/(2\pi i))$, only Ω is now a real matrix. A specific property of real bundles is that E and E^* are isomorphic: any linear form on E_x has the form $\langle \cdot, u \rangle$, where $u \in E_x$ and $\langle\ ,\ \rangle$ is the Euclidean inner product.Hence, the characteristic classes corresponding to the functions $f(z)$ and $f(-z)$ coincide. In particular, from this it follows that all the odd Chern classes c_{2k+1} are equal to zero. The class $(-1)^k c_{2k}$ is called the kth *Pontryagin class* p_k.

For even-dimensional oriented real bundles (that is, where $\dim E = 2k$), we can introduce a specific characteristic class $\chi(E) \in H^{2k}(M)$, called the *Euler class*. We consider oriented reference frames of the bundle, orthonormalized with respect to the Euclidean inner product $\langle\ ,\ \rangle$. The curvature matrix of the Euclidean connection in such a frame is skew-symmetric. We set $\chi(E) = \operatorname{Pf}(\Omega/(2\pi))$, where Pf is the Pfaffian of a skew-symmetric matrix. Since the transition functions take values in the group SO($2k$), the Pfaffian does not depend on the choice of frame. It can be proved that $\chi(E)$ is a closed form and that its cohomology class does not depend on the choice of connection.

It is obvious that the Euler class is multiplicative, that is, $\chi(E_1 \oplus E_2) = \chi(E_1)\chi(E_2)$. It is also clear that $\chi^2(E) = p_m$, since the Pfaffian satisfies $(\operatorname{Pf} A)^2 = \det A$.

By definition, the characteristic classes of the manifold M are those of the tangent bundle TM, or those of its cotangent bundle T^*M, which is isomorphic to TM. We abbreviate the notation $f(TM)$ (or $f(T^*M)$) for the characteristic classes of the manifold to $f(M)$.

3.2. Elements of K-Theory. By a *virtual bundle* over a (not necessarily compact) manifold M we understand a pair $\xi = \{E^0, E^1\}$ of complex vector bundles over M. We also allow for zero-dimensional bundles, denoted by 0. The symbol 1 denotes a trivial one-dimensional bundle. A virtual bundle is called *trivial* if E^0 and E^1 are isomorphic. Two virtual bundles $\xi_1 = \{E_1^0, E_1^1\}$ and $\xi_2 = \{E_2^0, E_2^1\}$ are called *isomorphic* if E_1^0 is isomorphic to E_2^0 and E_1^1 is isomorphic to E_2^1. The direct sum is defined in the obvious way, that is, $\xi_1 \oplus \xi_2 = \{E_1^0 \oplus E_2^0, E_1^1 \oplus E_2^1\}$. We also define the tensor product $\xi_1 \otimes \xi_2 = \{(E_1^0 \otimes E_2^0) \oplus (E_1^1 \otimes E_2^1), (E_1^1 \otimes E_2^0) \oplus (E_1^0 \otimes E_2^1)\}$. Here we assume that $E \oplus 0 = E$, $E \otimes 0 = 0$ and $E \otimes 1 = E$.

In the set of virtual bundles we introduce an equivalence relation by writing $\xi_1 \sim \xi_2$ if there are trivial bundles η_1 and η_2 such that $\xi_1 \oplus \eta_1$ is isomorphic to $\xi_2 \oplus \eta_2$. Such virtual bundles are called *stably isomorphic*. The equivalence classes form a commutative ring $K(M)$ with identity, where the addition $\xi_1 + \xi_2$ is defined by the direct sum and the multiplication $\xi_1 \xi_2$ by the tensor product. The inverse of $\xi = \{E^0, E^1\}$ is the virtual bundle $-\xi = \{E^1, E^0\}$, and the identity is the class $\{1, 0\}$.

To a manifold mapping $g : M \to N$ there corresponds a ring homomorphism $g^! : K(N) \to K(M)$ which maps the class of virtual bundles $\xi = \{E^0, E^1\}$ over N on to the class $g^! \xi = \{g^* E^0, g^* E^1\}$, where $g^* E$ is the bundle over M induced by g. Thus, the correspondence $M \to K(M)$ is a contravariant functor, called *Grothendieck's K-functor*.

The characteristic class ch can be extended to virtual bundles by setting $\operatorname{ch} \xi = \operatorname{ch} E^0 - \operatorname{ch} E^1 \in H^+(M)$, where $H^+(M)$ is the ring of even-dimensional cohomologies. From the properties of the Chern character it follows that the cohomology class $\operatorname{ch} \xi$ depends only on the class ξ in $K(M)$, and the mapping $\operatorname{ch} : K(M) \to H^+(M)$ is a ring homomorphism.

In index theory an important role is played by a modification of the K-functor, called a *K-functor with compact supports*. By a *virtual bundle with compact support* we understand a triple $\xi = \{E^0, E^1, a\}$, where E^0 and E^1 are bundles over M of the same dimension and $a : E^0 \to E^1$ is an isomorphism of bundles over $M \setminus X$, with X some compactum in M, called the *support* of ξ. Such a triple is called *trivial* if a can be extended to an isomorphism over the whole of M. The direct sum of triples is defined in the obvious way. Two triples $\xi_1 = \{E_1^0, E_1^1, a_1\}$ and $\xi_2 = \{E_2^0, E_2^1, a_2\}$ are called *isomorphic* if there are isomorphisms $\varphi^i : E_1^i \to E_2^i$ over M such that $\varphi^1 a_1 = a_2 \varphi^0$ over $M \setminus X$.

We introduce an equivalence relation by writing $\xi_1 \sim \xi_2$ if there are trivial triples η_1 and η_2 such that $\xi_1 \oplus \eta_1$ is isomorphic to $\xi_2 \oplus \eta_2$. The equivalence classes form a commutative semigroup $K^{\mathrm{comp}}(M)$ with respect to direct sum. In fact, $K^{\mathrm{comp}}(M)$ is a group, as is shown below.

Sometimes it is convenient to consider the function a in the triple $\xi = \{E^0, E^1, a\}$ to be defined everywhere on M so that it is an isomorphism outside X, but, generally speaking, not on X. In other words, the isomorphism a defined on $M \setminus X$ can be extended to the whole of M as a homomorphism. Such an extension is realized by means of a *truncating function* $\rho \geq 0$, which is equal to zero on X and to one outside some larger compactum. Then ρa yields the desired extension.

We mention three properties of virtual bundles with compact support and of their classes in $K^{\mathrm{comp}}(M)$.

1. *Stability.* Let $\xi(t) = \{E^0, E^1, a(t)\}$ be a smooth family depending on a parameter $t \in [0, 1]$, with support in a compactum X independent of t. Then the triples $\xi(0)$ and $\xi(t)$ are isomorphic for any t, that is, the class of ξ in $K^{\mathrm{comp}}(M)$ is invariant under homotopies. Indeed, let us define isomorphisms $\varphi^i(t) : E^i \to E^i$ $(i = 0, 1)$ by setting $\varphi^1(t) \equiv 1$ and taking $\varphi^0(t)$ to be a fundamental matrix of solutions for the system of ordinary differential equations $\dot{\varphi}^0 + \rho \dot{a}^{-1}(t) a(t) \varphi^0 = 0$, $\varphi^0(0) = 1$, where ρ is a truncating function. In view of its uniqueness, this solution coincides with $a^{-1}(t) a(0)$ outside the compactum where $\rho \equiv 1$, since $a(t) \varphi^0(t) = a(0)$. This proves that the triples $\xi(0)$ and $\xi(t)$ are isomorphic. We mention that if E^0 and E^1 are equipped with Hermitian inner products and a is a unitary isomorphism, then $\varphi^0(t)$ is also unitary.

2. *The logarithmic property.* Let $\xi = \{E^0, E^1, a\}$ and $\eta = \{E^1, E^2, b\}$ be virtual triples with compact support. Then $\zeta = \{E^0, E^2, ba\}$ defines an element $\xi + \eta$ in $K^{\mathrm{comp}}(M)$. Indeed, $\zeta \oplus \{E^1, E^1, 1\}$ and $\xi \oplus \eta$ are isomorphic, since the matrices

$$\begin{pmatrix} ba & 0 \\ 0 & 1 \end{pmatrix} = \begin{pmatrix} b & 0 \\ 0 & 1 \end{pmatrix} \begin{pmatrix} a & 0 \\ 0 & 1 \end{pmatrix} \quad \text{and} \quad \begin{pmatrix} a & 0 \\ 0 & b \end{pmatrix} = \begin{pmatrix} a & 0 \\ 0 & 1 \end{pmatrix} \begin{pmatrix} 1 & 0 \\ 0 & b \end{pmatrix}$$

are homotopic.

In particular, the inverse of $\xi = \{E^0, E^1, a\} \in K^{\mathrm{comp}}(M)$ is the triple $\{E^1, E^0, a^{-1}\}$, so that $K^{\mathrm{comp}}(M)$ is a group.

3. The group $K^{\mathrm{comp}}(M)$ is a module over $K(M)$. The product of $\eta = \{F^0, F^1\} \in K(M)$ and $\xi = \{E^0, E^1, a\} \in K^{\mathrm{comp}}(M)$, which gives the module structure, is defined as the class $F^0 \otimes \xi - F^1 \otimes \xi = \{F^0 \otimes E^0, F^0 \otimes E^1, 1 \otimes a\} - \{F^1 \otimes E^0, F^1 \otimes E^1, 1 \otimes a\}$ in $K^{\mathrm{comp}}(M)$.

We define the characteristic class $\mathrm{ch}\,\xi$ for $\xi \in K^{\mathrm{comp}}(M)$. Let $\xi = \{E^0, E^1, a\}$ be a virtual bundle with compact support, and ∂^0 and ∂^1 connections in E^0 and E^1, respectively. We recall that the covariant differential $(\partial b) u = \partial^1(bu) - b\partial^0 u$, where $u \in C^\infty(E^0)$, is defined for any homomorphism $b : E^0 \to E^1$. We claim that ∂^0 and ∂^1 can always be chosen so that $\partial a = 0$ outside a compactum. Indeed, starting with arbitrary connections ∂^0 and ∂^1, we introduce a connection $\tilde{\partial}^0$ acting on $u \in C^\infty(E^0)$ according to the formula $\tilde{\partial}^0 u = \partial^0 u + (\rho a^{-1} \partial a) u$, that is, the connection forms of ∂^0 and $\tilde{\partial}^0$ differ by $\Delta \Gamma = \rho a^{-1} \partial a$, where ρ is a truncating function. Then the covariant differential $\tilde{\partial} b$ defined by $\tilde{\partial}^0$ and ∂^1 satisfies $\tilde{\partial} b = \partial b - b(\rho a^{-1} \partial a)$, from which it follows that $\tilde{\partial} a = 0$ outside the compactum where $\rho \equiv 1$. We mention that if ∂^0 and ∂^1 are Hermitian connections and a is unitary, then $\tilde{\partial}^0$ is also Hermitian. The curvature $\tilde{\Omega}_0$ of $\tilde{\partial}^0$ is given by

$$\tilde{\Omega}_0 = \Omega_0 + \partial(\rho a^{-1} \partial a) + (\rho a^{-1} \partial a)^2 . \tag{1.10}$$

We define the form $\mathrm{ch}\,\xi$ by

$$\mathrm{ch}\,\xi = \mathrm{tr}\, e^{\tilde{\omega}_0} - \mathrm{tr}\, e^{\omega_1} \tag{1.11}$$

and claim that it has compact support. Indeed, differentiating the equality $\tilde{\partial}a = 0$, we find that $0 = \tilde{\partial}(\tilde{\partial}a) = \Omega_1 a - a\tilde{\Omega}_0$, from which it follows that $\tilde{\Omega}_0 = a^{-1}\Omega_1 a$ outside X, so that the traces of $e^{\tilde{\omega}_0}$ and e^{ω_1} coincide outside a compactum. Thus, we obtain the cohomology class $\operatorname{ch}\xi \in H^{+\operatorname{comp}}(M)$. It is not difficult to show that this class is invariant under homotopies of a, $\tilde{\partial}^0$ and ∂^1 such that $\tilde{\partial}a = 0$ outside a compactum.

It is clear that the group $H^{+\operatorname{comp}}(M)$ is a module over $H^+(M)$. From the properties of the Chern character it follows that $\operatorname{ch} : K^{\operatorname{comp}}(M) \to H^{+\operatorname{comp}}(M)$ is a module homomorphism.

3.3. The Thom Isomorphism. Let M be a compact manifold of dimension n and E an m-dimensional complex bundle over M. To avoid any possible confusion, we denote by N the total space of the bundle E, which is a noncompact $(n + 2m)$-dimensional manifold. The points of N are pairs (x, z), where $x \in M$ and $z \in E_x$. We define the embedding $i : M \to N$, $i(x) = (x, 0)$, and the projection $p : N \to M$, $p(x, z) = x$. Since N shrinks to M, the above embedding and projection induce mutually inverse isomorphisms $i^* : H(N) \to H(M)$ and $p^* : H(M) \to H(N)$, as well as isomorphisms $i^! : K(N) \to K(M)$ and $p^! : K(M) \to K(N)$. The *Thom isomorphism theorems* describe the structure of $K^{\operatorname{comp}}(N)$ as a module over $K(N) \approx K(M)$ and the structure of $H^{\operatorname{comp}}(N)$ as a module over $H(N) \approx H(M)$. They state that these modules are generated by one free generator. Our arguments require the construction of such generators and the relation between them.

First we consider the *Thom isomorphism in K-theory*. We denote the lifting of the bundles E and E^* to N by means of the projection p by the same symbols. The complex

$$0 \to \Lambda^0(E^*) \xrightarrow{\varepsilon(z)} \Lambda^1(E^*) \xrightarrow{\varepsilon(z)} \ldots \xrightarrow{\varepsilon(z)} \Lambda^n(E^*) \to 0 \qquad (1.12)$$

is defined at every point $(x, z) \in N$, where $\varepsilon(z)$ is the homomorphism of exterior multiplication of the p-form $u \in \Lambda^p(E_x^*)$ by the form $\langle \cdot, z \rangle$ (we assume that E is equipped with a Hermitian structure). It is clear that this complex is exact for $z \neq 0$. Just as in 2.3 we associated an elliptic operator with an elliptic complex, here with the complex (1.12) we associate the virtual bundle with compact support over N given by

$$\beta_E = \{\Lambda^+(E^*), \Lambda^-(E^*), b(z)\},$$

where $b(z) = \varepsilon(z) + \varepsilon^*(z)$ and $\varepsilon^*(z)$ is the dual homomorphism of $\varepsilon(z)$. We mention that $\varepsilon^*(z)$ coincides with the contraction homomorphism $i(z) : \Lambda^p \to \Lambda^{p-1}$, which is defined as follows. A form $u \in \Lambda^p(E_x^*)$ is an antisymmetric multilinear function $u(z_1, z_2, \ldots, z_p)$, where $z_i \in E_x$. Then

$$(i(z)u)(z_1, z_2, \ldots, z_{p-1}) = u(z, z_1, z_2, \ldots, z_{p-1}).$$

It is clear that the homomorphism $i(z)$ is antidifferentiation, that is, $i(z)(u \wedge v) = (i(z)u) \wedge v + (-1)^p u \wedge (i(z)v)$, where p is the degree of the form u. From this we obtain

$$i(z)\varepsilon(z) + \varepsilon(z)i(z) = |z|^2 \ . \tag{1.13}$$

Hence, $b^2(z) = i(z)\varepsilon(z) + \varepsilon(z)i(z)$ is the homomorphism of multiplication by $|z|^2$, analogous to the Laplace operator in 2.3.

From *Bott's periodicity theorem* (Atiyah (1967)), which is a fundamental assertion in K-theory, it follows that β_E is a generator of the module $K^{\mathrm{comp}}(N)$ over the ring $K(M)$. In other words, the homomorphism $i_! : K(M) \to K^{\mathrm{comp}}(N)$ defined by $i_!\xi = \beta_E(p^!\xi)$ is an isomorphism, called the *Thom isomorphism* in K-theory.

The element $\beta_E \in K^{\mathrm{comp}}(N)$ is caled the *Bott generator*. Sometimes it is convenient to write it in the form

$$\beta_E = \{\Lambda^+(E^*), \Lambda^-(E^*), b(\rho\nu)\} \ , \tag{1.14}$$

where $\nu = z/|z|$ is a unit vector and $\rho(z)$ a truncating function equal to 1 for $|z| \geq 1$.

We now go over to cohomologies. Let ψ be a closed form with compact support on N. In local coordinates, this form can be written as a sum of terms of the form $\psi_{i_1 \ldots i_k} \wedge dx^{i_1} \wedge \ldots \wedge dx^{i_k}$, where $\psi_{i_1 \ldots i_k}$ are forms on the fibre E_x. We define the operation p_* of *fibre integration* of a form by

$$p_*(\psi_{i_1 \ldots i_k} \wedge dx^{i_1} \wedge \ldots \wedge dx^{i_k}) = \left(\int_{E_x} \psi_{i_1 \ldots i_k}\right) dx^{i_1} \wedge \ldots \wedge dx^{i_k} \ ,$$

where the integral over E_x is taken of the component of the top degree $2m$ with respect to dz^i, $d\bar{z}^i$. The orientation of E_x is the standard orienation of a complex space, defined by the form $i^m dz^1 \wedge d\bar{z}^1 \wedge \ldots \wedge dz^m \wedge d\bar{z}^m$. It is clear that this operation is independent of the choice of local coordinates $(x, z^1, \bar{z}^1, \ldots, z^m, \bar{z}^m)$ on N, and that it defines a homomorphism $p_* : H^{\mathrm{comp}}(N) \to H(M)$ which reduces the degree of the form to $2m$. *Thom's theorem* states that p_* is an isomorphism. We denote by $i_* : H(M) \to H^{\mathrm{comp}}(N)$ the inverse isomorphism. The image of $1 \in H(M)$ under this mapping is called the *Thom generator* $U_E = i_*1 \in H^{2m\,\mathrm{comp}}(N)$. We indicate the explicit construction of U_E.

Let ∂ be a Hermitian connection in the bundle E and Γ the connection form in a local reference frame. This also defines a connection in the lifting of E to the manifold N, with the same connection form Γ. The latter is called the lifting of ∂, and we denote it by the same symbol. The bundle E over N contains the section $\nu = z/|z|$, which is defined for $z \neq 0$. Starting with ∂, we construct a connection $\tilde{\partial}$ in the bundle E over N such that $\tilde{\partial}\nu = 0$ for $|z| \geq 1$. This can be given, for example, by

$$\tilde{\partial}u = \partial u + \rho(-\partial\nu\langle u, \nu\rangle + \nu\langle u, \partial\nu\rangle + \langle\partial\nu, \nu\rangle u) \ . \tag{1.15}$$

Differentiating the equality $\tilde{\partial}\nu = 0$, we find that $\tilde{\Omega}\nu = 0$, from which it follows that $\det\tilde{\omega} = 0$ for $|z| \geq 1$. In this way, the mth Chern form $\det\tilde{\omega}$ defined by $\tilde{\partial}$ yields a $2m$-dimensional cohomology class with compact support. It can be proved that $p_* \det\tilde{\omega} = 1$ (see, for example, Berline and Vergne (1985) and Bott (1967b)). Consequently, this form is the Thom generator U_E.

We now compute $p_* \operatorname{ch} \beta_E$. With the connection $\tilde{\partial}$ in the bundle E over N we associate connections in E^* and in the exterior powers $\Lambda^k(E^*)$. Here $\tilde{\partial}b(\nu) = 0$, since $\tilde{\partial}\nu = 0$. The form $\operatorname{ch} \beta_E$ is given by (1.11), where $\tilde{\omega}_0$ and ω_1 are the curvatures of $\tilde{\partial}$ in $\Lambda^+(E^*)$ and $\Lambda^-(E^*)$, respectively. From (1.9) we obtain

$$\operatorname{ch} \beta_E = \operatorname{ch} \Lambda^+(E^*) - \operatorname{ch} \Lambda^-(E^*) = \det\tilde{\omega} \cdot \det \frac{1 - e^{-\tilde{\omega}}}{\tilde{\omega}} \; . \tag{1.16}$$

The first factor is a form with compact support on N which defines the class $U_E \in H^{\mathrm{comp}}(N)$. In the second factor we replace the curvature $\tilde{\omega}$ of $\tilde{\partial}$ by the curvature ω of ∂, which does not change its cohomlogy class in $H(N)$, consequently, nor the class of all products in $H^{\mathrm{comp}}(N)$. Since the form ω does not contain the differentials dz^i and $d\bar{z}^i$, fibre integration yields

$$p_* \operatorname{ch} \beta_E = \mathcal{T}^{-1}(E) \; . \tag{1.17}$$

We make some remarks about the Thom homomorphism in the case of real bundles. Let E be a real oriented $2m$-dimensional bundle with an Euclidean structure over a compact manifold M, and let N be the space of the bundle. The points of N have the form (x, y), where $x \in M$ and $y \in E_x$. Let ∂ be an Euclidean connection on E, as well as on its lifting to N, $\tilde{\partial}$ the Euclidean connection on the lifting of E satisfying the condition $\tilde{\partial}\nu = 0$ outside a compactum, where $\nu = y/|y|$ is a section of the bundle E over N which is defined for $y \neq 0$. Such a connection is defined by the same formula (1.15) as in the complex case, but here $\langle \partial\nu, \nu \rangle = 0$. As before, the homomorphism p_* of fibre integration is an isomorphism, and the inverse isomorphism i_* is given by multiplication by the Thom class U_E, which in the real case is defined by the form $\operatorname{Pf}(\tilde{\Omega}/(2\pi))$, where $\tilde{\Omega}$ is the curvature of $\tilde{\partial}$. It can be verified immediately that $p_* U_E = 1$ and that $i^* U_E = \chi(E)$ is the Euler class of E.

In K-theory the situation is more complicated. As in the complex case, the element β_E is defined by (1.14), but it is not a generator of $K^{\mathrm{comp}}(N)$. The equality (1.16) remains valid as before, but fibre integration yields a different result. Indeed,

$$\det\bar{\omega} = (-1)^m (\operatorname{Pf}(\bar{\Omega}/(2\pi)))^2 \sim (-1)^m U_E \chi(E)$$

in $H^{\mathrm{comp}}(N)$. Thus, instead of (1.17) we obtain

$$p_* \operatorname{ch} \beta_E = (-1)^m \chi(E) \mathcal{T}^{-1}(E) \; . \tag{1.18}$$

§4. The Atiyah-Singer Formula

Let $A : C^\infty(E^0) \to C^\infty(E^1)$ be an elliptic PDO on a compact n-dimensional manifold M and $a : E^0 \to E^1$ its principal symbol, which defines an isomorphism of the bundles E^0 and E^1 (more precisely, of their liftings to T^*M) over $T^*M \setminus 0$. Thus, the elliptic operator defines an element $d(A) \in K^{\mathrm{comp}}(T^*M)$ given by the virtual bundle $\{E^0, E^1, a\}$ with compact support. This is called the *distinguishing element of the elliptic operator.*

Theorem 1.4. *The index of the elliptic operator A is computed by means of the formula*

$$\mathrm{ind}\, A = \int\limits_{T^*M} \mathrm{ch}\, d(A) \mathcal{T}(M)\,, \qquad (1.19)$$

where the orientation of T^*M is given by the form $d\xi_1 \wedge dx^1 \wedge \ldots \wedge d\xi_n \wedge dx^n$.

The expression on the right-hand side is called the *topological index.*

The Todd class $\mathcal{T}(M)$ is defined by the form $\det(\omega/(1 - e^{-\omega}))$, where ω is expressed in terms of the curvature tensor R^i_{jkl} of the Riemannian metric by

$$\omega = -\frac{1}{2\pi i}\left(\frac{1}{2} R^i_{jkl} dx^k \wedge dx^l\right).$$

As already mentioned in the Introduction, there are many ways of proving this theorem. Since it is not possible to dwell on them in detail, we confine ourselves only to a brief outline of the main methods. One of them, based on the Thom isomorphism, is examined in more detail in Chap. 3 in connection with the proof of a more general index theorem.

The left- and right-hand sides in (1.19) are stable homotopic invariants of the symbol of the elliptic operator on $S(M)$. Naturally, we would want to use deformations, direct sums, and tensor multiplication to bring the operator to a simpler form, for which (1.19) could be verified directly. Unfortunately, the properties of the index that we listed earlier are not sufficient for us to be able to do this. For example, in the first proof Atiyah and Singer used, in addition, the invariance of the index with respect to bordisms, which for the analytic index was proved with considerable difficulty (Palais (1965)). The second proof was based on the Thom isomorphism in K-theory. If N is a normal bundle of dimension m under the embedding of M in $\mathbb{R}^{n+m}$, then T^*N may be regarded as a complex m-dimensional bundle over T^*M. By means of the Thom isomorphism in K-theory, with $d(A) \in K^{\mathrm{comp}}(T^*M)$ we associate an element $\alpha \in K^{\mathrm{comp}}(T^*N)$ that has the same topological index. But, once again, we encounter a fundamental difficulty, which consists in interpreting α as a class of the symbol of an elliptic operator on the compactification of N and proving that its analytic index coincides with the index of A (see Atiyah and Singer (1968a,b)).

We also mention the analytic proofs. Using Theorem 1.1 or Theorem 1.2, it is not difficult to obtain an expression for the analytic index in terms of the so-called full symbol of the operator and its derivatives with respect to local coordinates. However, bringing that expression to the invariant form (1.19) is far from simple. This can be done for operators having some specific property of invariance, by means of the operation of group averaging. Among them are classical operators such as $d + d^*$, the signature operator, and the Dirac operator. To extend the theorem to arbitrary operators, we need to use a result of K-theory which states that ch defines an isomorphism of the cohomologies of $H^+(M)$ with real coefficients in the ring $K(M) \otimes \mathbb{R}$. Such a proof is given in Atiyah, Bott and Patody (1973). The idea of group averaging is used in Chapter 3, but without the additional use of this K-theory result, which is not applicable there.

We make a few remarks concerning the Atiyah-Singer formula.

1. We know that the index depends only on the principal symbol on $S(M)$; therefore, it is natural to want to transform the integral in (1.19) into one over $S(M)$. With this aim, we find an explicit expression for the form $\operatorname{ch} d(A)$. Let ∂^0 and ∂^1 be connections on E^0 and E^1, respectively, as well as on their liftings to T^*M, and let $\tilde{\partial}^0 = \partial^0 + (\rho a^{-1}\partial a)$. Then $\operatorname{ch} d(A)$ is computed by means of (1.11). We consider a homotopy of the connection $\tilde{\partial}_t^0 = \partial^0 + t(\rho a^{-1}\partial a)$ which links ∂^0 and $\tilde{\partial}^0$. Let $\omega_0(t)$ be the curvature of $\tilde{\partial}^0$ (divided by $-2\pi i$). Then (1.11) can be written as

$$\operatorname{ch} d(A) = \operatorname{tr} e^{\omega_0} - \operatorname{tr} e^{\omega_1} + \int \frac{d}{dt}(\operatorname{tr} e^{\tilde{\omega}_0(t)})\, dt\ .$$

Since the support of $\operatorname{ch} d(A)$ is contained in $B(M)$ (the bundle of unit balls $|\xi| \leq 1$), then we can integrate over $B(M)$ in (1.19). The first two terms in the expression of $\operatorname{ch} d(A)$ yield zero, since these forms do not contain $d\xi_i$. Using the formula for the variation of curvature (see 3.2), we can rewrite the third term as the differential

$$-\frac{1}{2\pi i}d\int_0^1 \rho\operatorname{tr} a^{-1}\partial a e^{\tilde{\omega}_0(t)}dt\ .$$

Using Stokes's formula, we obtain

$$\operatorname{ind} A = -\frac{1}{2\pi i}\int_{S(M)} T(M)\int_0^1 \operatorname{tr} a^{-1}\partial a e^{\tilde{\omega}_0(t)}dt\ . \tag{1.20}$$

From this and (1.10) it follows that

$$\tilde{\omega}_0(t) = \omega_0 - \frac{1}{2\pi i}\{t\partial(a^{-1}\partial a) + t^2(a^{-1}\partial a)^2\}\ .$$

Taking into account that $\partial a^{-1} = -a^{-1}\partial a a^{-1}$ and $\partial^2 a = \Omega_1 a - a\Omega_0$, we can rewrite the last formula as

$$\tilde{\omega}_0(t) = (1 - t)\omega_0 + ta^{-1}\omega_1 a + \frac{1}{2\pi i}t(1 - t)(a^{-1}\partial a)^2 \,. \tag{1.21}$$

2. From (1.20) and (1.21) we deduce a simple corollary: if E^0 and E^1 are of dimension 1 and $\dim M > 2$, then $\operatorname{ind} A = 0$ for any elliptic operator. Indeed, $(a^{-1}\partial a)^2 = 0$ for one-dimensional bundles, and the integral with respect to t can be computed explicitly; we obtain

$$\operatorname{ind} A = -\frac{1}{2\pi i} \int\limits_{S(M)} T(M)\frac{e^{\omega_1} - e^{\omega_0}}{\omega_1 - \omega_0} a^{-1}\partial a \,.$$

If $\dim M > 2$, then the dimension of the fibre of $S(M)$ is greater than 1. The integral over $S(M)$ is equal to zero, since the differentials $d\xi_i$ are contained only in the form $a^{-1}\partial a$ of degree one.

In the case $\dim M = 2$, the integral $(1/(2\pi i)) \int\limits_{|\xi|=1} a^{-1}\partial a = k$ over the fibre of $S(M)$ is a constant; hence, for an oriented manifold we obtain

$$\operatorname{ind} A = k \int\limits_{M} \frac{\omega_1 + \omega_0}{2} \,. \tag{1.22}$$

Here we have used the fact that $T(M) = 1$ for two-dimensional manifolds, since the dimensions of the Pontryagin classes are multiples of 4. The change of sign is connected with the change of orientation in fibre integration.

3. The index is equal to zero for any elliptic differential operator on a manifold of odd dimension. Indeed, the form in the integrand in (1.20) remains invariant under the antipodal mapping $(x, \xi) \mapsto (x, -\xi)$ of $S(M)$, while the orientation of $S(M)$ changes to the opposite one. We mention that for PDOs the index may be non-zero, since the form $a^{-1}\partial a$ may change under the antipodal mapping.

4. Suppose that the bundles E^0 and E^1 are trivial (the case of a system). Then in (1.21) only the last term is non-zero, and the integral with respect to t can be computed. If $T(M) = 1$, then we arrive at

$$\operatorname{ind} A = -\frac{1}{(2\pi i)^n} \frac{(n-1)!}{(2n-1)!} \int\limits_{S(M)} \operatorname{tr}(a^{-1}da)^{2n-1} \,. \tag{1.23}$$

We mention without proof that for matrices of order $n < \dim M$ this integral is equal to zero, while for $n = \dim M$ it admits the following topological inerpretation: let $\deg a$ be the degree of the continuous mapping into the sphere

$$S(M) \xrightarrow{a} \operatorname{GL}(n, \mathbb{C}) \to S^{2n-1} \,,$$

where the last mapping is defined by the first column of the matrix a; then (see Atiyah and Singer (1968b))

$$\operatorname{ind} A = -\frac{1}{(n-1)!} \deg a \,.$$

§5. Examples

5.1. The Gauss-Bonnet Theorem. Let M be a compact oriented manifold of dimension $2n$, Λ the bundle of exterior differential forms, and $\Lambda^{\pm}$ the bundles of the forms of even and odd degrees, respectively. The operator of exterior differentiation $d : C^{\infty}(\Lambda^k) \to C^{\infty}(\Lambda^{k+1})$ defines the *de Rham complex* of M. The principal symbol of d (up to a factor i) is given by exterior multiplication by the covector ξ, so that the sequence of symbols

$$0 \to \Lambda^0 \xrightarrow{\xi\wedge} \Lambda^1 \xrightarrow{\xi\wedge} \ldots \xrightarrow{\xi\wedge} \Lambda^{2n} \to 0$$

is exact for $\xi \neq 0$. With the de Rham complex we associate in the standard way (see 2.3) the elliptic operator $A = d + d^* : C^{\infty}(\Lambda^+) \to C^{\infty}(\Lambda^-)$. The index of this operator is called the *Euler characteristic of the manifold M*. We compute the index by means of the Atiyah-Singer formula. The distinguishing element $d(A) \in K^{\mathrm{comp}}(T^*M)$ is in fact the element β_E for the real bundle $E = T^*M$ (see 3.3). Using (1.18), we obtain

$$\mathrm{ind}\, A = \int\limits_{T^*M} (-1)^n U_E \chi(E) T^{-1}(E) T(E) = \int\limits_{M} \chi(E)$$

(the change of sign is caused by the change in orientation in the fibre integration). Therefore, the Euler characteristic of the manifold is equal to the integral of the Euler class. This is called the *Gauss-Bonnet formula*.

5.2. The Riemann-Roch Theorem. Let M be a complex compact manifold of complex dimension n. The complexification $\mathbb{C}\otimes TM$ of the real tangent bundle can be written as the direct sum of the holomorphic tangent bundle $T_{\mathbb{C}}M$ with local reference frame $\partial/\partial z^i$ and the antiholomorphic one $\bar{T}_{\mathbb{C}}M$ with local reference frame $\partial/\partial z^i$. We denote by $\Lambda^{p,q}$ the bundle of exterior differential forms of degree p in dz^i and of degree q in $d\bar{z}^i$. In particular, $\Lambda^{0,1}$ is the dual bundle of $\bar{T}_{\mathbb{C}}M$. All of TM, T^*M, $T_{\mathbb{C}}M$ and $\bar{T}_{\mathbb{C}}^*M$ are isomorphic as real bundles. To simplify the notation, we write $\bar{T}M$ instead of $\bar{T}_{\mathbb{C}}M$ and identify it with T^*M as a manifold.

Let E be a holomorphic bundle over M; this means that the transition matrix functions are holomorphic. Then the Cauchy-Riemann operator d'' is correctly defined on the sections of E, and we obtain the *Dolbeault complex*

$$0 \to C^{\infty}(E) \xrightarrow{d''} C^{\infty}(E \otimes \Lambda^{0,1}) \xrightarrow{d''} \ldots \xrightarrow{d''} C^{\infty}(E \otimes \Lambda^{0,n}) \to 0 \, .$$

This complex is elliptic, since for any $v \neq 0 \in \bar{T}_x M$ the sequence of symbols

$$0 \to E \xrightarrow{\zeta\wedge} E \otimes \Lambda^{0,1} \xrightarrow{\zeta\wedge} \ldots \xrightarrow{\zeta\wedge} E \otimes \Lambda^{0,n} \to 0 \, ,$$

where $\zeta = \langle \cdot\, , v \rangle \in \Lambda_x^{0,1}$, is exact. *Hirzebruch's theorem* (Chern (1956)), which generalizes the classical Riemann-Roch theorem, states that the Euler characteristic χ of this complex can be expressed by means of the formula

$$\chi = \int_M \operatorname{ch} E \mathcal{T}(T_{\mathbb{C}}M) \, .$$

To prove this, we apply the Atiyah-Singer theorem to the elliptic operator $A = d'' + (d'')^*$ corresponding to the Dolbeault complex (see 2.3). From the definition of the Thom isomorphism in K-theory and the comparison of the symbol sequence with (1.12) we see that the distinguishing element $d(A) \in K^{\mathrm{comp}}(\bar{T}M)$ is in fact the image of $[E] \in K(M)$ under the Thom isomorphism $i_! : K(M) \to K(\bar{T}M)$. By (1.17),

$$p_* \operatorname{ch} d(A) = \operatorname{ch} E p_* \beta_{\bar{T}M} = \operatorname{ch} E \mathcal{T}^{-1}(\bar{T}M) \, .$$

Taking also into account that

$$\mathcal{T}(M) = \mathcal{T}(\mathbb{C} \otimes TM) = \mathcal{T}(T_{\mathbb{C}}M)\mathcal{T}(\bar{T}_{\mathbb{C}}M)$$

and performing the fibre integration in the Atiyah-Singer formula, we find that

$$\chi = \operatorname{ind} A = \int_{\bar{T}M} \operatorname{ch} d(A)\mathcal{T}(T_{\mathbb{C}}M)\mathcal{T}(\bar{T}_{\mathbb{C}}M) = \int_M \operatorname{ch} E \mathcal{T}(T_{\mathbb{C}}M) \, .$$

In the classical case, when $\dim_{\mathbb{C}} M = 1$ and the one-dimensional holomorphic bundle E is generated by the divisor $D = \sum n_i x_i$ on the Riemann surface M (see Forster (1977) and Springer (1957)), the Dolbeault complex contains two terms, that is, d'' is the Cauchy-Riemann operator with index

$$\operatorname{ind} d'' = \dim \operatorname{Ker} d'' - \dim \operatorname{Coker} d'' = \int_M c_1(E) + \tfrac{1}{2}c_1(T_{\mathbb{C}}M) = \deg D + 1 - g \, ,$$

where $\deg D = \sum n_i$ is the degree of the divisor D and g the genus of the Riemann surface, equal to half of the dimension of $H^1(M)$. Here we have used the fact that, by the Gauss-Bonnet theorem, the integral

$$\int_M c_1(T_{\mathbb{C}}M) = \int_M \chi(TM)$$

is equal to the Euler characteristic $2 - 2g$ of M.

We mention that $\operatorname{Ker} d''$ is the space of holomorphic sections of E, that is, the space of meromorphic functions f such that $z^{n_i} f$ are holomorphic in the neighbourhood of the point x_i, where z is the local parameter at x_i.

The classical *Riemann-Roch theorem* may serve as an illustration of the application of the index theorem to the proof of existence of solutions to elliptic equations. Thus, if $\deg D > g - 1$, then this theorem yields the *Riemann inequality* $\dim \operatorname{Ker} d'' \geq \deg D + 1 - g > 0$, from which, in particular, we deduce the existence of meromorphic functions with poles of order not exceeding prescribed vales at given points. For $\deg D > 2g - 2$ it can be proved that Coker d'' is empty and Riemann's inequality becomes an equality that determines the dimension of the space of solutions.

5.3. Spinor Structure and the Dirac Operator. This example is very important because here we encounter for the first time objects with which we deal constantly in Chap. 3. We refer to algebra bundles and connections on them associated with vector bundles equipped with additional structures. We also illustrate an application of index theory based on integer values of the index, which plays an important role in Chap. 3.

By a *Clifford algebra* C_n we understand an associative algebra with identity over $\mathbb{C}$, generated by elements $e_1, e_2, \ldots, e_n$ such that $e_i e_j + e_j e_i = -2\delta_{ij}$. Any element $a \in C_n$ can be represented uniquely in the form

$$a = \sum a_{i_1 \ldots i_n} e_1^{i_1} e_2^{i_2} \ldots e_n^{i_n} , \tag{1.24}$$

where the indices i_k take the values 0 and 1 and the summation extends over all such collections of indices. The dimension of C_n as a linear space over $\mathbb{C}$ is 2^n. In C_n we can define the operations of involution, induced by the action on the generators $e_i^* = -e_i$, and trace, given by $\operatorname{tr} a = c a_{00\ldots 0}$, where c is a normalizing constant. It is easy to verify that $\operatorname{tr} ab = \operatorname{tr} ba$. Consequently, C_n is a C^*-algebra with norm $\|a\|^2 = \operatorname{tr} aa^*$.

Let $\alpha \in so(n)$ be a real skew-symmetric matrix. We associate with it the element $a = \varphi(\alpha) = \frac{1}{4}\alpha_{ij} e_i e_j$ (here and in what follows we adopt the convention of summation over repeated indices). We denote by $\operatorname{spin}(n)$ the real subspace in C_n consisting of the elements $\varphi(\alpha)$. Using commutation relations, we easily find that for $a = \varphi(\alpha)$

$$[a, e_i] = a e_i - e_i a = \alpha_{ij} e_j \tag{1.25}$$

and

$$[\varphi(\alpha), \varphi(\beta)] = \varphi([\alpha, \beta]) ,$$

where on the left-hand side we take the commutator in the Clifford algebra and on the right-hand side the commutator in the matrix algebra. Hence, $\operatorname{spin}(n)$ is isomorphic to $so(n)$ as a Lie algebra. The elements of the form e^a, where $a = \operatorname{spin}(n)$ and the exponential is computed in the Clifford algebra as a power series, generate a group $\operatorname{Spin}(n)$ called the *spinor group*. From (1.25) and the well-known formula

$$e^a e_i e^{-a} = \sum_{k=0}^{\infty} \frac{1}{k!} \underbrace{[a, [a, \ldots, [a, e_i]] \ldots]}_{k}$$

it follows that $e^a e_i e^{-a} = (e^\alpha)_{ij} e_j$ for $a = \varphi(\alpha)$. The mapping $\psi : e^{\varphi(\alpha)} \mapsto e^\alpha$ defines a group homomorphism $\psi : \operatorname{Spin}(n) \to SO(n)$ which, as is easily seen, satisfies $\psi^{-1}(1) = \pm 1$. Consequently, ψ is a double covering.

A change $e_i' = g_{ij} e_j$ of generators by means of an orthogonal matrix does not change the relations, the involution, and the trace; therefore, we can speak of the Clifford algebra $C(E)$ of an Euclidean n-dimensional space E generated by the elements of an orthonormal basis for E.

In what follows we consider the case when n is even, that is, $n = 2m$. We denote by $e_1, \ldots, e_m, e'_1, \ldots, e'_m$ the generators of the Clifford algebra and introduce new generators $z_k = \frac{1}{2}(e_k + ie'_k)$, $z_k^* = \frac{1}{2}(-e_k + ie'_k)$. The latter satisfy the *canonical anticommutation relations*

$$z_i z_j + z_j z_i = z_i^* z_j^* + z_j^* z_i^* = 0 , \qquad z_i^* z_j + z_j z_i^* = \delta_{ij} . \tag{1.26}$$

In particular, $z_i^2 = (z_i^*)^2 = 0$. It is easy to see that the elements $z_i^* z_i$ are idempotent, that is, $(z_i^* z_i)^2 = z_i^* z_i$. We introduce the element $p_0 = z_1^* z_1 z_2^* z_2 \ldots z_m^* z_m$, which is also idempotent and satisfies $z_i^* p_0 = 0$. Using the terminology of physics, we call p_0 a *vacuum projection*, the z_i^* *annihilation operators*, and the z_i *creation operators*. Any element $a \in C_{2m}$ can be written uniquely in a Wick normal form, that is, as a non-commutative polynomial in z_i and z_i^* where all the creation operators are on the left of the annihilation operators.

Let S be a left ideal of the algebra C_{2m}, generated by a vacuum projection. Any element $s \in S$ has the form ap_0, where $a \in C_{2m}$, and can be written uniquely as $F(z_1, z_2, \ldots, z_m)p_0$, where F is a polynomial in the creation operators which is at most linear in each variable; hence, $\dim S = 2^m$. S is called a *spinor space* and its elements are called *spinors*. The algebra C_{2m} acts in S as left multiplication, that is, if $s = ap_0$ and $b \in C_{2m}$, then $bs = bap_0$. This is called *Clifford multiplication*. C_{2m}, regarded as an algebra of linear transformations of S, is isomorphic to the whole matrix algebra $\mathrm{Hom}(S, S)$ of order 2^n. This is the unique irreducible representation of C_{2m}. We choose the normalizing factor in the definition of trace to be equal to 2^m so that $\mathrm{tr}\, p_0 = 1$, and equip S with the inner product $\langle s_1, s_2 \rangle = \mathrm{tr}\, p_0 a_2^* a_1 p_0$. Then the involution in C_{2m} corresponds to the Hermitian duality of linear transformations in S, and the trace in C_{2m} coincides with the trace of a linear transformation. This construction of the spinor representation is a particular case of the GNS (Gel'fand-Naimark-Segal) construction in the theory of C^*-algebra (see Emch (1972)).

Now let E be a real oriented Euclidean $2m$-dimensional bundle over a manifold M. It defines the bundle $C(E)$ of Clifford algebras, whose fibre at a point x is the Clifford algebra of the space E_x. A local orthonormal oriented reference frame $e_1, e_2, \ldots, e_{2m}$ defines a system of generators in every fibre $C(E_x)$. The sections of the bundle $C(E)$ are given locally by (1.24), where $a_{i_1 i_2 \ldots i_{2m}}$ are complex function of x. Let ∂ be a Euclidean connection in E with local connection form Γ, which takes values in so$(2m)$. We associate with it a connection in $C(E)$, also denoted by ∂, which acts on the generators e_i just as on the basis sections of E, that is,

$$\partial e_i = \Gamma_{ij} e_j = [\varphi(\Gamma), e_i] ,$$

and can be uniquely extended to any sections by means of Leibniz's rule $\partial(ab) = (\partial a)b + a\partial b$, so that for the sections (1.24)

$$\partial a = da + [\varphi(\Gamma), a] \,, \tag{1.27}$$

where d is the exterior differentiation of the coefficients. Since $\varphi : \mathrm{so}(2m) \to \mathrm{spin}(2m)$ is an isomorphism, it follows that $\partial^2 a = [\varphi(\Omega), a]$, where Ω is the curvature of ∂.

Next, let $S(E)$ be a 2^m-dimensional complex bundle over M, in whose fibres $S(E_x)$ the algebra $C(E_x)$ acts by means of Clifford multiplication, that is, $S(E)$ is a spinor bundle and $C(E)$ is isomorphic to $\mathrm{Hom}(S(E), S(E))$. Any connection ∇ in $S(E)$ also induces a connection in $C(E) = \mathrm{Hom}(S, S)$ by $(\nabla a)s = \nabla(as) - a\nabla s$. The curvature Ω_∇ of ∇ in $S(E)$ is a 2-form with values in $C(E)$. We say that a bundle E with a Euclidean connection ∂ admits a *spinor structure* if here is a bundle $S(E)$ with a connection ∇ such that $\Omega_\nabla = \varphi(\Omega)$ and ∇ induces the connection ∂ in $C(E)$.

We also indicate an equivalent definition in terms of transition functions, although we do not need it in what follows. Let $\{U_i\}$ be a covering of M by means of neighbourhoods for which any intersection is contractible. Let $g_{UV} \in \mathrm{SO}(2m)$ be the transition function of E over $U \cap V$. Then $g_{UV} g_{VW} g_{WU} = 1$ in $U \cap V \cap W$. Since $\psi : \mathrm{Spin}(2m) \to \mathrm{SO}(2m)$ is a double covering, it follows that we can choose a single-valued pre-image $\psi^{-1}(g_{UV}) \in \mathrm{Spin}(2m)$ over the intersections $U \cap V$. The bundle E has a spinor structure if these pre-images can be chosen so that $\psi^{-1}(g_{UV})\psi^{-1}(g_{VW})\psi^{-1}(g_{WU}) = 1$ (generally speaking, this product is equal to ± 1).

We analyze the instructive case when E has a complex structure, that is, it is a real form of some complex Hermitian bundle F. This means that we can choose local generators $z_1, \ldots, z_m, z_1^*, \ldots, z_m^*$ in $C(E)$ from among the creation and annihilation operators, so that the former change under transitions by means of a unitary matrix, that is, $z_i' = u_{ij} z_j$, $u = (u_{ij}) \in U(m)$. Then the vacuum projection $p_0 = z_1^* z_1 z_2^* z_2 \ldots z_m^* z_m$ is independent of the choice of generators and defines a global section of the bundle $C(E)$ for which $\partial p_0 = 0$, where ∂ is the real form of the Hermitian connection in F, as well as the associated connection in $C(E)$. The above construction of the spinor space carries over to bundles, and we obtain a bundle $S(E)$ with action $C(E)$. Any section of $S(E)$ has the form $s = ap_0$, where $a \in C(E)$. We introduce a connection in $S(E)$ in a natural way: for $s = ap_0$ we set

$$\partial s = \partial(ap_0)p_0 = (\partial a)p_0 \,. \tag{1.28}$$

However, the curvature of this connection does not, generally speaking, coincide with $\varphi(\Omega)$. Indeed,

$$\partial^2 s = [\varphi(\Omega), a]p_0 = \varphi(\Omega)ap_0 - a\varphi(\Omega)p_0 = (\varphi(\Omega) - \mathrm{tr}\, p_0\varphi(\Omega)p_0)s \,.$$

The last expression is obtained as follows. The form $\varphi(\Omega)$ commutes with p_0 since $\partial^2 p_0 = 0$. Consequently, $\varphi(\Omega)p_0 = \varphi(\Omega)p_0 p_0 = p_0\varphi(\Omega)p_0$. Since p_0 is a one-dimensional projecton in $S(E)$, it follows that $p_0\varphi(\Omega)p_0 = p_0\, \mathrm{tr}\, p_0\varphi(\Omega)p_0$, which yields the necessary expression. We call $\varphi(\Omega) - \mathrm{tr}\, p_0\varphi(\Omega)p_0$ *Wick (normal) curvature* and denote it by $\varphi_n(\Omega)$, as opposed to $\varphi(\Omega)$, which we call *Weyl curvature*. This terminology is motivated by the fact that for $\varphi(\Omega)$ the constant term is equal to zero in the form (1.24), while for $\varphi_n(\Omega)$ the constant term is equal to zero in the Wick normal form. The equivalent normalization conditions are $\operatorname{tr}\varphi(\Omega) = 0$ and $\operatorname{tr} p_0\varphi_n(\Omega)p_0 = 0$. All this happens because in the algebra $C(E)$ both the connection form $\varphi(\Gamma)$ and the curvature form $\varphi(\Omega)$ are not uniquely defined, but only up to scalar forms, while the connection in $S(E)$ defines the curvature uniquely.

In the case where $\varphi_n(\Omega) \neq \varphi(\Omega)$, we may try to correct the situation by tensoring $S(E)$ by a one-dimensional complex bundle L and defining a connection ∇ in $L \otimes S(E)$ in the usual way, namely, $\nabla = \partial_L \otimes 1 + 1 \otimes \partial$. This implies that $\Omega_\nabla = \varphi_n(\Omega) + \Omega_L$. If L is chosen so that $\Omega_L = \operatorname{tr} p_0\varphi(\Omega)p_0$, then the problem is solved. To do this, it is necessary and sufficient that the form $-(\operatorname{tr} p_0\varphi(\Omega)p_0)/(2\pi i)$ should be integer-valued. It can be shown that this form defines a cohomology class $\frac{1}{2}c_1(F)$, and that a spinor structure exists if and only if this cohomology class is integer-valued.

We extend our considerations to a real oriented even-dimensional bundle that admits a spinor structure. In this case we denote the connection ∇ in $S(E)$ also by ∂. The element $r = i^{-m}e_1 e_2 \ldots e_{2m}$ does not vary if we change the basis by means of a matrix $g \in \mathrm{SO}(2m)$; therefore, it defines a global section of $C(E)$ for which $\partial r = 0$. Its square is equal to 1. Consequently, $S(E)$ can be written as the direct sum $S^+ \oplus S^-$ of the eigenspaces of r corresponding to the eigenvalues ± 1 (even and odd spinors), and the connection ∂ in $S(E)$ preserves this decomposition. It is easy to see that $e_i r = -r e_i$; hence, multiplication by e_i maps $S^\mp$ into $S^\pm$.

Let M be an even-dimensional oriented compact manifold. We say that it admits a spinor structure if its tangent (or cotangent) bundle TM admits a spinor structure. Therefore, the tangent vectors e_i at $x \in M$ may also be regarded as elements of $C(T_x M)$ acting by means of Clifford multiplication on $S(T_x M)$. Let $\partial_i = i(e_i)\partial$ be the covariant differentiation of the sections of $S(TM)$ along the vector e_i. We define the *Dirac operator* $D : C^\infty(S) \to C^\infty(S)$ by $Ds = e_i\partial_i s$ (we adopt the convention of summation over repeated indices). Here e_i are the elements of an orthonormal basis, which act by means of Clifford multiplication. It is clear that D does not depend on the choice of basis. The principal symbol of D is $a(x,\xi) = e_i\xi_i$, where ξ_i are (up to a factor i) the coordinates of the covector ξ with respect to the basis e_i. Its square is $-\xi_i\xi_i = -|\xi|^2$; consequently, D is an elliptic operator. The square of the Dirac operator is called the *Laplace operator* Δ and the sections $s \in \operatorname{Ker}\Delta$ are called *harmonic spinors*. It is obvious that D maps $C^\infty(S^\pm)$ into $C^\infty(S^\mp)$; hence, the space H of harmonic spinors can be written as the direct sum of the subspaces $H^\pm$ of even and odd harmonic spinors.

We consider the restriction of the Dirac operator $D : C^\infty(S^+) \to C^\infty(S^-)$ and compute its index, which, by definition, is

$$\operatorname{ind} D = \dim H^+ - \dim H^- = \operatorname{tr} r|_H \,.$$

We apply the Atiyah-Singer formula. The distinguishing element $d(D) \in K^{\text{comp}}(T^*M)$ is given by a virtual bundle with compact support, that is, $d(D) = \{S^+, S^-, \rho\nu\}$, where $\nu = \xi_i e_i/|\xi|$ and ρ is a truncating function on T^*M, regarded as a section of the bundle $C(E)$ over T^*M. To avoid confusion, we denote by E both the tangent bundle and its lifting to T^*M. Let ∂ be a Riemannian connection in E, as well as its lifting to the bundle E over T^*M. We denote by $\tilde{\partial}$ a connection in the lifting of E to T^*M such that $\tilde{\partial}\nu = 0$ outside a compactum (see (1.15)); let $\tilde{\Omega}$ be the curvature of $\tilde{\partial}$. We also denote by $\tilde{\partial}$ the associated connections in the bundles $C(E)$ and $S(E)$ lifted to T^*M. Then $\varphi(\tilde{\Omega})$ is the curvature of $\tilde{\partial}$ on $S(E)$; we denote by $\varphi^{\pm}(D)$ its restriction to $S^{\pm}(E)$. The cohomology class $\mathrm{ch}\, d(D)$ is given by the form with compact support

$$\mathrm{ch}\, d(D) = \mathrm{tr}\, e^{-\varphi^+(\tilde{\Omega})/(2\pi i)} - \mathrm{tr}\, e^{-\varphi^-(\tilde{\Omega})/(2\pi i)}$$

$$= \mathrm{tr}\, r e^{\varphi(\tilde{\Omega})/(2\pi i)} = i^{-m}\, \mathrm{tr}\, e_1 e_2 \ldots e_{2m} e^{\tilde{\omega}_{ij} e_i e_j/4} \, ,$$

where $\tilde{\omega} = -\tilde{\Omega}/(2\pi i)$. This expression is computed by means of the following standard method. We denote it by $f(\tilde{\Omega})$ and compute it under the assumption that $\tilde{\Omega}$ is a number matrix in $\mathrm{so}(2m)$. Then we can choose a basis with respect to which $\tilde{\Omega}$ is block-diagonal with blocks of the form $\begin{pmatrix} 0 & \lambda_k \\ -\lambda_k & 0 \end{pmatrix}$, $\lambda_k \in \mathbb{R}$. Since $(ie_1 e_2)^2 = 1$ yields $e^{-\lambda e_1 e_2/(4\pi i)} = \mathrm{ch}(\lambda/(4\pi)) + ie_1 e_2 \,\mathrm{sh}(\lambda/(4\pi))$, it follows that

$$f(\tilde{\Omega}) = (-1)^m 2^m \prod_{k=1}^{m} \mathrm{sh}\, \frac{\lambda_k}{4\pi} = (-1)^m \prod_{k=1}^{m} \frac{\lambda_k}{2\pi} \prod_{k=1}^{m} \frac{\mathrm{sh}(\lambda_k/(4\pi))}{\lambda_k/(4\pi)} \, ,$$

where the factor 2^m is a normalizing constant in the definition of $\mathrm{tr}\, a$ in the Clifford algebra. This expression can be rewritten in the invariant form

$$f(\tilde{\Omega}) = (-1)^m \mathrm{Pf}\, \frac{\tilde{\Omega}}{2\pi} \left(\det \frac{\mathrm{sh}(\tilde{\omega}/2)}{\tilde{\omega}/2} \right)^{1/2} \tag{1.29}$$

and is an analytic function of the elements of $\tilde{\Omega}$. Consequently, (1.29) also holds for matrices $\tilde{\Omega}$ with elements in the commutative algebra of even-dimensioal differential forms. Similar to (1.18), we find that in $H^{+\text{comp}}(T^*M)$ the form $\mathrm{ch}\, d(D)$ is cohomologic to $(-1)^m U_E A^{-1}(E)$, where U_E is the Thom class of E and $A^{-1}(E) = (\det((\mathrm{sh}(\omega/2))/(\omega/2)))^{1/2}$ is the inverse of the A-class of the bundle E over M. For real bundles we have $\mathcal{T}(E) = A^2(E)$, since ω is a skew-symmetric matrix, which implies that $\det e^{\omega/2} = e^{(\mathrm{tr}\,\omega)/2} = 1$, which, in turn, yields

$$\det \frac{\omega}{1 - e^{-\omega}} = \det \frac{\omega/2}{\mathrm{sh}(\omega/2)} \det e^{\omega/2} = \det \frac{\omega/2}{\mathrm{sh}(\omega/2)} \, . \tag{1.30}$$

By the Atiyah-Singer formula,

$$\operatorname{ind} D = \int\limits_{T^*M} (-1)^m U_E A^{-1}(E) A^2(E) = \int\limits_{M} A(E) \, ,$$

where the factor $(-1)^m$ has been dropped because of the change of orientation in fibre integration.

The number on the right-hand side, called the A-*genus* of the manifold, is defined independently of whether M admits a spinor structure or not. However, generally speaking, this number is not an integer. Thus, for example, for $M = \mathbf{CP}^2$ the A-genus is equal to $-\frac{1}{8}$. Since the index of the Dirac operator must be an integer, this means that a spinor structure exists only if the A-genus is an integer.

In conclusion, we mention that a necessary and sufficient condition for the existence of a spinor structure is that the second Stieffel-Whitney class (see Husemöller (1966)) should be equal to zero.

Chapter 2
Generalisations

§1. The Atiyah-Bott Fixed Point Theorem

Let $g : M \to M$ be a smooth mapping of a compact manifold and E a vector bundle over M. The mapping g defines an *induced bundle* g^*E whose fibre $(g^*E)_x$ at a point x is the fibre $E_{g(x)}$ of E at the point $g(x)$, as well as the action $g^* : C^\infty(M, E) \to C^\infty(M, g^*E)$ on the sections, according to the formula $(g^*u)(x) = u(g(x))$. In addition, let $\Phi : g^*E \to E$ be a bundle homomorphism. Then the composition $T_g = \Phi \circ g^*$ defines the action $T_g : C^\infty(M, E) \to C^\infty(M, E)$ by

$$(T_g u)(x) = \Phi(x)u(g(x)) \, .$$

The mapping T_g obtained in this way is called a *geometric endomorphism* of the section space.

Let E^0 and E^1 be two bundles, and $T_g^0 = \Phi^0 \circ g^*$ and $T_g^1 = \Phi^1 \circ g^*$ geometric endomorphisms with one and the same mapping g. The PDO $A : C^\infty(M, E^0) \to C^\infty(M, E^1)$ is called g-*invariant* if it commutes with the operators T_g, that is,

$$A T_g^0 = T_g^1 A \, . \tag{2.1}$$

For an elliptic g-invariant PDO the spaces $\operatorname{Ker} A$ and $\operatorname{Coker} A$ are finite-dimensional and, by (2.1), T_g^0 and T_g^1 act on $\operatorname{Ker} A$ and $\operatorname{Coker} A$, respectively. By the *Lefschetz number* of a map g for a g-invariant elliptic operator A we understand the number

$$L_A(g) = \operatorname{tr} T_g^0|_{\operatorname{Ker} A} - \operatorname{tr} T_g^1|_{\operatorname{Coker} A} \,.$$

If g, T_g^0 and T_g^1 are identities, then the Lefschetz number is in fact the index of A.

The above concepts can be generalized to the case of elliptic complexes. By a *geometric endomorphism* of an elliptic complex

$$0 \to C^\infty(M, E_0) \xrightarrow{A_0} C^\infty(M, E^1) \xrightarrow{A_1} \ldots \xrightarrow{A_{k-1}} C^\infty(M, E^k) \to 0$$

we understand a collection $T_g^p = \Phi^p \circ g^* : C^\infty(M, E^p) \to C^\infty(M, E^p)$ of geometric endomorphisms of the section space which satisfy the commutativity condition

$$A_p T_g^p = T_g^{p+1} A_p \,.$$

In this case T_g^p acts in the cohomologies H^p of the complex, and the Lefschetz number is defined as

$$L(g) = \sum_{p=0}^{k} (-1)^p \operatorname{tr} T_g^p|_{H^p} \,.$$

A classical example is the de Rham complex. If $g : M \to M$ is any smooth mapping, then $T_g^p : C^\infty(M, \Lambda^p) \to C^\infty(M, \Lambda^p)$ is the pre-image of a p-form under g and the corresponding homomorphisms Φ^p are the exterior powers $\Lambda^p(dg)^*$, where dg is the differential of the mapping and $(dg)^*$ the dual mapping in the cotangent spaces.

Similar to (1.3), for the Lefschetz number we easily derive the formula

$$L_A(g) = \operatorname{tr}(T_g^0 - R T_g^1 A) - \operatorname{tr}(T_g^1 - A R T_g^1) \,, \tag{2.2}$$

where R is a parametrix of the operator A. Taking (2.1) into account, we can write the first term as $\operatorname{tr}(1 - RA)T_g^0$; however, the form (2.2) is so advantageous, that in it A does not have to satisfy (2.1) exactly, but only up to a trace class operator. Moreover, the addition of trace class terms to A and R does not influence the difference of the traces.

The *Atiyah-Bott theorem*, which generalizes Lefschetz's classical fixed point theorem, considers a mapping g that has finitely many non-degenerate fixed points. A fixed point x is called *non-degenerate* xif 1 is not an eigenvalue of the differential $dg(x) : T_x M \to T_x M$ at this point.

Theorem 2.1 (Atiyah and Bott (1967)). *If A is an elliptic g-invariant PDO on a compact manifold and the mapping g has finitely many non-degenerate fixed points, then*

$$L_A(g) = \sum_x \frac{\operatorname{tr} \Phi^0(x) - \operatorname{tr} \Phi^1(x)}{|\det(1 - dg(x))|} \,, \tag{2.3}$$

where the summation extends over all the fixed points.

We mention the curious fact that this formula does not depend on A.

For elliptic complexes, the analogue of (2.3) has the form

$$L(g) = \sum_x \frac{\sum_{p=0}^{k} (-1)^p \operatorname{tr} \Phi^p(x)}{|\det(1 - dg(x))|} \ .$$

In particular, for the de Rham complex the sum in the summand is equal to $\det(1 - (dg(x))^*) = \det(1 - dg(x))$, so that each term is equal to ± 1, depending on whether the orientation of the tangent space at a fixed point of the mapping $dg(x)$ is preserved or not. Therefore, the right-hand side is equal to the algebraic sum of the fixed points.

The proof of Theorem 2.1 is considerably shorter than that of the index theorem. We briefly outline the proof, making use of a so-called microlocal partition of unity and the asymptotic computation of the integral by the method of stationary phase. This technique is often used in the theory of PDOs and Fourier integral operators (see Shubin (1978), Hörmander (1971) and Treves (1980)).

Let $\{\rho_i(x)\}$ be a partition of unity on M such that all its functions are equal either to 0 or to 1 in the neighbourhood of the fixed points. Also, let $\psi_0(\xi)$ be a function with compact support on T^*M and equal to 1 in the neighbourhood of $\xi = 0$, and let $\psi_\infty = 1 - \psi_0$. We define operators ψ_{ij} ($j = 0, \infty$) in terms of local coordinates on M by setting $\psi_{ij}u = F^{-1}\psi_j(h\xi)F\rho_i(x)u(x)$, where F is the Fourier transformation and h a positive number. The ψ_{ij} yield a partition of the identity operator; moreover, the ψ_{i0} are infinitely smoothing operators, that is, they are trace class operators in any Sobolev space. Taking into account that $\operatorname{tr} RT_g^1\psi_{i0}A = \operatorname{tr} ART_g^1\psi_{i0}$, we can rewrite (2.2) in the form

$$\begin{aligned}
L_A(g) = &\sum_i \{\operatorname{tr} T_g^0\psi_{i0} - \operatorname{tr} T_g^1\psi_{i0}\} \\
&+ \sum_i \{\operatorname{tr}(T_g^0 - RT_g^1 A)\psi_{i\infty} - \operatorname{tr}(T_g^1 - ART_g^1)\psi_{i\infty}\} \\
&- \sum_i \operatorname{tr} RT_g^1[A, \psi_{i0}] \ .
\end{aligned} \tag{2.4}$$

$T_g^0\psi_{i0}$ can be rewritten in terms of local coordinates as the iterated integral

$$T_g^0\psi_{i0}u = (2\pi h)^{-n} \iint e^{(i/h)\xi(g(x)-y)}\Phi^0(x)\psi_0(\xi)\rho_i(y)u(y)\,dy\,d\xi \ ,$$

and its trace is

$$\operatorname{tr} T_g^0\psi_{i0} = (2\pi h)^{-n} \iint e^{(i/h)\xi(g(x)-x)} \operatorname{tr} \Phi^0(x)\psi_0(\xi)\rho_i(x)\,d\xi\,dx \ . \tag{2.5}$$

As $h \to 0$, the integral (2.5) can be computed by the metod of stationary phase; the stationary points are those where $\xi = 0$ and $g(x) - x = 0$. The contribution of the fixed point x in the dominant term, which is independent of h, is

$$\frac{\operatorname{tr} \Phi^0(x)}{|\det(1 - dg(x))|}.$$

The $T_g^1 \psi_{i0}$ make a similar contribution, and their difference yields the right-hand side of (2.3). The remaining terms in (2.4) can also be represented as integrals with rapidly oscillating exponentials whose exponents do not have critical points because the function $1 - \psi_0(\xi)$ is equal to zero in the neighbourhood of $\xi = 0$. Consequently, they tend to zero faster than any power of h. Since, in fact, the Lefschetz number does not depend on h, we arrive at the Atiyah-Bott formula.

A similar metod can also be applied to the index theorem. However, there we need the subsequent terms in the asymptotic expansions, and the resulting formula contains not only the principal symbols, but also their derivatives of higher order and lower terms. In view of the Atiyah-Singer theorem, we may assert that all the non-invariant terms must somehow cancel out to yield the invariant expression occurring in the formula of the theorem. This cancellation process can be traced only in very special cases (see Fedosov (1974), Atiyah, Bott and Patody (1973), Bismut (1984) and Hörmander (1979)).

The Atiyah-Bott theorem has numerous generalisations. Thus, if the diffeomorphism g and its liftings T_g^0 and T_g^1 run through a transformation group G and A is invariant with respect to G, then the Lefschetz number is a function on the group, equal to the difference of the characters of two finite-dimensional representations of G in $\operatorname{Ker} A$ and $\operatorname{Coker} A$. This function is denoted by $\operatorname{ind}_G A$ and is called the *G-index* of the operator A. For finite and compact groups G we obtain a general formula for the G-index (Atiyah and Segal (1968)), which contains both the Atiyah-Bott formula (2.3) and the Atiyah-Singer formula (1.19) as special cases. The investigations concerning the G-index have stimulated the development of the corresponding framework: equivariant K-theory (Atiyah (1967)), equivariant cohomologies (Berline and Vergne (1985), Atiyah and Bott (1984)), and localisation formulae (Berline and Vergne (1985), Bott (1967b)). We also mention the research concerning the index of so-called transversally elliptic operators (Atiyah (1974)), where the G-index is a generalized function on the group G, equal to the difference of two generalized characters of infinite-dimensional representations of G.

An interesting generalisation of the Atiyah-Bott formula to manifolds with boundary was obtained in Brenner and Shubin (1981). The fixed points of the mapping g on the boundary are classified as stable or unstable, with only the former contributing to the Lefschetz number.

§2. The Index of a Family of Elliptic Operators

Let $A(x) : E^0 \to E^1$ be a family of bounded Fredholm operators in Hilbert spaces, parametrized by the points of a compact manifold M and continuous with respect to the operator norm. With this a family we associate an element $\operatorname{ind} A(x) \in K(M)$ in the following way. As was proved in Atiyah (1967), there is a subspace $L^0 \subset E^0$ of finite codimension such that $\operatorname{Ker} A \cap L^0$ consists of zero alone, the image $L_x^1 = A(x)L^0 \subset E^1$ of L^0 under the operator $A(x)$ has finite codimension in E^1, and the factor-space E^1/L_x^1 is a locally trivial vector bundle over M. The element $\operatorname{ind} A(x) \in K(M)$ is given by the virtual bundle $\{E^0/L^0, E^1/L_x^1\}$ and depends only on the homotopic class of the mapping of M in the space of Fredholm operators. If $\operatorname{Ker} A(x)$ and $\operatorname{Coker} A(x)$ form locally trivial bundles, then $\operatorname{ind} A(x)$ coincides with the class $\{\operatorname{Ker} A(x), \operatorname{Coker} A(x)\}$ in $K(M)$.

We derive a formula for $\operatorname{ch}(\operatorname{ind} A(x))$, similar to (1.3) and (2.2). Moreover, in view of the applications discussed in Chap. 3, we construct a more general index theory for Fredholm families on a locally compact manifold M, which may be regarded as infinite-dimensional analogues of virtual bundles with compact support (see Sect. 3.2, Chap. 1). We also state the Atiyah-Singer formula for a family of elliptic PDOs, which expresses $\operatorname{ch}(\operatorname{ind} A(x))$ in terms of topological invariants of the elliptic family.

Let M be a locally compact manifold and $\mathcal{E}$ a locally trivial *Hilbert bundle* over M. This means that the fibre of the bundle is a Hilbert space E and the transition functions $f_{ij}(x)$ take values in a structure group which either coincides with $\operatorname{GL}(E)$, or is a subgroup of it and depends smoothly on x in the operator norm. (We mention that an infinite-dimensional Hilbert bundle with the structure group $\operatorname{GL}(E)$ over a compact manifold is trivial (Atiyah (1967)).) We also need the concept of connection in $\mathcal{E}$. Just as in the finite-dimensional case (see the definition in 3.1, Chap. 1), a *connection* associates with a section $u \in C^\infty(\mathcal{E})$ its *covariant differential* $\partial u \in C^\infty(\mathcal{E} \otimes \Lambda^1)$, which is given locally in the form $du + \Gamma u$, where Γ is a 1-form with values in the space $\operatorname{Hom}(E, E)$ of bounded operators in the Hilbert space E. The *curvature* Ω of a connection ∂ is the operator-valued 2-form defined by $\partial^2 u = \Omega u$. We set $\omega = -\Omega/(2\pi i)$.

A *Fredholm family* Ξ is given by a collection $\{\mathcal{E}^0, \mathcal{E}^1, A, R\}$, where $\mathcal{E}^0$ and $\mathcal{E}^1$ are Hilbert bundles over M, $A = A(x) : \mathcal{E}_x^0 \to \mathcal{E}_x^1$ is a bounded Fredholm operator in the fibres, and $R = R(x) : \mathcal{E}_x^1 \to \mathcal{E}_x^0$ its parametrix. We assume that $A(x)$ and $R(x)$ depend smoothly on x in the operator norm, and that $1 - R(x)A(x)$ and $1 - A(x)R(x)$ are smooth in the trace norm. The set $X \subset M$, where $1 - R(x)A(x)$ and $1 - A(x)R(x)$ are non-zero, is called the *support of the family* Ξ. We denote by $F_X(M)$ the set of Fredholm families with support in X, and by $F^{\mathrm{comp}}(M)$ the set of families with compact support. The notation $T(x) \sim 0 \,(\operatorname{mod} X)$ is used for a trace class operator that depends smoothly on x in the trace norm and is identically zero outside X. If X is compact, then we simply write $T(x) \sim 0$.

A family $\Xi \in F_X(M)$ or $\Xi \in F^{\mathrm{comp}}(M)$ is called *trivial* if $1 - RA = 0$ and $1 - AR = 0$ on the whole of M, that is, its support is empty. Two families Ξ_1, $\Xi_2 \in F_X(M)$ are called *isomorphic* if there are fibre isomorphisms $U^i = U^i(x) : \mathcal{E}^i_{1x} \to \mathcal{E}^i_{2x}$ such that the diagram

$$
\begin{array}{ccc}
\mathcal{E}^0_1 & \underset{\overset{A_1}{\longrightarrow}}{\underset{R_1}{\longleftarrow}} & \mathcal{E}^1_1 \\[2mm]
U^0 \downarrow & & \downarrow U^1 \\[2mm]
\mathcal{E}^0_2 & \underset{\overset{A_2}{\longrightarrow}}{\underset{R_2}{\longleftarrow}} & \mathcal{E}^1_2
\end{array}
$$

is *almost commutative*. This means that $U^1 A_1 - A_2 U^0 \sim 0$ and $U^0 R_1 - R_2 U^1 \sim 0 \,(\mathrm{mod}\, X)$. Two families are called *stably isomorphic* if there are trivial families Z_1 and Z_2 such that $\Xi_1 \oplus Z_1$ and $\Xi_2 \oplus Z_2$ are isomorphic.

Let $\Xi(t) = \{\mathcal{E}^0, \mathcal{E}^1, A(x,t), R(x,t)\} \in F_X(M)$, $t \in [0,1]$, be a smooth homotopy of Fredholm families. Then $\Xi(t)$ is isomorphic to $\Xi(0)$ for any t. Here we can take U^0 to be equal to 1 and define $U^0(t)$ as the solution of the differential equation $\dot{U}^0 + R\dot{A}U^0 = 0$, $U^0(0) = 1$. Indeed, the relation $A(t)U^0(t) - A(0) \sim 0 \,(\mathrm{mod}\, X)$, which is evident for $t = 0$, holds for all t since $\dot{A}U^0 - A\dot{U}^0 = \dot{A}U^0 - AR\dot{A}U^0 = (1 - RA)\dot{A}U^0 \sim 0 \,(\mathrm{mod}\, X)$.

As was mentioned at the beginning of this section, for infinite-dimensional Hilbert bundles over a compact manifold M there is a family of subspaces $L^0_x \subset \mathcal{E}^0_x$ of finite codimension such that $\mathrm{Ker}\, A(x) \cap L^0_x = 0$, $L^1_x = A(x)L^0_x$ has finite codimension in $\mathcal{E}^1_x$, and the factor spaces $\mathcal{E}^0_x/L^0_x$ and $\mathcal{E}^1_x/L^1_x$ form locally trivial bundles. From this it follows that any Fredholm family Ξ over a compact manifold is isomorphic to the direct sum of the finite-dimensional family $\{\mathcal{E}^0_x/L^0_x, \mathcal{E}^1_x/L^1_x, 0, 0\}$ and a trivial infinite dimensional family defined by the bundles L^0_x and L^1_x and the invertible operator $A(x) : L^0_x \to L^1_x$.

Let $\Xi_1 \in F_{X_1}(M)$ and $\Xi_2 \in F_{X_2}(M)$, and suppose that $X_1 \cap X_2$ is a precompact set. We define the *tensor product* $\Xi_1 \otimes \Xi_2 \in F^{\mathrm{comp}}(M)$ to be the collection $\{(\mathcal{E}^0_1 \otimes \mathcal{E}^0_2) \oplus (\mathcal{E}^1_1 \otimes \mathcal{E}^1_2), (\mathcal{E}^1_1 \otimes \mathcal{E}^0_2) \oplus (\mathcal{E}^0_1 \otimes \mathcal{E}^1_2), \mathcal{A}, \mathcal{R}\}$, where

$$
\mathcal{A} = \begin{pmatrix} A_1 \otimes (1 - R_2 A_2) & -1 \otimes R_2 \\ 1 \otimes A_2 & R_1 \otimes 1 \end{pmatrix},
$$
$$
\mathcal{R} = \begin{pmatrix} R_1 \otimes 1 & 1 \otimes R_2 \\ -1 \otimes A_2 & A_1 \otimes (1 - A_2 R_2) \end{pmatrix}.
\tag{2.6}
$$

It is easy to see that $\Xi_1 \otimes \Xi_2$ is a Fredholm family with support in $X_1 \cap X_2$, and that $\Xi_1 \otimes \Xi_2$ and $\Xi_2 \otimes \Xi_1$ are homotopic. The homotopy is given by

$$
\mathcal{A}(t) = \begin{pmatrix} A_1 \otimes (1 - t R_2 A_2) & -1 \otimes R_2 \\ (1 - (1-t)R_1 A_1) \otimes A_2 & R_1 \otimes 1 \end{pmatrix}
$$

and

$$\mathcal{R}(t) = \begin{pmatrix} R_1 \otimes 1 & 1 \otimes R_2 \\ -(1 - (1-t)A_1 R_1) \otimes A_2 & A_1 \otimes (1 - tA_2 R_2) \end{pmatrix} .$$

Indeed, using matrix multiplication, we obtain

$$1 - \mathcal{R}(t)\mathcal{A}(t) = \begin{pmatrix} (1 - R_1 A_1) \otimes (1 - R_2 A_2) & 0 \\ 0 & (1 - A_1 R_1) \otimes (1 - A_2 R_2) \end{pmatrix} \sim 0 ,$$

$$1 - \mathcal{A}(t)\mathcal{R}(t) = \begin{pmatrix} (1 - A_1 R_1) \otimes (1 - R_2 A_2) & 0 \\ 0 & (1 - R_1 A_1) \otimes (1 - A_2 R_2) \end{pmatrix} \sim 0 .$$

Therefore, tensor multiplication is, up to an isomorphism, commutative and distributive over direct sum.[1]

For a Fredholm family $\Xi \in F^{\mathrm{comp}}(M)$ we define the cohomology class $\mathrm{ch}\,\Xi \in H^{+\mathrm{comp}}(M)$, similar to the character of virtual bundles with compact support. Let ∂_0 and ∂_1 be connections on the Hilbert bundles $\mathcal{E}^0$ and $\mathcal{E}^1$. The covariant differential ∂A is defined by means of the relation $(\partial A)u = \partial_1(Au) - A(\partial_0 u)$ for any section $u \in C^\infty(\mathcal{E}^0)$; ∂R is defined analogously. We claim that ∂_0 and ∂_1 can be chosen so that

$$\partial A \sim 0 , \qquad \partial R \sim 0 , \qquad \Omega_1 A - A\Omega_0 \sim 0 \tag{2.7}$$

(the second and third relations follow from the fact that $\partial A \sim 0$). Indeed, if ∂_0 and ∂_1 are arbitrary, then we define $\tilde{\partial}_0$ by setting $\tilde{\partial}_0 u = \partial_0 u + R\partial Au$. In this case, $\tilde{A} = \partial A - AR\partial A = (1 - AR)\partial A \sim 0$.

Assuming that ∂_0 and ∂_1 are chosen so that (2.7) holds, we set

$$\mathrm{ch}\,\Xi = \mathrm{tr}(e^{\omega_0} - Re^{\omega_1} A) - \mathrm{tr}\, e^{\omega_1}(1 - AR) . \tag{2.8}$$

From (2.7) it follows that $e^{\omega_1} A - Ae^{\omega_0} \sim 0$, so that the operators under the trace symbol are trace class operators and the form (2.8) has compact support. Since $\mathrm{tr}\, Re^{\omega_1} A = \mathrm{tr}\, ARe^{\omega_1}$ for a finite-dimensional family, it follows that here (2.8) changes into (1.11), which defines the character of a virtual bundle with compact support. The formula (2.8) is also meaningful for $\Xi \in F_X(M)$, in which case it defines a form with support in X.

Theorem 2.2. *The form $\mathrm{ch}\,\Xi$ is closed and has the following properties.*

1) *The cohomology class $\mathrm{ch}\,\Xi \in H^{+\mathrm{comp}}(M)$ is invariant under homotopies of the family and of the connections ∂_0 and ∂_1 if the condition (2.7) is satisfied.*

2) *If $\Xi = \xi \oplus Z$, where ξ is finite-dimensional and Z trivial, then the classes $\mathrm{ch}\,\Xi$ and $\mathrm{ch}\,\xi$ in $H^{+\mathrm{comp}}(M)$ coincide.*

3) *If $\Xi_1 \in F_{X_1}(M)$, $\Xi_2 \in F_{X_2}(M)$ and the set $X_1 \cap X_2$ is precompact, then the cohomology classes $\mathrm{ch}\,\Xi_1 \otimes \Xi_2$ and $\mathrm{ch}\,\Xi_1 \, \mathrm{ch}\,\Xi_2$ in $H^{+\mathrm{comp}}(M)$ coincide.*

---

[1] It can be shown that a Fredholm operator $\mathcal{A}$ defined by (2.6) is homotopic to a Fredholm operator defined by (1.4). The advantage of (2.6) over (1.4) consists in the fact that the operator character of A_1, R_1, A_2 and R_2 is not essential in (2.6). This observation is important in Chapter 3.

Proof. Using Bianchi's identity $\partial_i \omega_i = 0$, we find that

$$d \operatorname{ch} \Xi = \operatorname{tr} e^{\omega_1} \partial A R + \operatorname{tr} e^{\omega_1} A \partial R - \operatorname{tr} \partial R e^{\omega_1} A - \operatorname{tr} R e^{\omega_1} \partial A = 0 \, ,$$

since $\partial A \sim 0$, $\partial R \sim 0$, and the factors under the trace can be permuted cyclically.

We prove the property 1). The homotopy $\Xi(t) = \{\mathcal{E}^0, \mathcal{E}^1, A(t), R(t)\}$ defines a Fredholm family on $M \times [0,1]$, and the homotopy of the connections ∂_{0t} and ∂_{1t} defines the connections $\mathcal{D}_0 u(x,t) = \partial_{0t} u + dt(\dot{u} + R\dot{A}u)$ and $\mathcal{D}_1 u(x,t) = \partial_{1t} u + dt\dot{u}$ in the liftings of the bundles $\mathcal{E}^0$ and $\mathcal{E}^1$ to $M \times [0,1]$. Then $\mathcal{D}A = \partial_t A + dt(\dot{A} - AR\dot{A}) \sim 0$ on $M \times [0,1]$, since $\partial_t A \sim 0$ and $(1 - AR) \sim 0$. The form $\operatorname{ch} \Xi$ on $M \times [0,1]$, constructed in terms of $\mathcal{D}_0$ and $\mathcal{D}_1$, is closed and has compact support on $M \times [0,1]$. From this it follows that its restrictions to $M \times \{0\}$ and $M \times \{1\}$ belong to one and the same cohomology class with compact support. Indeed, if $a + dt \wedge b$ is a closed form with compact support on $M \times [0,1]$, where a does not contain dt, then $\dot{a} - db = 0$, where d is the exterior differential on M, so that $\dot{a} = 0$ in $H^{+\operatorname{comp}}(M)$.

We now prove 2). If the connections ∂_0 and ∂_1 preserve direct sum decomposition, then the statement is obvious. In the general case, let $p_0(x)$ and $p_1(x)$ be finite-dimensional projections splitting Ξ into a direct sum. Then the connections $\tilde{\partial}_0$ and $\tilde{\partial}_1$ defined by $\tilde{\partial}_i u = \partial_i u + (2p^i - 1)\partial_i p^i u$ are such that $\tilde{\partial}_i p^i = 0$, that is, they preserve decomposition, and, as before, the condition $\partial A \sim 0$ is satisfied. The assertion now follows from the fact that the class $\operatorname{ch} \Xi$ in $H^{+\operatorname{comp}}(M)$ does not depend on the choice of connection.

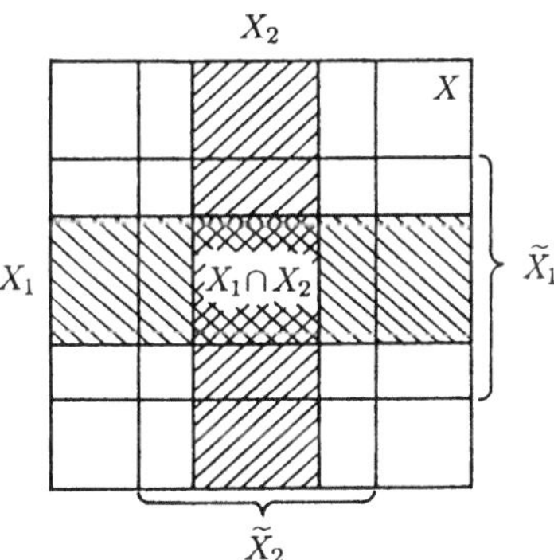

We prove 3). Let ∂_1^0 and ∂_1^1 be connections in $\mathcal{E}_1^0$ and $\mathcal{E}_1^1$, respectively, satisfying $\partial A_1 \sim 0 \pmod{X_1}$; similarly, let ∂_2^0 and ∂_2^1 be connections in $\mathcal{E}_2^0$ and $\mathcal{E}_2^1$ satisfying $\partial A_2 \sim 0 \pmod{X_2}$. In the tensor products $\mathcal{E}_j^i \otimes \mathcal{E}_l^k$ ($i, k = 0, 1$; $j, l = 1, 2$) we can introduce connections in a natural way by $\partial_j^i \otimes 1 + 1 \otimes \partial_l^k$. We denote by $\mathcal{D}_0$ and $\mathcal{D}_1$ the corresponding connections in $(\mathcal{E}_1^0 \otimes \mathcal{E}_2^0) \oplus (\mathcal{E}_1^1 \otimes \mathcal{E}_2^1)$ and $(\mathcal{E}_1^1 \otimes \mathcal{E}_2^0) \oplus (\mathcal{E}_1^0 \otimes \mathcal{E}_2^1)$, respectively, and let $\tilde{\mathcal{D}}_0$ be defined, as usual, by $\tilde{\mathcal{D}}_0 u = \mathcal{D}_0 u + \mathcal{R}\mathcal{D}\mathcal{A}u$. Then $\tilde{\mathcal{D}}\mathcal{A} \sim 0 \pmod{X_1 \cap X_2}$. The forms $\operatorname{ch} \Xi_1$ and $\operatorname{ch} \Xi_2$ have support in X_1 and X_2, respectively, and $\operatorname{ch} \Xi_1 \otimes \Xi_2$ and $\operatorname{ch} \Xi_1 \operatorname{ch} \Xi_2$ have support in $X_1 \cap X_2$. Let X be a compact neighbourhood of $X_1 \cap X_2$ (see the figure). We claim that $\mathrm{ch}\,\Xi_1 \otimes \Xi_2 - \mathrm{ch}\,\Xi_1\,\mathrm{ch}\,\Xi_2$ in X is the differential of a form with support in $X_1 \cap X_2$. This would prove 3). Over the compactum X, the families Ξ_1 and Ξ_2 can be partitioned into direct sums $\xi_1 \oplus Z_1$ and $\xi_2 \oplus Z_2$, where ξ_i are finite-dimensional families with support in X_i and Z_i are trivial families. Since tensor multiplication by a trivial family yields a trivial family, it follows that the cohomology classes of the forms $\mathrm{ch}\,\Xi_1 \otimes \Xi_2$ and $\mathrm{ch}\,\xi_1 \otimes \xi_2$ in $H(X, X \setminus (X_1 \cap X_2))$ coincide. So do the classes of the forms $\mathrm{ch}\,\Xi_i$ and $\mathrm{ch}\,\xi_i$ in $H(X, X \setminus X_i)$. Therefore, it suffices to prove the assertion for finite-dimensional families.

In the finite-dimensional case (see (1.11)), $\mathrm{ch}\,\xi_1 \otimes \xi_2 = \mathrm{tr}\,e^{\tilde{\omega}_0} - \mathrm{tr}\,e^{\omega_1}$, where $\tilde{\Omega}_0$ and Ω_1 are the curvatures of $\tilde{\mathcal{D}}_0$ and $\mathcal{D}_1$, respectively. Then (1.10) yields

$$\tilde{\Omega}_0 = \Omega_0 + \mathcal{D}(\mathcal{R}\mathcal{D}\mathcal{A}) + (\mathcal{R}\mathcal{D}\mathcal{A})^2 , \qquad (2.9)$$

where Ω_0 is the curvature of $\mathcal{D}_0$ and $\mathcal{R}$ and $\mathcal{A}$ are given by (2.6). Direct computation shows that

$$\mathcal{R}\mathcal{D}\mathcal{A} \sim \begin{pmatrix} -1 \otimes \partial R_2 A_2 & \partial R_1 \otimes R_2 - R_1 \otimes \partial R_2 \\ A_1 \otimes (\partial A_2 + A_2 \partial R_2 A_2) & 1 \otimes A_2 \partial R_2 \end{pmatrix} (\mathrm{mod}\, X_1 \cap X_2) .$$

Let φ_i ($i = 1,2$) be truncating functions on X, equal to 0 in X_i and to 1 outside some neighbourhood $\tilde{X}_i \supset X_i$, where $\tilde{X}_1 \cap \tilde{X}_2 \subset X$ (see the figure). Since the situation is finite-dimensional, we replace A_i and R_i by $\tilde{A}_i = \varphi_i A_i$ and $\tilde{R}_i = \varphi_i R_i$, respectively, and construct $\tilde{A}$ and $\tilde{R}$ by means of (2.6). This change leads to a variation in the curvature of $\tilde{\mathcal{D}}_0$, but does not affect the cohomology class of $\mathrm{ch}\,\xi_1 \otimes \xi_2$ in $H(X, X \setminus (\tilde{X}_1 \cap \tilde{X}_2))$. Since $\partial A_i = 0$ and $\partial R_i = 0$ outside X_i, it follows that $\partial \tilde{A}_i = A_i d\varphi_i$ and $\partial \tilde{R}_i = R_i d\varphi_i$. Hence,

$$\mathcal{R}\mathcal{D}\mathcal{A} \sim \tilde{\mathcal{R}}\mathcal{D}\tilde{\mathcal{A}} \sim \begin{pmatrix} -\varphi_2 d\varphi_2 & R_1 \otimes R_2(\varphi_2 d\varphi_1 - \varphi_1 d\varphi_2) \\ A_1 \otimes A_2 \varphi_1(d\varphi_2 + \varphi_2^2 d\varphi_2) & \varphi_2 d\varphi_2 \end{pmatrix} .$$

From this we see that the matrix $S = \mathcal{D}(\tilde{\mathcal{R}}\mathcal{D}\tilde{\mathcal{A}}) + (\tilde{\mathcal{R}}\mathcal{D}\tilde{\mathcal{A}})^2$ contains only the product of differentials $d\varphi_1 \wedge d\varphi_2$, so the products of any two of its entries are equal to zero. Consequently,

$$\mathrm{ch}\,\xi_1 \otimes \xi_2 = \mathrm{tr}\,e^{\omega_0} - \mathrm{tr}\,e^{\omega_1} + \mathrm{tr}\,e^{\omega_0} S.$$

The first two terms yield $\mathrm{ch}\,\xi_1\,\mathrm{ch}\,\xi_2$, while the third one is cohomologic to 0 in $H(X, X \setminus (\tilde{X}_1 \cap \tilde{X}_2))$. Indeed, e^{ω_0} is a diagonal matrix, and, computing its elements, we find that

$$\mathrm{tr}\,e^{\omega_0} S = (\mathrm{tr}\,e^{\omega_1^0}\,\mathrm{tr}\,e^{\omega_2^0} - \mathrm{tr}\,e^{\omega_1^1}\,\mathrm{tr}\,e^{\omega_2^1})\varphi_1 d\varphi_2 \wedge \varphi_2(d\varphi_2 + \varphi_2^2 d\varphi_2) .$$

The factor in parentheses is a closed form, while the remaining ones can be rewriten as $-\frac{1}{4}d(1 - \varphi_1^2) \wedge d(\varphi_2^2 + \frac{1}{4}\varphi_2^4)$, from which it follows that $\mathrm{tr}\,e^{\omega_0} S$ is cohomologic to 0 in $H(X, X \setminus (\tilde{X}_1 \cap \tilde{X}_2))$. The theorem is proved.

A somewhat more general definition, which enables us to consider the case of zero-dimensional bundles $\mathcal{E}^0$ or $\mathcal{E}^1$, can be given by taking a Fredholm family to be a collection $\{P^0, P^1, A, R\}$, where P^i are projections in $\mathcal{E}^i$, that is, $(P^i)^2 = P^i$, and A and R satisfy the conditions

$$P^1 A = AP^0 = A\,, \qquad P^0 R = RP^1 = R\,, \tag{2.10}$$

$$P^0 - RA \sim 0\,, \qquad P^1 - AR \sim 0\,. \tag{2.11}$$

This means that we consider the Fredholm family in the sub-bundles $P^0\mathcal{E}^0$ and $P^1\mathcal{E}^1$ instead of $\mathcal{E}^0$ and $\mathcal{E}^1$. We leave the generalisation of Theorem 2.2 in this case to the reader.

We conclude this section by formulating the *Atiyah-Singer theorem on the index of a family of elliptic operators* (Atiyah and Singer (1971a)). Let N be a locally trivial bundle manifold with basis X, fibre M, and the group of diffeomorphisms of M as structure group (X and M are compact manifolds). Next, let E^0 and E^1 be vector bundles over N. For a point $x \in X$ we denote by E^i_x the restrictions of the E^i to the fibre M_x. Now let $A(x) : C^\infty(E^0_x) \to C^\infty(E^1_x)$ be a smooth family of elliptic operators of order m (the smoothness is understood in the norm of the operators from H^s to H^{s-m} in Sobolev spaces) with principal symbol $a(x, y, \eta)$, where $(y, \eta) \in T^*M_x \setminus 0$. We denote by T^*M the bundle over N whose fibre at the point $(x, y) \in N$ is the cotangent space to M_x at $y \in M_x$ (the cotangent bundle along the fibres). The triple $\xi = \{E^0, E^1, a\}$ defines the distinguishfing element $d(A) \in K^{\mathrm{comp}}(T^*M)$. Let $\mathcal{T}(T^*M)$ be the Todd class of the (complexified) bundle T^*M.

Theorem 2.3. *The cohomology class of* $\mathrm{ch\,ind}\,A(x) \in H^+(X)$ *is given by the form*

$$\int\limits_{T^*M_x} \mathrm{ch}\,d(A)\mathcal{T}(T^*M)\,, \tag{2.12}$$

where the integral is taken along the fibres.

§3. The Index of Almost Periodic Operators

We begin with a simple example. Let $A = a(x, D)$ be a PDO in the space of vector functions on $\mathbb{R}^n$, that is, an operator acting on vector functions $u(x) \in C^\infty_0$ according to the formula

$$(Au)(x) = (2\pi)^{-n} \int e^{i(x,\xi)} a(x, \xi) \hat{u}(\xi)\, d\xi\,, \tag{2.13}$$

where $\hat{u}$ is the Fourier transform of u. The function $a(x, \xi)$ is called the *complete symbol* of A in $\mathbb{R}^n$. We assume that $a(x, \xi)$ belongs to the *class S^m* of symbols, defined by the condition

$$\|\partial^\alpha_\xi \partial^\beta_x a(x, \xi)\| \le C_{\alpha\beta X}(1 + |\xi|^2)^{(m-|\alpha|)/2} \tag{2.14}$$

for any compactum $X \subset \mathbb{R}^n$ and any multi-indices α and β, where $\| \ \|$ is some matrix norm. We say that A is of *order* m if there is a function $a_m(x, \xi)$, $x \in \mathbb{R}^n$, $\xi \in \mathbb{R}^n \setminus 0$, positively homogeneous of degree m with respect to ξ and such that $a(x, \xi) - \rho(\xi) a_m(x, \xi) \in S^{m-1}$, where $\rho(\xi)$ is a truncating function equal to 1 outside a neighbourhood of $\xi = 0$. The function $a_m(x, \xi)$ is called the *principal symbol* of the operator. The operator is called *elliptic* if its principal symbol is invertible for $\xi \neq 0$. By comparison with operators on a manifold, those in $\mathbb{R}^n$ are constructed more simply, since they are defined globally by means of the principal symbol, while for operators on a manifold the representation (2.13) is only local. Clearly, the formula (1.5) given in Chapter 1 for the principal symbol remains valid for operators in $\mathbb{R}^n$.

We assume that the complete symbol of the elliptic operator is a periodic function in x, that is, it is invariant with respect to shifts on the integral lattice $\mathbb{Z}^n$ in $\mathbb{R}^n$. If we consider A in the Sobolev spaces $H^s(\mathbb{R}^n)$, then the operator and its formal adjoint have infinite-dimensional kernels (because $\mathbb{R}^n$ is not compact), so that A is not a Fredholm operator and its index is not defined. However, we can define the index of A in a generalized sense, by setting it equal to the index of the same operator in the space of periodic functions, that is, when A is regarded as an operator on the torus $\mathbf{T}^n = \mathbb{R}^n / \mathbb{Z}^n$.

We can also look at this example from another angle. Let R be a parametrix of an operator A, *invariant* with respect to shifts on the above lattice, that is, an operator such that the order of $1 - RA$ and $1 - AR$ is strictly less than $-n$. Since $\mathbb{R}^n$ is not compact, the latter are not trace class operators in $L^2(\mathbb{R}^n)$. Nevertheless, we can define for them a *generalized trace* $\tilde{\mathrm{tr}}$ by setting it equal to the trace of these operators on the torus. For an operator T with a smooth kernel $T(x, y)$ invariant with respect to simultaneous shifts in x and y on the lattice, the trace $\tilde{\mathrm{tr}}$ is defined by

$$\tilde{\mathrm{tr}}\, T = \int_{T^n} T(x, x)\, dx = \langle T(x, x) \rangle \, ,$$

where $\langle \ \rangle$ denotes the mean value of a periodic function in $\mathbb{R}^n$. In this notation, the generalized index of a periodic operator A in $\mathbb{R}^n$ is given by the formula

$$\mathrm{ind}\, A = \tilde{\mathrm{tr}}(1 - RA) - \tilde{\mathrm{tr}}(1 - AR) \, .$$

We now consider the more general case of operators with almost periodic coefficients in $\mathbb{R}^n$. The theory of such operators is discussed in detail in Shubin (1979), and their index theory in Coburn, Moyer and Singer (1973) and Shubin (1979). Let $a(x, D)$ be a PDO in $\mathbb{R}^n$ with complete symbol $a(x, \xi) \in S^m$, which for any ξ is a *uniform almost periodic function* of x. This means that the set $a(x + \omega, \xi)$ obtained by means of shifts on all the vectors $\omega \in \mathbb{R}^n$ is precompact in the space of continuous bounded functions equipped with the topology of uniform convergence. Such an operator is called an *almost periodic* (a.p.) PDO. The *ellipticity* of an a.p. PDO of order m means that its principal symbol $a_m(x, \xi)$ is invertible for $\xi \neq 0$ and $\|a_m^{-1}(x, \xi)\| \leq C|\xi|^{-m}$ uniformly with respect to x. For an elliptic a.p. PDO there is a *parametrix* $R = r(x, D)$, which is an a.p. operator such that the order of $1 - RA$ and $1 - AR$ is strictly lower than $-n$. An a.p. operator T of order lower than $-n$ has a continuous kernel $T(x, y)$ with the property that $T(x, x + z)$ is a uniform a.p. function of x for any $z \in \mathbb{R}^n$. We define the *generalized trace* of T by

$$\operatorname{tr} T = \langle \operatorname{tr} T(x, x) \rangle \, .$$

Here $\operatorname{tr} T(x, x)$ is the matrix trace and $\langle f \rangle$ the mean value over $\mathbb{R}^n$ of the a.p. function $f(x)$, defined by

$$\langle f \rangle = \lim_{l \to \infty} \frac{1}{|Q_l|} \int_{Q_l} f(x) \, dx \, ,$$

where Q_l is the cube $|x^i| \le l/2$ and $|Q_l|$ its volume. The *index of an elliptic a.p. PDO* can now be defined by

$$\operatorname{ind} A = \operatorname{tr}(1 - RA) - \operatorname{tr}(AR) \, . \tag{2.15}$$

In fact, the original definition of the index is different and the formula (2.15) is obtained as a theorem. However, without going into details, we adopt (2.15) as the definition.

It can be proved that the index of an elliptic a.p. operator has the usual properties (stability in the class of a.p. operators and the logarithmic property), but that now it is not necessarily an integer. We can also derive a formula for the index, analogous to (1.22) (Shubin (1979))

$$\operatorname{ind} A = -\frac{(n-1)!}{(2n-1)!} \frac{1}{(2\pi i)^n} \left\langle \int_{|x|=1} \operatorname{tr}(a_m^{-1} da_m)^{2n-1} \right\rangle \, ,$$

and a similar formula for families of elliptic a.p. PDOs.

There are many different generalisations of the above concepts. We mention the index theory for elliptic operators on a non-compact manifold, which are invariant with respect to a discrete group of motions with a compact fundamental domain—a generalisation of periodic PDOs (see Atiyah (1976) and Atiyah and Schmid (1977)). A further generalisation with important applications in group representation theory is considered in Connes and Moscovici (1982). Another one is the index theory for homogeneous stochastic PDOs (Fedosov and Shubin (1978)), and, finally, K-theory in von Neumann algebras (Breuer (1968, 1969)).

The examples considered in this chapter may serve as illustrations of some general construction scheme for index theory. In all the cases we consider operator algebras with a trace such that $\operatorname{tr} AB = \operatorname{tr} BA$. For operators with a parametrix, the existence of a trace enables us to introduce the index by means of a formula analogous to (1.3). The requirement that the elements of the algebra should be operators often turns out to be unessential. A fairly advanced index theory is also possible for abstract algebras with trace (Fedosov (1978)). In the next chapter this scheme is used for the algebra of quantum observables in deformation quantisation.

§4. The Index of Boundary Value Problems

A classical differential operator $\mathcal{A} = \begin{pmatrix} A \\ \gamma'T \end{pmatrix}$ on a compact manifold M_+ with boundary ∂M_+ (this is also referred to as a boundary value problem) associates with a function $u \in C^\infty(M_+)$ a pair of functions $f = Au \in C^\infty(M_+)$ and $g = \gamma'Tu \in C^\infty(\partial M_+)$. Here A and T are differential operators with smooth coefficients, the coefficients of A depend on $x \in M_+$, those of T on $x' \in \partial M_+$, and γ' is the operator of restriction of functions to the boundary. In the general case, u, f and g are sections of bundles over M_+ and ∂M_+. The operator $\gamma'T$, called a *trace*, or *boundary*, *operator*, prescribes boundary conditions which u must satisfy, and is usually given by a column vector $\begin{pmatrix} \gamma'T_1 \\ \vdots \\ \gamma'T_k \end{pmatrix}$, where the operators $T_1, \ldots, T_k$ are, generally speaking, of different orders $m_1, m_2, \ldots, m_k$.

$\mathcal{A}$ is called *elliptic* if the principal symbol $a(x, \xi)$ of A is invertible for all $x \in M_+$, $\xi \neq 0$, and, in addition, at all boundary points the principal symbols $a(x', \xi)$ and $t(x', \xi) = \begin{pmatrix} t_1(x', \xi) \\ \vdots \\ t_k(x', \xi) \end{pmatrix}$ of A and T, respectively, satisfy the Shapiro-Lopatinskij condition (see below). As in the case of manifolds without boundary, an elliptic operator is a Fredholm operator in the corresponding function spaces, and its index depends only on the principal symbols $a(x, \xi)$ and $t(x', \xi)$ and remains invariant under ellipticity-preserving deformations.

The classical boundary value problems form a very narrow class of operators on manifolds with boundary, which is inadequate for an index theory. In Boutet de Monvel (1971), an algebra of PDOs on a manifold with boundary which contains classical boundary value problems and parametrices of elliptic boundary value problems was introduced and studied, and an index theory was also constructed.

A *Boutet de Monvel operator* is given by a matrix

$$\mathcal{A} = \begin{pmatrix} \gamma^+ A + \gamma'B & K \\ \gamma'T & Q \end{pmatrix} : \begin{array}{ccc} C^\infty(M_+, E^0) & & C^\infty(M_+, E^1) \\ \oplus & \to & \oplus \\ C^\infty(\partial M_+, F^0) & & C^\infty(\partial M_+, F^1) \end{array}, \tag{2.16}$$

where E^0 and E^1 are bundles over M_+ and F^0 and F^1 bundles over ∂M_+. The operator γ^+A is defined as follows. We assume that M_+ is embedded in a larger manifold M of the same dimension, and that the PDO A acts on functions defined on M. We extend $u \in C^\infty(M_+)$ by zero outside M_+, and to this extension we apply A and then the operator γ^+ of restriction to M_+. A natural condition for the PDO A is that $\gamma^+ A u$ should belong to $C^\infty(M_+)$ for any $u \in C^\infty(M_+)$. This smoothness property is guaranteed by the so-called *transmission condition*. In particular, the order of A can be only a (positive or negative) integer. As for the remaining operators, $\gamma' B$ is called the Green's operator, $\gamma' T$ the trace (boundary) operator, K the potential (coboundary) operator, and Q a PDO on the boundary ∂M_+. For classical boundary value problems, A and T are differential operators and $\gamma' B$, K and Q are zero.

With an operator $\mathcal{A}$ of the form (2.16) we associate the *principal interior symbol* $\sigma_i(\mathcal{A}) = a(x, \xi)$, which is the principal symbol of the PDO A on $T^*(M_+) \setminus 0$, and the *principal boundary symbol* $\sigma_b(\mathcal{A}) = a_b(x', \xi')$, which is an operator-valued function on $T^*(\partial M_+) \setminus 0$. At a fixed point $(x', \xi') \in T^*(\partial M_+) \setminus 0$ the symbol a_b is given by

$$
a_b = \begin{pmatrix} \Pi^+ a(\xi_n) + \Pi'_{\eta_n} b(\xi_n, \eta_n) & k(\xi_n) \\ \Pi' t(\xi_n) & q \end{pmatrix} : \begin{matrix} E^0_{x'} \otimes H^+ \\ \oplus \\ E^0_{x'} \end{matrix} \rightarrow \begin{matrix} E^1_{x'} \otimes H^+ \\ \oplus \\ E^1_{x'} \end{matrix} . \tag{2.17}
$$

Here $H^+ = \{F[\theta_+(x^n)u(x^n)], u(x^n) \in S(\mathbb{R})\}$ is the space of Fourier images of functions of the form $\theta_+(x^n)u(x^n)$, where $u(x^n)$ belongs to the Schwarz space $S(\mathbb{R})$ and $\theta_+(x^n)$ is the characteristic function of the semi-axis $\mathbb{R}_+(x_n \geq 0)$. We also introduce the space $H^- = \{F[\theta_-(x^n)u(x^n)], u(x^n) \in S(\mathbb{R})\} \oplus H'$, where H' is the space of polynomials in ξ_n, and set $H = H^+ \oplus H^-$. Let $\Pi^\pm$ be the projections in H on the direct terms $H^\pm$ (we mention that H^+ is the Fourier image of the operator of multiplication by the characteristic function $\theta_+(x^n)$). We introduce a functional Π' on H, which is equal to 0 on H^- and to $\lim_{\xi_n \to \infty} i\xi_n \hat{u}(\xi_n)$ for $\hat{u}(\xi_n) \in H^+$ (the Fourier image of the functional $\gamma' u(x) = \lim_{x^n \to 0+} u(x^n)$). For absolutely integrable functions $\hat{u}(\xi_n) \in H$, the functional Π' has the form $\Pi' \hat{u} = (1/(2\pi)) \int_{-\infty}^{\infty} \hat{u}(\xi_n) \, d\xi_n$. In computations it is convenient to assume that all $\hat{u}(\xi_n) \in H$ are rational functions without poles on the real axis (these functions are dense in the corresponding spaces). Then $\Pi^\pm$ and Π' are given by integrals over a closed contour Γ in the complex plane, which circumscribes all the poles of $\hat{u}(\xi_n)$ in the upper half-plane; that is,

$$
(\Pi^\pm u)(\xi_n) = \frac{1}{2\pi i} \int_\Gamma \frac{\hat{u}(\eta)}{\eta - \xi_n} \, d\eta \, ,
$$

where ξ_n lies outside Γ for Π^+ and inside Γ for Π^-, and

$$
\Pi' \hat{u} = \frac{1}{2\pi} \int_\Gamma \hat{u}(\xi_n) \, d\xi_n \, .
$$

Consequently, $\Pi^+\hat{u}(\xi_n) \in H^+$ is a rational function that has no poles in the lower half-plane and tends to 0 as $\xi_n \to \infty$, and $\Pi^-\hat{u}(\xi_n)$ is a rational function with no poles in the upper half-plane. The elements of the matrix (2.17) satisfy the following conditions. In the neighbourhood of the boundary, the interior symbol $a(x,\xi)$ can be rewritten in the form $a(x', x^n, \xi', \xi_n)$ and the function $a(\xi_n)$ in (2.17) is equal to $a(x', 0, \xi', \xi_n)$, where $(x', \xi') \in T^*(\partial M_+)$ are regarded as parameters. The *transmission condition* mentioned above means that $a(\xi_n) \in H$. The function $b(\xi_n, \eta_n)$ (the *Green's symbol*) belongs to H^+ with respect to ξ_n and to H^- with respect to η_n, $k(\xi_n) \in H^+$ is the potential symbol, $t(\xi_n) \in H^-$ the trace symbol, and q a homomorphism. Consequently, the operator (2.17) acts on the vector-valued function $\hat{u}(\xi_n) \in E^0_{x'} \otimes H^+$ and the vector $v \in F^0_{x'}$ according to the formula

$$a_b \begin{pmatrix} \hat{u}(\xi_n) \\ v \end{pmatrix} = \begin{pmatrix} \Pi^+ a(\xi_n)\hat{u}(\xi_n) + \Pi'_{\eta_n} b(\xi_n, \eta_n)\hat{u}(\eta_n) + k(\xi_n)v \\ \Pi' t(\xi_n)\hat{u}(\xi_n) + qv \end{pmatrix} .$$

The operators a_b of the form (2.17) form an algebra, that is, the product of two such operators is an operator of the same form. We denote this algebra by $\mathfrak{A}_b$.

$\mathcal{A}$ is called *elliptic* if its interior and boundary symbols are invertible. We mention that the boundary symbol is an operator in an infinite-dimensional space of functions, and its invertibility is understood in the operator sense. For classical boundary value problems, the condition of invertibility of the boundary symbol is known as the *Shapiro-Lopatinskij condition*.

We define the trace tr' on $\mathfrak{A}_b$ (in the case when $E^0 = E^1$ and $F^0 = F^1$). Regarding (2.17) as an operator in the space $(E \otimes H^+) \oplus F$, it is easy to see that it is not a trace class operator; consequently, it does not have an operator trace except when $\Pi^+ a(\xi_n) = 0$. In the later case, the trace exists and is equal to

$$\Pi' \, \mathrm{tr}\, b(\xi_n, \xi_n) + \mathrm{tr}\, q . \tag{2.18}$$

where tr denotes the endomorphism trace (of a matrix).

We define the boundary trace $\mathrm{tr}'\, a_b$ for the operator (2.17) by means of (2.18) even if $\Pi^+ a(\xi_n) \neq 0$. Then the trace tr' exists for any operator $a_b \in \mathfrak{A}_b$; however, in contrast to the operator trace, it is not equal to zero on commutators.

Lemma 2.1. *For any a_1, $a_2 \in \mathfrak{A}_b$ we have*

$$\mathrm{tr}'\, [a_1, a_2] = -i\Pi' \, \mathrm{tr}\, \frac{\partial a_1(\xi_n)}{\partial \xi_n} a_2(\xi_n) = i\Pi' \, \mathrm{tr}\, a_1(\xi_n)\frac{\partial a_2(\xi_n)}{\partial \xi_n} . \tag{2.19}$$

Proof. From the explicit formulae for the multiplication of operators in $\mathfrak{A}_b$ it follows that

$$\mathrm{tr}'\, [a_1, a_2] = \Pi' (\mathrm{tr}\, g_{12}(\xi, \xi) - \mathrm{tr}\, g_{21}(\xi, \xi)) ,$$

where g_{12} and g_{21} are the Green's symbols defined by

$$g_{12}(\xi, \eta) = -\Pi_\xi^+ \Pi_\eta^- \frac{(a_1^+(\xi) - a_1^+(\eta))(a_2^-(\xi) - a_2^-(\eta))}{i(\xi - \eta)} \,,$$

with g_{21} obtained by permuting the subscripts. Here $a^\pm(\xi) = \Pi^\pm a(\xi)$. Since $((a_1^+(\xi) - a_1^+(\eta))/(i(\xi - \eta)))a_2^-(\xi) \in H^+$ as a function of η for ξ fixed, by removing the parentheses we obtain

$$g_{12} = \Pi_\eta^- \frac{a_1^+(\xi) - a_1^+(\eta)}{i(\xi - \eta)} a_2^-(\eta) = \frac{1}{2\pi i} \int_\Gamma \frac{a_1^+(\xi) - a_1^+(\zeta)}{i(\xi - \zeta)(\zeta - \eta)} a_2^-(\zeta)\, d\zeta \,,$$

where Γ contains all the poles of $a_1^+(\xi)$ but not those of $a_2^-(\zeta)$ (there is no singularity at $\zeta = \xi$). The last equality holds if a_1 and a_2 are rational, which does not imply any loss of generality. Next,

$$\Pi' \operatorname{tr} g_{12}(\xi, \xi) = -\frac{1}{(2\pi i)^2} \int_{\Gamma_1} d\xi \int_{\Gamma_2} \operatorname{tr} \frac{a_1^+(\xi) - a_1^+(\zeta)}{(\xi - \zeta)^2} a_2^-(\zeta)\, d\zeta \,,$$

where Γ_1 lies inside Γ_2. Computing first the integral over Γ_1 by means of the residues outside Γ_1 (the residue at ∞ is 0), we find that

$$\Pi' \operatorname{tr} g_{12}(\xi, \xi) = \frac{1}{2\pi i} \int_{\Gamma_2} \operatorname{tr} \frac{\partial a_1^+(\zeta)}{\partial \zeta} a_2^-(\zeta)\, d\zeta$$

$$= -i\Pi' \operatorname{tr} \frac{\partial a_1^+(\zeta)}{\partial \zeta} a_2^-(\zeta) = i\Pi' \operatorname{tr} a_1^+(\zeta) \frac{\partial a^-(\zeta)}{\partial \zeta} \,.$$

A similar expression is also derived for $\Pi' \operatorname{tr} g_{21}(\xi, \xi)$. Finally,

$$\operatorname{tr}'(a_1 a_2 - a_2 a_1) = -i\Pi' \operatorname{tr}\left(\frac{\partial a_1^+(\xi)}{\partial \xi} a_2^-(\xi) + \frac{\partial a_1^-(\xi)}{\partial \xi} a_2^+(\xi) \right)$$

$$= -i\Pi' \operatorname{tr} \frac{\partial a_1(\xi)}{\partial \xi} a_1(\xi) \,,$$

since Π' is equal to zero on terms of the form $(\partial a_1^+(\xi)/\partial \xi)\, a_2^+(\xi)$ and $(\partial a_1^-(\xi)/\partial \xi)\, a_2^-(\xi)$.

The formula for the index of an elliptic boundary value problem can be written in terms of the character of Fredholm families (see Chap. 2, Sec. 2). We extend the principal interior symbol $a(x, \xi)$ and its inverse $a^{-1}(x, \xi)$ from $T^*(M_+) \setminus 0$ to the whole cotangent bundle $T^*(M_+)$ by multiplying them by a truncating function $\rho(\xi)$ equal to zero in the neighborhood of the zero section. We denote by $a(x, \xi)$ and $r(x, \xi)$ the functions obtained in this way. They define a finite-dimensional family with compact support $d_i(A) = \{E^0, E^1, a, r\} \in K^{\mathrm{comp}}(TM_+)$ on $T^*(M_+)$, which is the distinguishing element of the elliptic operator A.

The principal boundary symbols a_b and a_b^{-1} of $\mathcal{A}$ and its parametrix also define a Fredholm family with compact support on $T^*(\partial M_+)$, namely,

$$d_b(\mathcal{A}) = \{(E^0 \otimes H^+) \oplus F^0, (E^1 \otimes H^+) \oplus F^1, a_b(x', \xi'), r_b(x', \xi')\} \,,$$

but, in contrast to d_i, this family is infinite-dimensional (since H^+ is an infinite-dimensional space). Here a_b and r_b are smooth extensions of the principal boundary symbols of $\mathcal{A}$ and its parametrix on the whole of $T^*(\partial M_+)$. The only condition that both these extensions must satisfy is that $a(\xi_n)$ in (2.17) should coincide with $a(x', 0, \xi', \xi_n)$, which is the restriction of the interior (extended) symbol to the boundary, and, similarly, $r(\xi_n) = r(xf', 0, \xi', \xi_n)$.

We define a connection on all bundles: ∂_i on the bundles E^i ($i = 0, 1$) over M_+, and ∂'_i on the bundles $(E^i \otimes H^+) \oplus F^i$ over ∂M_+. Here ∂'_i acts trivially on the cofactor H^+ and coincides with ∂_i on the cofactor E^i. Both ∂_i and ∂'_i define connections on the liftings of the E^i to $T^*(M_+)$ and on those of the $(E^i \otimes H^+) \oplus F^i$, respectively, which are denoted by the same symbols ∂_i and ∂'_i. As was described in 2.2 (see also 1.3.2), we define connections $\tilde{\partial}_i$ on the liftings of the E^i to $T^*(M_+)$ by setting $\tilde{\partial}_1 = \partial_1$ and $\tilde{\partial}_0 u = \partial_0 u + (r\partial a)u$ (here ∂a is the covariant differential of the homomorphism $a : E^0 \to E^1$ with respect to the connections ∂_0 and ∂_1). Similarly, we set $\tilde{\partial}_1 = \partial_1$ and $\tilde{\partial}'_0 u = \partial'_0 u + (r_b \partial' a_b)u$. The curvatures of these connections (divided by $-2\pi i$) are denoted by ω_i, $\tilde{\omega}_i$, ω'_i and $\tilde{\omega}'_i$. Consequently,

$$\tilde{\omega}_0 = \omega_0 - \frac{1}{2\pi i}\left(\partial(r\partial a) + (r\partial a)^2\right), \tag{2.20}$$

$$\tilde{\omega}'_0 = \omega'_0 - \frac{1}{2\pi i}\left(\partial'(r_b\partial' a_b) + (r_b\partial' a_b)^2\right). \tag{2.21}$$

Denoting by $\tilde{\partial}$ and $\tilde{\partial}'$ the covariant differentials of homomorphisms with respect to $\tilde{\partial}_i$ and ∂'_i, we find that (see (2.7))

$$\tilde{\partial} a \sim \tilde{\partial} r \sim \tilde{\omega}_1 a - a\tilde{\omega}_0 \sim 0 \,,$$
$$\tilde{\partial}' a_b \sim \tilde{\partial}' r_b \sim \tilde{\omega}'_1 a_b - a_b \tilde{\omega}'_0 \sim 0 \,,$$

where $\sim$ denotes equality outside some compact.

We introduce the differential forms with compact support

$$\mathrm{ch}\, d_i(\mathcal{A}) = \mathrm{tr}\, e^{\tilde{\omega}_0} - \mathrm{tr}\, e^{\omega_1} \,, \tag{2.22}$$

$$\mathrm{ch}'\, d_b(\mathcal{A}) = \mathrm{tr}'\,(e^{\tilde{\omega}'_0} - e^{\tilde{\omega}'_0} r_b a_b) - \mathrm{tr}'\,(e^{\omega'_1} - a_b e^{\tilde{\omega}'_0} r_b)\,. \tag{2.23}$$

(see the formula (2.8) for the character of Fredholm families), where tr' is the trace in the algebra $\mathfrak{A}_b$ defined by (2.18). We mention that in (2.23) we cannot pass to the shorter form (2.22), but must use the full formula for the character of an infinite-dimensional Fredholm family. This is due to the fact that tr' does not annihilate the commutators: $\mathrm{tr}'\,[a_b, e^{\tilde{\omega}'_0} r_b] \neq 0$. There is also a difference between (2.23) and the formula (2.8) for the character of a family. Specifically, (2.8) can be written in various guises that reduce to the same result (among them also one of the form (2.23)), since the operator trace annihilates commutators. Since tr' does not have this property, the form (2.23) becomes essential.

Theorem 2.4. *The index of the boundary value problem is expressed by the formula*

$$\operatorname{ind} \mathcal{A} = \int\limits_{T^*(M_+)} \operatorname{ch} d_i(\mathcal{A})\mathcal{T}(M_+) + \int\limits_{T^*(\partial M_+)} \operatorname{ch}' d_b(\mathcal{A})\mathcal{T}(M_+)\,, \qquad (2.24)$$

where $\mathcal{T}(M_+)$ is the differential form given by the Todd class of the tangent bundle.

To prove this assertion, in view of Boutet de Monvel's results (Boutet de Monvel (1971); see also Rempel and Schulze (1982)) it suffices to verify two properties of the functional (2.24): 1) invariance with respect to homotopies, and 2) if the symbols $a(x,\xi)$, $a^{-1}(x,\xi)$, $a_b(x',\xi')$ and $a_b^{-1}(x',\xi')$ can be extended to the whole cotangent bundles $T^*(M_+)$ and $T^*(\partial M_+)$ so that $d_b(\mathcal{A})$ is a trivial family and $d_i(\mathcal{A})$ is a trivial family in the neighbourhood of the boundary, then (2.24) must reduce to the Atiyah-Singer formula (1.19) for a manifold without boundary.

We need to verify only the first property, since the second one is evident. A homotopy defines Fredholm families d_i and d_b on the manifolds $T^*(\partial M_+) \times [0,1]$ and $T^*(\partial M_+) \times [0,1]$, respectively, as well as connections ∂_i, $\tilde{\partial}_i$, ∂_i' and $\tilde{\partial}_i'$ on the liftings of the bundles E^i and $(E^i \otimes H^+) \oplus F^i$ to these manifolds.

We compute the change in the first term in (2.24). The cotangent bundle $T^*(M_+)$ restricted to ∂M_+ splits into the direct sum of $T^*(\partial M_+)$ and the trivial one-dimensional bundle corresponding to the coordinate ξ_n : $T^*(M_+)|_{\partial M_+} = T^*(\partial M_+) \oplus 1$. Since the form $\operatorname{ch} d_i(\mathcal{A})$ is closed on $T^*(M_+) \times [0,1]$, it follows that the integral over the boundary of this manifold is equal to zero, that is,

$$\int\limits_{T^*(M_+)\times\{1\}} - \int\limits_{T^*(M_+)\times\{0\}} = - \int\limits_{\{T^*(\partial M_+)\oplus 1\}\times[0,1]} \operatorname{ch} d_i(\mathcal{A})\mathcal{T}(M_+)$$

$$= - \int\limits_{T^*(\partial M_+)\times[0,1]} \int\limits_{-\infty}^{\infty} d\xi_n\, i\!\left(\frac{\partial}{\partial\xi_n}\right) \operatorname{ch} d_i(\mathcal{A})\mathcal{T}(M_+)\,,$$

where $i(\partial/\partial\xi_n)$ denotes the operator of the interior product of a form and the vector field $\partial/\partial\xi_n$. By (2.22) and (2.20),

$$i\!\left(\frac{\partial}{\partial\xi_n}\right)\operatorname{ch} d_i(\mathcal{A}) = \operatorname{tr} e^{\tilde{\omega}_0} i\!\left(\frac{\partial}{\partial\xi_n}\right)\tilde{\omega}_0 = -\frac{1}{2\pi i}\operatorname{tr} e^{\tilde{\omega}_0}\left(\frac{\partial}{\partial\xi_n}(r\partial a) - \tilde{\partial}\!\left(r\frac{\partial a}{\partial\xi_n}\right)\right),$$

where ξ_n is regarded as a parameter and ∂ and $\tilde{\partial}$ are connections on $T^*(\partial M_+) \times [0,1]$. Thus, the change in the first term is equal to

$$\int\limits_{T^*(\partial M_+)\times[0,1]} \frac{1}{2\pi i} \int\limits_{-\infty}^{\infty} d\xi_n \mathcal{T}(M_+)\operatorname{tr} e^{\tilde{\omega}_0}\left(\frac{\partial}{\partial\xi_n}(r\partial a) - \tilde{\partial}\!\left(r\frac{\partial a}{\partial\xi_n}\right)\right).$$

We now compute the change in the second term

$$\int_{T^*(\partial M_+)\times\{1\}} - \int_{T^*(\partial M_+)\times\{0\}} \mathrm{ch}'\, d_b(\mathcal{A})\mathcal{T}(M_+) = \int_{T^*(\partial M_+)\times[0,1]} d\mathrm{ch}'\, d_b(\mathcal{A})\mathcal{T}(M_+)\,.$$

The form $\mathrm{ch}'\, d_b(\mathcal{A})$ is not closed, since tr' does not annihilate commutators. We have

$$d\,\mathrm{ch}'\, d_b(\mathcal{A}) = d\,\mathrm{tr}'\, e^{\tilde\omega'_0} - d\,\mathrm{tr}'\, e^{\omega'_1} + d\,\mathrm{tr}[a_b, e^{\tilde\omega'_0}r]\,.$$

Next,

$$d\,\mathrm{tr}'\, e^{\omega'_1} = \mathrm{tr}'\, \partial' e^{\omega'_1} = 0\,,$$
$$d\,\mathrm{tr}'\, e^{\tilde\omega'_0} = \mathrm{tr}'\, \partial'_0 e^{\tilde\omega'_0} = \mathrm{tr}\, \tilde\partial'_0 e^{\tilde\omega'_0} - \mathrm{tr}'\, [r_b\partial a_b, e^{\tilde\omega'_0}] = -\,\mathrm{tr}'\, [r_b\partial a_b, e^{\tilde\omega'_0}]\,,$$

since, by Bianchi's identity, $\partial'_1 e^{\omega'_1} = \tilde\partial'_0 e^{\tilde\omega'_0} = 0$. By Lemma 2.1,

$$d\,\mathrm{ch}'\, d_b(\mathcal{A}) = i\varPi'\,\mathrm{tr}\,\frac{\partial}{\partial\xi_n}(r\partial a)e^{\tilde\omega_0} - id\varPi'\,\mathrm{tr}\,\frac{\partial a}{\partial\xi_n}e^{\tilde\omega_0}r$$

$$= \frac{i}{2\pi}\int_{-\infty}^{\infty} d\xi_n\,\mathrm{tr}\left(\frac{\partial}{\partial\xi_n}(r\partial a) - \tilde\partial\left(r\frac{\partial a}{\partial\xi_n}\right)\right)e^{\tilde\omega_0}\,.$$

Here we have used the fact that, by Bianchi's identity,

$$d\,\mathrm{tr}\,\frac{\partial a}{\partial\xi_n}e^{\tilde\omega_0}r = \mathrm{tr}\,\tilde\partial\left(r\frac{\partial a}{\partial\xi_n}e^{\tilde\omega_0}\right) = \mathrm{tr}\,\tilde\partial\left(r\frac{\partial a}{\partial\xi_n}\right)e^{\tilde\omega_0}\,.$$

From this we see that the total change is equal to zero.

Similar arguments enable us to transform (2.24) into integrals over $S(M_+)$ and $S(\partial M_+)$, just as in the case of a manifold without boundary. Omitting the details, we state the final result. On $S(M_+)\times[0,1]$ and $S(\partial M_+)\times[0,1]$ we introduce the forms

$$\tilde\omega_0 = \omega_0 - \frac{1}{2\pi i}(\partial(ta^{-1}\partial a) + (ta^{-1}\partial a)^2)\,,$$
$$\tilde\omega'_0 = \omega'_0 - \frac{1}{2\pi i}(\partial'(ta_b^{-1}\partial' a_b) + (ta_b^{-1}\partial a_b)^2)\,.$$

Then

$$\mathrm{ind}\,\mathcal{A} = \int_{S(M_+)\times[0,1]} \mathrm{tr}\, e^{\tilde\omega_0}\mathcal{T}(M_+)$$

$$+ \int_{S(\partial M_+)\times[0,1]} \{(1-t)\,\mathrm{tr}'\, e^{\omega'_0} + t\,\mathrm{tr}'\, a_b e^{\omega'_0}a_b^{-1}\}\mathcal{T}(M_+)\,,\quad (2.25)$$

where the orientation of the manifold $S(M_+)\times[0,1]$ is given by the form $\prod_{i=1}^{n} d(t\xi_i)\wedge dx^i$, with a similar remark for $S(\partial M_+)\times[0,1]$. We mention that the first term in (2.25) is the same as for a manifold without boundary.

The formulae (2.24) and (2.25) can be generalized in a natural way to the case of a family of elliptic operators on a manifold with boundary. Here the result is fully analogous to Theorem 2.3.

In many cases, the method of reduction to the boundary turns out to be useful in the computation of the index. We illustrate it on the example of classical boundary value problems. Let

$$\mathcal{A}_1 = \begin{pmatrix} \gamma^+ A \\ \gamma' T_1 \end{pmatrix}, \qquad \mathcal{A}_2 = \begin{pmatrix} \gamma^+ A \\ \gamma' T_2 \end{pmatrix}$$

be two elliptic boundary value problems differing only by the boundary conditions, and let $\mathcal{R}_1 = (\gamma^+ R + \gamma' G_1, K_1)$ be a parametrix of $\mathcal{A}_1$. Then

$$\mathcal{A}_2 \mathcal{R}_1 = \begin{pmatrix} 1 & 0 \\ * & \gamma' T_2 K_1 \end{pmatrix}$$

up to infinitely smoothing operators. $S = \gamma' T_2 K_1$ is an elliptic PDO on the boundary. By the logarithmic property of the index,

$$\operatorname{ind} \mathcal{A}_2 = \operatorname{ind} \mathcal{A}_1 + \operatorname{ind} \mathcal{A}_2 \mathcal{R}_1 = \operatorname{ind} S + \operatorname{ind} \mathcal{A}_1. \tag{2.26}$$

Consequently, if the index of $\mathcal{A}_1$ is known, then the computation of the index of $\mathcal{A}_2$ reduces to that of the index of the PDO S on the boundary. The equality (2.26) is called the *Agranovich-Dynin formula*.

We illustrate the application of (2.25) and (2.26) on a specific example. Let $\mathcal{A}$ be a differential operator on the unit disk $M_+ : x_1^2 + x_2^2 \leq 1$, with symbol

$$a(x, \xi) = b(x)c(\xi) = \begin{pmatrix} i & \mu(x^1 + ix^2) \\ -\mu(x^1 - ix^2) & -i \end{pmatrix} \begin{pmatrix} \xi_1 + i\xi_2 & 0 \\ 0 & \xi_1 - i\xi_2 \end{pmatrix},$$

where μ is a real number. The boundary conditions are given by means of the operator $T = (\alpha(x'), \beta(x'))$, where α and β are complex functions on the circle, nowhere equal to zero.

For the two-dimensional manifold M_+, after integration with respect to $t \in [0, 1]$ the formula (2.25) becomes

$$\operatorname{ind} \mathcal{A} = -\frac{1}{6(2\pi i)^2} \int\limits_{S(M_+)} \operatorname{tr}(a^{-1}da)^3$$

$$-\frac{1}{2(2\pi i)} \int\limits_{S(\partial M_+)} \operatorname{tr}' a_b^{-1} da_b + \operatorname{tr}' da_b a_b^{-1}. \tag{2.27}$$

Simple calculations show that the first term is equal to $\mu^2/(1 + \mu^2)$. We compute the boundary term first in the particular case when $T_1 = (1, 1)$. In the neighbourhood of the boundary we choose x^1 to be the polar angle and x^2 the distance to the boundary. For $x^2 = 0$

$$a(x^1, 0, \xi_1, \xi_2) = b(x^1)c(\xi_1, \xi_2) = \begin{pmatrix} -ie^{ix^1} & -\mu \\ \mu & ie^{-ix^1} \end{pmatrix} \begin{pmatrix} \xi_2 - i\xi_1 & 0 \\ 0 & \xi_2 + i\xi_1 \end{pmatrix} .$$

The boundary symbol of $\mathcal{A}$ is

$$a_b = \begin{pmatrix} \Pi^+ a(\xi_2) \\ \Pi'(1,1) \end{pmatrix}$$

(the dependence on x^1 and ξ_1 is not indicated). This operator is invertible in the algebra $\mathfrak{A}_b$ for $\xi_1 \neq 0$, and

$$a_b^{-1} = (\Pi^+ a^{-1}(\xi_2) + k(\xi_2)\Pi'_\eta(1,1)(a^{-1}(\eta))^-, k(\xi_2)) ,$$

where $k(\xi_2)$ is a two-element column with entries in H^+ satisfying the conditions

$$\Pi^+ c(\xi_2)k(\xi_2) = 0 , \qquad \Pi'(1,1)k(\xi_2) = 1 .$$

The explicit expression of $k(\xi_2)$ is

$$k(\xi_2) = -i \begin{pmatrix} (\xi_2 - i\xi_1)^{-1} & \theta(\xi_1) \\ (\xi_2 + i\xi_1)^{-1} & \theta(-\xi_1) \end{pmatrix} ,$$

where θ is the Heaviside function. Hence,

$$da_b a_b^{-1} = \begin{pmatrix} \Pi^+ da(\xi_2)a^{-1}(\xi_2) & 0 \\ 0 & 0 \end{pmatrix} ,$$

since $da_b = (db(x^1))c(\xi_1, \xi_2)$ and $\Pi^+ ck = 0$. In particular, $da_b b_b^-$ does not contain Green's symbols, so $\mathrm{tr}'\, da_b a_b^{-1} = 0$. By Lemma 2.1, from this it follows that

$$\mathrm{tr}'\, a_b^{-1} da_b = \mathrm{tr}'\, [a_b^{-1}, da_b] = \frac{i}{2\pi} \int_\Gamma d\xi_2\, \mathrm{tr}\, a^{-1} d\left(\frac{\partial a}{\partial \xi_2}\right)$$

$$= -\mathrm{tr} \begin{pmatrix} \theta(\xi_1) & 0 \\ 0 & \theta(-\xi_1) \end{pmatrix} b^{-1}(x^1)db(x^1) = -\frac{idx^1}{1+\mu^2} \operatorname{sgn} \xi_1 .$$

Therefore, the boundary term in (2.27) is equal to $1/(1+\mu^2)$, and $\mathrm{ind}\, \mathcal{A}_1 = 1$ under the boundary condition $T_1 = (1,1)$.

In the case of the boundary condition $T = (\alpha(x^1), \beta(x^1))$ we use the formula (2.26). The symbol of the operator $S = \Pi'TK$ is

$$s = \alpha(x^1)\theta(\xi_1) + \beta(x^1)\theta(-\xi_1) ,$$

hence,

$$\mathrm{ind}\, S' = \frac{1}{2\pi}(\Delta \arg \alpha(x^1) - \Delta \arg \beta(x^1)) ,$$

from which it follows that

$$\mathrm{ind}\, \mathcal{A} = 1 + \frac{1}{2\pi}(\Delta \arg \alpha(x^1) - \Delta \arg \beta(x^1)) .$$

§5. The Index of Toeplitz Operators

In this section we explain Boutet de Monvel's results (Boutet de Monvel (1979); see also Boutet de Monvel and Guillemin (1989)).

First we recall the definition of a strictly pseudoconvex domain. Let $\mathbb{C}^n$ be an n-dimensional complex space with coordinates $z_j = x_j + iy_j$, and Ω a bounded domain in $\mathbb{C}^n$ with a smooth boundary $\partial\Omega$. We assume that Ω is given by $\rho < 0$, where $\rho = \rho(z, \bar{z})$ is a smooth real function such that $d\rho \neq 0$ on $\partial\Omega$. A complex vector field X defined on the boundary and in its neighbourhood is called a *holomorphic (antiholomorphic) tangent* vector field if it has the form

$$X = \sum_{j=1}^{n} a_j(z, \bar{z}) \frac{\partial}{\partial z_j}$$

$\left(\sum_{j=1}^{n} a_j(z, \bar{z})(\partial/\partial \bar{z}_j) \right)$, where a_j are complex functions and $X\rho|_{\partial\Omega} = 0$. If $X = \sum_{j=1}^{n} a_j(\partial/\partial z_j)$ is a holomorphic tangent field, then $\bar{X} = \sum_{j=1}^{n} \bar{a}_j(\partial/\partial \bar{z}_j)$ is an antiholomorphic tangent field, and vice versa.

The Hermitian form

$$L(X, Y) = \sum_{i,j=1}^{n} \frac{\partial^2 \rho}{\partial z_i \partial \bar{z}_j} a_i \bar{b}_j$$

considered on holomorphic (or antiholomorphic) vector fields X and Y is called the *Levy form*. The domain Ω is called *strictly pseudoconvex* if for any holomorphic tangent vector field X the quadratic form $L(X, X) > 0$ at all the points of the boundary.

We define the *tangent Cauchy-Riemann operator* d_b'' on $\partial\Omega$.

We extend $u \in C^\infty(\partial\Omega)$ smoothly to the neighbourhood of the boundary, apply the Cauchy-Riemann operator to it to get $d''u$ (d' and d'' are the holomorphic and antiholomorphic parts of the differential, that is, $d = d' + d''$), restrict this form to the boundary, and consider it only on antiholomorphic tangent vector fields. Therefore, for any antiholomorphic tangent vector field $X = \sum_{j=1}^{n} a_j(\partial/\partial \bar{z}_j)$ we set

$$\langle d_b''u, X \rangle = \langle d''u, X \rangle|_{\partial\Omega} = Xu|_{\partial\Omega} = \sum_{j=1}^{n} a_j \frac{\partial u}{\partial \bar{z}_j}\bigg|_{\partial\Omega}.$$

The result does not depend on the method of extension of u, since any other extension has the form $u + \rho v$ and $\rho|_{\partial\Omega} = X\rho|_{\partial\Omega} = 0$. A fundamental result in the theory of functions of several complex variables states that for a strictly pseudoconvex domain, the function $u \in C^{\infty}(\partial\Omega)$ admits a holomorphic extension to Ω if and only if $d_b'' u = 0$ (u satisfies the tangent Cauchy-Riemann equations).

The above definitions are local and invariant with respect to holomorphic changes of variables. In the neighbourhood of a point $z_0 \in \partial\Omega$ (which can be shifted to 0 by means of a linear change of variables) where $\partial\rho/\partial\bar{z}_n \neq 0$, the function ρ can be reduced by means of a holomorphic change of variables to the form

$$\rho = z_n + \bar{z}_n + c z_n \bar{z}_n + \sum_{j=1}^{n-1} z_j \bar{z}_j + O(|z|^3) \ . \tag{2.28}$$

Antiholomorphic tangent vector fields are generated by the operators $D_j = \partial/\partial\bar{z}_j - b_j(\partial/\partial\bar{z}_n)$, where $b_j = (\partial\rho/\partial\bar{z}_j)/(\partial\rho/\partial\bar{z}_n) = z_j + O(|z|^2)$, and for any field $X = \sum_{j=1}^{n-1} a_j D_j$ the Levy form at z_0 has the form $\sum_{j=1}^{n-1} a_j \bar{a}_j > 0$.

As local coordinates on $\partial\Omega$ we may take the $n-1$ complex coordinates z_j, $j = 1, 2, \ldots, n-1$ (or the real coordinates x_j, y_j) and y_n. Then the tangent Cauchy-Riemann operator has the form $d_b'' u = \sum_{j=1}^{n-1} D_j u \, d\bar{z}_j$, where $D_j u = \partial u/\partial\bar{z}_j - (i/2)b_j(\partial u/\partial y_n)$.

In the cotangent bundle $T^*(\partial\Omega)$ in the neighbourhod of z_0 we introduce dual coordinates ξ_j, η_j and η_n. Then the symbols of the operators D_j are $\sigma_j = \sigma(D_j) = \frac{1}{2}i(\zeta_j - ib_j\eta_n)$, where $\zeta_j = \xi_j + i\eta_j$. By the *characteristic set* of the tangent Cauchy-Riemann operator we understand a set Σ in $T^*(\partial\Omega) \setminus 0$ on which all $\sigma_j = 0$. In the fibre over z_0 we have $\zeta_j = ib_j\eta_n = 0$ since $\sigma_j = 0$ for $z = z_0$, so that Σ_{z_0} consists of two half-lines $\Sigma_{z_0}^+$: $\zeta_j = 0$, $\eta_n > 0$ and $\Sigma_{z_0}^-$: $\zeta_j = 0$, $\eta_n < 0$. More invariantly, $\Sigma_{z_0}^+$ is the half-line defined by the covector $dy_n = (1/(2i))(d'\rho - d''\rho)|_{z_0} = \alpha(z_0)$. Hence, the set Σ consists of two bundles of half-lines Σ^+ and Σ^-. The bundle Σ^+ is a cone over $\partial\Omega$: the points of Σ^+ have the form $\lambda\alpha$ ($\lambda > 0$), where $\alpha = (1/(2i))(d'\rho - d''\rho)|_{\partial\Omega} = (1/i)d'\rho|_{\partial\Omega} = -(1/i)d''\rho|_{\partial\Omega}$. Since the Levy form is non-degenerate, α is a *contact form* on $\partial\Omega$. This means that the $(2n-1)$-form $\nu = \alpha \wedge (d\alpha)^{n-1}$ never vanishes. In turn, from this it follows that the 2-form $\omega = d(\lambda\alpha)$ on Σ^+ is *symplectic*, that is, it is a closed non-degenerate 2-form, since $\omega^n = n\lambda^{n-1}d\lambda \wedge \nu \neq 0$. In addition, ω is the restriction to Σ^+ of the standard sympectic form

$$\sum_{j=1}^{n-1} \mathrm{Re}\, d\zeta_j \wedge d\bar{z}_j + d\eta_n \wedge dy_n = \sum_{j=1}^{n-1} d\xi_j \wedge dx_j + \sum_{j=1}^{n} d\eta_j \wedge dy_j \ . \tag{2.29}$$

All these assertions are easily verified locally at z_0 by means of the representation (2.28).

In a conical neighbourhood V of the point $(z_0, \alpha(z_0)) \in T^*((\partial\Omega) \setminus 0)$ we perform a canonical transformation (a diffeomorphism that preserves the symplectic form (2.29)) to introduce new coordinates (t, q, τ, p), where $t \in \mathbb{R}^n$,

$q \in \mathbb{R}^{n-1}$, and τ and p are dual variables so that $\sigma_j = \frac{1}{2}(ip_j + q_j|\tau|)$ $(j = 1, 2, \ldots, n-1)$. In the neighbourhood of z_0 we have

$$\sigma_j = \tfrac{1}{2}i(\xi_j + y_j + O(|z|^2)\eta_n) + \tfrac{1}{2}(x_j + O(|z|^2)\eta_n - \eta_j) = \tfrac{1}{2}ip_j + \tfrac{1}{2}|\tau|q_j \;.$$

Since the Poisson brackets[2] are preserved by a canonical transformation, for $z = z_0$ we must have

$$\{\operatorname{Im}\sigma_j, \operatorname{Re}\sigma_k\} = \tfrac{1}{2}\delta_{jk}\eta_k = \tfrac{1}{4}\delta_{jk}|\tau| = \left\{\tfrac{1}{2}p_j, \tfrac{1}{2}|\tau|q_k\right\} \;,$$

which is satisfied if we set $|\tau| = 2\eta_n$.

In the case when the b_j are equal to the z_j without remainder terms, it is not difficult to write the explicit form of the generating function $S(x, y, \tau, p)$ of such a transformation, which is homogeneous of degree 1 in τ and p. In a conical neighbourhood of the point $x = y = p = 0$, $\tau = (0, \ldots, 0, 2)$ we can take, for example,

$$S = \frac{|\tau|}{2}y_n + \sum_{j=1}^{n-1}\left(p_j x_j + p_j \frac{\tau_j}{|\tau|} - \tau_j y_j - \frac{|\tau|}{2}x_j y_j\right) \;.$$

Then the usual formulae

$$q_j = \frac{\partial S}{\partial p_j}\;, \qquad t_i = \frac{\partial S}{\partial \tau_i}\;, \qquad \xi_j = \frac{\partial S}{\partial x_j}\;, \qquad \eta_i = \frac{\partial S}{\partial y_i}$$

$(j = 1, 2, \ldots; i = 1, 2, \ldots, n)$ yield a non-degenerate canonical transformation φ of a conical neighbourhood $U \subset T^*(\mathbb{R}^{2n-1}) \setminus 0$ of the point $t_j = \tau_j = q_j = p_j = 0$ $(j = 1, 2, \ldots, n-1)$, $t_n = 0$, $\tau_n = 2$, on to a conical neighbourhood $V \subset T^*(\partial\Omega)\setminus 0$ of the point $(z_0, \alpha(z_0))$. Under this transformation, the submanifold $\Sigma^+ \cap V$ is mapped on to $\Sigma_0^+ \cap U \subset T^*(\mathbb{R}^{2n-1}) \setminus 0$, where Σ_0^+ is defined by the equations $p_j = q_j = 0$, that is, it coincides with $T^*(\mathbb{R}^n)$. The symplectic form $\omega = d(\lambda\alpha)$ on Σ^+ is mapped on to a standard symplectic form on $T^*(\mathbb{R}^n)$, namely,

$$\varphi^*\omega = \sum_{i=1}^{n} d\tau_i \wedge dt_i \;.$$

With the canonical transformation φ we associate in a standard manner a Fourier integral operator F such that $D_j F \sim F D_j^0$, where $D_j^0 = \frac{1}{2}(\partial/\partial q_j + q_j|D_t|)$ is an operator in $\mathbb{R}^{2n-1}$ with symbol $\frac{1}{2}(ip_j + |\tau|q_j)$. The symbol $\sim$ denotes microlocal equality in the conical neighbourhood V up to infinitely smoothing operators. In summary, in a conical neighbourhood of a point of the manifold Σ^+ the tangent Cauchy-Riemann equation $d_b'' u = 0$ is microlocally equivalent to the system $D_j^0 = 0$. A similar assertion also holds for the points of Σ^-, but with the operators $(D_j^0)^* = \frac{1}{2}(-\partial/\partial x_j + q_j|D_t|)$.

---

[2] Poisson brackets on a symplectic manifold are discussed in the introduction to Chap. 3.

We now define Toeplitz operators on $\partial\Omega$. Let S be the *Szegö projection*, which is the orthogonal projection in $L^2(\partial\Omega)$ with respect to the measure defined by the form ν, on the subspace of solutions of the tangent Cauchy-Riemann equation. With a classical PDO Q on $\partial\Omega$ we associate the *Toeplitz operator* $T_Q = SQS : C^\infty(\partial\Omega) \to C^\infty(\partial\Omega)$. By definition, the *order of the Toeplitz operator* is assumed to be the same as the order of Q. By the *principal symbol* $\sigma(T_Q)$ of the Toeplitz operator we understand the restriction $\sigma(Q)|_{\Sigma^+}$ of the principal symbol of Q to Σ^+.

The following assertions hold.

1) The Toeplitz operators form an algebra, and

$$\sigma(T_{Q_1} \cdot T_{Q_2}) = \sigma(T_{Q_1})\sigma(T_{Q_2}) \ .$$

2) $\sigma([T_{Q_1}, T_{Q_2}]) = -i\{\sigma(T_{Q_1}), \sigma(T_{Q_2})\}$, where the Poisson brackets are considered on Σ^+ with respect to the symplectic form ω.

3) If Q is a PDO of order m and $\sigma(Q)|_{\Sigma} = 0$, then T_Q is an operator of order $m - 1$, that is, there is a PDO Q_1 of order $m - 1$ such that $T_Q = T_{Q_1}$ up to infinitely smoothing operators.

These properties easily follow from the above microlocal model of the operator d_b''. We clarify them, replacing, for simplicity, d_b'' by the system of operators D_j^0 and the Szegö projection S by its *microlocal model* S_0, which is the orthogonal projection in $L^2(\mathbb{R}^{2n-1})$ on the subspace of solutions of the equations $D_j^0 u = 0$.

We introduce the *Hermite operator*

$$(H_0 f)(t, q) = (2\pi)^{-n} \int e^{it\tau - |q||\tau|/2} \left(\frac{|\tau|}{2\pi}\right)^{(n-1)/4} \hat{f}(\tau)\, d\tau \ ,$$

where $\hat{f}$ is the Fourier transform of $f(t)$. It can be verified immediately that for any $f(t) \in C^\infty(\mathbb{R}^n)$ the function $u(t, q) = H_0 f$ satisfies the system of equations $D_j^0 u = 0$ and the operator H_0 maps $L^2(\mathbb{R}^n)$ isometrically on to the subspace of solutions of this system in $L^2(\mathbb{R}^{2n-1})$. Consequently, $H_0^* H_0 = 1$ in $L^2(\mathbb{R}^n)$, and $H_0 H_0^* = S_0$ is an orthogonal projection in $L^2(\mathbb{R}^{2n-1})$, which is the analogue of the Szegö projection. For the analogues of Toeplitz operators we have

$$T_Q^0 = H_0 H_0^* Q H_0 H_0^* = H_0 P_0 H_0^* \ ,$$

where $P_0 = H_0^* Q H_0 : C^\infty(\mathbb{R}^n) \to C^\infty(\mathbb{R}^n)$. It can be verified immediately that P_0 is a PDO on $\mathbb{R}^n$ and that $\sigma(P_0) = \sigma(Q)|_{\Sigma_0^+}$, where $\Sigma_0^+ = T^*(\mathbb{R}^n)\backslash 0 \subset T^*(\mathbb{R}^{2n-1}) \setminus 0$. Hence, the correspondence $T_Q^0 \mapsto P_0 = H_0^* P_0 H_0$ defines an isomorphism between the algebra of Toeplitz operators T_Q^0 and the algebra of PDOs on $\mathbb{R}^n$, which implies the properties 1)–3).

We mention that the microlocal model of the operator d_b'' at the points of Σ^- leads to the system $(D_j^0)^* u = \partial u/\partial q_j + y_j |D_t| u = 0$, which does not have solutions in L^2, since the corresponding analogue of the Szegö projection is equal to zero. This clarifies why the manifold Σ^- is not essential in the definition of the symbol of a Toeplitz operator. We also mention the analogy between the D_j^0 and the annihilation operators, and between the $(D_j^0)^*$ and the creation operators (see Chap. 3, Sect. 1).

In view of the properties 1)–3), the elliptic theory can be generalized in the obvious way to Toeplitz operators. T_Q is called elliptic if $\sigma(T_Q) \neq 0$ everywhere on Σ^+. Ellipticity implies the existence of a parametrix, that is, a Toeplitz operator R such that $S - RT_Q$ and $S - T_Q R$ are infinitely smoothing operators. This means that T_Q, regarded on the subspace of values of the projection S, is a Fredholm operator, and we obtain the index formula

$$\operatorname{ind} T_Q = \operatorname{tr}(S - RT_Q) - \operatorname{tr}(S - T_Q R) \,.$$

The *index theorem for Toeplitz operators* is analogous to the Atiyah-Singer index theorem for elliptic PDOs. In its simplest version it reads as follows.

Theorem 2.5. *Let T_Q be an elliptic matrix of Toeplitz operators on the boundary of a strictly pseudoconvex domain $\Omega \subset \mathbb{C}^n$, and let $\sigma(z)$ be a matrix-valued function on $\partial\Omega$ equal to the symbol $\sigma(T_Q)$ at the point $(z, \alpha(z)) \in \Sigma^+$. Then*

$$\operatorname{ind} T_Q = -\frac{1}{(2\pi i)^n} \frac{(n-1)!}{(2n-1)!} \int\limits_{\partial\Omega} (\sigma^{-1} d\sigma)^{2n-1} \,. \tag{2.30}$$

The concepts of tangent Cauchy-Riemann operator and Szegö projection of Toeplitz operators can be generalized in a natural way to the case of holomorphic bundles over a stricty pseudoconvex domain Ω in a Stein manifold. Let E^0 and E^1 be holomorphic vector bundles over Ω, and $\sigma(z)$ a function on $\partial\Omega$ with values in $\operatorname{Hom}(E^-|_{\partial\Omega}, E^1|_{\partial\Omega})$ and equal to the symbol of the elliptic Toeplitz operator

$$T_Q : C^\infty(\partial\Omega, E^0|_{\partial\Omega}) \to C^\infty(\partial\Omega, E^1|_{\partial\Omega}) \,.$$

The function $\sigma(z)$ can be extended to a homomorphism of the bundles E^0 and E^1 over Ω by means of a truncating function ρ and defines the distinguishing element $d(T_Q)$, which is the virtual bundle $\{E^0, E^1, \rho\sigma\} \in K^{\mathrm{comp}}(\Omega)$ with compact support in Ω.

Theorem 2.6. *We have*

$$\operatorname{ind} T_Q = \int\limits_\Omega \operatorname{ch} d(T_Q) \mathcal{T}(\Omega) \,,$$

where $\mathcal{T}(\Omega)$ is the Todd class of the complex tangent bundle of the manifold Ω.

Chapter 3
Deformation Quantisation and the Index

A manifold M is called *symplectic* if there is a *symplectic form* on it, that is, a closed non-degenerate 2-form $\sigma = \frac{1}{2}\sigma_{ij}dx^i \wedge dx^j$. Such a manifold is necessarily even-dimensional: $\dim M = 2n$. The *Poisson bracket* operation is defined on the space $C^\infty(M)$ by $\{f,g\} = \sigma^{ij}(\partial f/\partial x^i)(\partial g/\partial x^j)$, where σ^{ij} are the elements of the inverse of σ_{ij}, and defines a Lie algebra structure on $C^\infty(M)$. In addition, on a space of functions there is also a commutative algebra structure.

We denote by $Z = C^\infty(M)[[h]]$ the set of formal power series

$$a(x,h) = \sum_{k=0}^\infty h^k a_k(x)$$

in the formal variable h, with coefficients $a_k(x) \in C^\infty(M)$. By a *deformation quantisation* on a symplectic manifold M we understand the structure of an associative algebra with identity over $\mathbb{C}$ on the set Z, with the usual linear operations and an associative product $a * b$ such that

1) the coefficients of the series $a * b$ are expressed polynomially in terms of those of a and b and their partial derivatives with respect to local coordinates;

2) $a * b = a_0(x)b_0(x) + \cdots$;

3) $[a,b] = a*b - b*a = -ih\{a_0(x), b_0(x)\} + \cdots$, where the dots denote terms containing higher powers of h. This algebra is called the *algebra of quantum observables*. The condition 2) means that it is a deformation of the commutative algebra of functions, while 3) expresses the so-called *correspondence principle*.

Deformation quantisation has been considered in detail in Bayen, Flato, Fronsdal, Lichnerowicz and Sternheimer (1978) (see also the survey Karasev and Maslov (1984)). In the former, a construction of quantisation is proposed, and conditions on M are obtained under which the construction is possible. These conditions are difficult to prove in the general case; a sufficient condition is that $H_3(M)$ should be trivial. Later it was proved in Lecomte and De Wilde (1983) that the $*$-product exists for any symplectic manifold.

We propose another quantisation construction, which is possible on any symplectic manifold. Then in the algebra of quantum observables we can introduce a trace, which takes values in the set of formal Laurent series in h with finitely many negative powers. Using the trace, we introduce the concept of index by means of (1.8). We can prove an index formula analogous to the Atiyah-Singer one. Here the index is a polynomial in $1/h$ whose degree does not exceed n. From this we derive necessary conditions for the existence of an operator representation of the algebra of quantum observables: h can take only numerical values for which any index is an integer. In 5.3, Chap. 1 similar considerations yielded a necessary condition for the existence of a spinor structure. In specific examples, the conditions for the index to be an integer coincide with the well-known quantisation conditions obtained in operator constructions.

§1. The Algebra of Quantum Observables

1.1. The Weyl Symbols. The material in this article has much in common with that in 5.3, Chap. 1. Using the terminology of physics, we can say that the algebra of Weyl symbols is the boson analogue of the Clifford algebra.

Let Σ be a real symplectic space of dimension $2n$, and σ a non-degenerate skew-symmetric bilinear form on Σ, called a *symplectic form*. Also, let g be a symmetric positive definite form given on Σ.

We introduce the *algebra of Weyl symbols* $S^\infty(\Sigma)$. Let $S^m = S^m(\Sigma)$ be the space of complex-valued functions $a(y) \in C^\infty(\Sigma)$ which for any multi-index $\gamma = (\gamma_1, \gamma_2, \ldots, \gamma_{2n})$ satisfy the estimates

$$\left| \frac{\partial^\gamma a}{\partial y^\gamma} \right| \le C_\gamma (1 + |y|^2)^{(m-|\gamma|)/2} ,$$

where $|y|^2 = g(y, y)$. These estimates are analogous to (2.14); the difference is that here we do not distinguish between the variables x and ξ, and the pair (x, ξ) is denoted by the single symbol y. We denote by S^∞ and $S^{-\infty}$ the union and intersection of all the S^m, $m \in \mathbb{R}$, respectively. In particular, $S^{-\infty}$ is the Schwarz space.

For a fixed $\lambda > 0$ we define multiplication in S^∞ by the formula

$$(a \circ b)(y) = \frac{(\pi\lambda)^{-2n}}{(2n)!} \int_{\Sigma \times \Sigma} \exp\left(\frac{2i}{\lambda} \sigma(t, \tau) \right) a(y + t) b(y + \tau) \sigma^{2n}(dt, d\tau) , \quad (3.1)$$

where $\sigma^{2n}(dt, d\tau)$ is the exterior power of the 2-form $\sigma_{ij} dt^i \wedge d\tau^j$ (the orientation on $\Sigma \times \Sigma$ is given by the form σ^{2n}). The integral is understood as the limit as $\varepsilon \to 0+$ of the same integral with the two extra terms $-\varepsilon g(t, t) - \varepsilon g(\tau, \tau)$ in the index of the exponential. If we expand $a(y + t) b(y + \tau)$ in a formal Taylor series in t and τ and integrate term by term, then we arrive at

$$(a \circ b)(y) \sim \sum_{k=0}^{\infty} \frac{1}{k!} \left(-\frac{i\lambda}{2} \right)^k \sigma^{i_1 j_1} \sigma^{i_2 j_2} \ldots \sigma^{i_k j_k}$$

$$\times \frac{\partial^k a(y)}{\partial y^{i_1} \partial y^{i_2} \ldots \partial y^{i_k}} \frac{\partial^k b(y)}{\partial y^{j_1} \partial y^{j_2} \ldots \partial y^{j_k}} ,$$

which we abbreviate to

$$\sum_{k=0}^{\infty} \frac{1}{k!} \left(-\frac{i\lambda}{2} \right)^k \left\langle \bigotimes^k \sigma^{-1}, \partial^k a \otimes \partial^k b \right\rangle . \quad (3.2)$$

Here y^i, $i = 1, 2, \ldots, 2n$, are the coordinates of the vector y with respect to a fixed basis, σ^{ij} the elements of the inverse σ^{-1} of the matrix σ_{ij} of the symplectic form σ and the convention of summation from 1 to $2n$ over repeated indices is understood. The expansion is asymptotic in the sense that for $a \in S^{m_1}$ and $b \in S^{m_2}$ the difference between $a \circ b$ and the Nth partial sum of (3.2) belongs to $S^{m_1+m_2-2N}$. For polynomials, the expansion (3.2) is an exact equality. In particular,

$$y^i \circ y^j = y^i y^j - \frac{i\lambda}{2}\sigma^{ij} \, ,$$

from which we obtain the commutation relations

$$[y^i, y^j] = y^i \circ y^j - y^j \circ y^i = -i\lambda\sigma^{ij} \, . \tag{3.3}$$

Multiplication induces on S^∞ a structure of associative algebra with identity (the function identically equal to 1) and an involution given by complex conjugation. As is easily seen, the centre of the algebra consists only of constants.

The real symplectic transformations s of Σ generate automorphisms A_s of the algebra $S^\infty : (A_s a)(y) = a(s^{-1}y)$.

For the symbols $a(y) \in S^m$, $m < -2n$, we define the *trace*

$$\operatorname{tr} a(y) = \frac{(2\pi\lambda)^{-n}}{n!} \int\limits_\Sigma a(y)\sigma^n \, ,$$

where σ^n is the exterior power of the 2-form $\sigma = \frac{1}{2}\sigma_{ij}dy^i \wedge dy^j$ (the orientation of Σ is given by the form σ^n). Using the multiplication formula (3.1), we can show that $\operatorname{tr} a \circ b = \operatorname{tr} a(y)b(y)$, where $a(y)b(y)$ is the usual multiplication of functions. From this we deduce that the trace of the product does not depend on the order of the two factors. The symbols in $S^{-\infty}$ form an ideal in S^∞ on which the trace is certainly defined. We call it the *trace ideal*.

We now introduce the dependence on the parameter $x \in M$. Let M be a manifold and Σ a symplectic bundle over M. Then we can define a locally trivial bundle $S^\infty(\Sigma)$ over M whose fibre at x is the algebra $S^\infty(\Sigma_x)$, and the transition functions are given by the automorphisms $A_{s(x)}$, where $s(x)$ is the transition function of Σ, which takes values in the group $\operatorname{Sp}(2n)$ of symplectic linear transformations. The sections of $S^\infty(\Sigma)$ are given by smooth functions $a(x, y)$, $x \in M$, $y \in \Sigma_x$, which for x fixed belong to $S^\infty(\Sigma_x)$ as functions of y. It is clear that the fibre multiplication $\circ$ is invariantly defined by (3.1), since σ does not change under symplectic transformations.

Let ∂ be a *symplectic connection* on the bundle Σ, that is, a connection that preserves the symplectic form σ. This means that

$$d\sigma(u_1, u_2) = \sigma(\partial u_1, u_2) + \sigma(u_1, \partial u_2)$$

for any sections u_1 and u_2 of Σ. A local reference frame $e_1, e_2, \ldots, e_{2n}$ of Σ is called *symplectic* if the coefficients $\sigma_{ij} = \sigma(e_i, e_j)$ are constant. In a local symplectic frame the connection ∂ is given by a matrix 1-form Γ with elements Γ_j^i, and the condition that ∂ is symplectic means that the forms $\Gamma_{ij} = \sigma_{ik}\Gamma_j^k$ are symmetric with respect to i and j.

With the connection ∂ in Σ we associate a connection (denoted by the same symbol) on the algebra bundle $S^\infty(\Sigma)$, which in a local symplectic frame is defined by a formula similar to (1.27), namely,

$$\partial a = da + [\Gamma_w, a] \,, \tag{3.4}$$

where d is the differential with respect to x and $\Gamma_w = (1/(2\lambda))\Gamma_{ij}y^i y^j$ a local 1-form with values in $S^\infty(\Sigma)$, called a *Weyl connection form*. For local linear forms $y^i = \sigma^{ik}\sigma(e_k, y)$ regarded as local sections of the bundle $S^\infty(\Sigma)$, the covariant differential ∂y^i defined by (3.4) coincides with $-\Gamma_j^i y^j$, which is the covariant differential of the form y^i regarded as a section of the dual bundle of Σ. For any section $a \in C^\infty(S^\infty(\Sigma))$ we have $\partial^2 a = [\Omega_w, a]$, where

$$\Omega_w = d\Gamma_w + \Gamma_w^2 = \frac{i}{2\lambda}\Omega_{ij}y^i y^j \tag{3.5}$$

is a global 2-form with values in $S^\infty(\Sigma)$, called the *Weyl curvature* of the connection ∂. Here $\Omega_{ij} = \sigma_{ik}\Omega_j^k$, where Ω_j^k is the curvature of the symplectic connection on Σ. We mention that the form Γ_w (as well as Ω_w) is not defined uniquely, but up to a term with values in the centre; this non-uniqueness is eliminated by means of the condition $\Gamma_w|_{y=0} = 0$ ($\Omega_w|_{y=0} = 0$), which is called the *Weyl normalisation condition*.

We now consider the question of operator representations of $S^\infty(\Sigma)$. There is a well-known representation of the algebra $S^\infty(\Sigma)$ by means of Weyl PDOs in $\mathbb{R}^n$ (Hörmander (1979)), but we need another one, namely, a representation in the Fock space.

On any symplectic bundle there is a Hermitian structure (Lichnerowicz (1955)). This means that there is an n-dimensional complex bundle E over M equipped with a Hermitian inner product $\langle \ , \ \rangle$, such that Σ is isomorphic to the realification of E, with $\sigma(u, v) = \operatorname{Im}\langle u, v\rangle$ and $g(u, v) = \operatorname{Re}\langle u, v\rangle$. Thus, on Σ there is an operator J of complex structure, corresponding to multiplication by i in the complex bundle E. Obviously,

$$g(u, v) = \sigma(Ju, v) = \sigma(u, Jv) \,, \qquad \sigma(u, v) = g(u, Jv) = -g(Ju, v) \,,$$

and $J^2 = -1$.

Let z^α and $\bar{z}^\alpha$ ($\alpha = 1, 2, \ldots, n$) be the complex coordinates of the vector $y \in \Sigma_x$ in an orthonormal reference frame e_α of E. Considering them as local sections of the bundle $S^\infty(\Sigma)$, we find that

$$z^\alpha \circ z^\beta = z^\alpha z^\beta \,, \qquad \bar{z}^\alpha \circ z^\beta = \bar{z}^\alpha z^\beta + \lambda\delta^{\alpha\beta} \,,$$

from which we obtain the canonical commutation relations

$$[z^\alpha, z^\beta] = [\bar{z}^\beta, \bar{z}^\alpha] = 0 \,, \qquad [\bar{z}^\alpha, z^\beta] = 2\lambda\delta^{\alpha\beta} \tag{3.6}$$

(cf. (1.26)).

The multiplication formula (3.1) yields

$$\exp\left(-\frac{t_1}{\lambda}|y|^2\right) \circ \exp\left(-\frac{t_2}{\lambda}|y|^2\right) = (1+t_1 t_2)^{-n}\exp\left(-\frac{t_1+t_2}{1+t_1 t_2}|y|^2\right)$$

for $t_1 > 0$ and $t_2 > 0$, where $|y|^2 = g(y,y) = z^\alpha \bar{z}^\alpha$. Expanding in powers of t_1, we arrive at the equality

$$|y|^2 \circ \exp\left(-\frac{t}{\lambda}|y|^2\right) = \{(1-t^2)|y|^2 + n\lambda t\}\exp\left(-\frac{t}{\lambda}|y|^2\right) . \tag{3.7}$$

For $t_1 = t_2 = 1$ we also find that the function $p_0(y) = 2^n \exp(-(1/\lambda)|y|^2) \in S^{-\infty}$ is such that $p_0 \circ p_0 = p_0$, $\operatorname{tr} p_0 = 1$ and $\bar{z}^\alpha \circ p_0 = p_0 \circ z^\alpha = 0$. Using the terminology of physics, we call p_0 the *vacuum projection*, and z^α and $\bar{z}^\alpha$ *creation* and *annihilation operators*, respectively. Under symplectic transformations of Σ corresponding to unitary transformations of the space E, the vacuum projection remains unchanged so that p_0 defines a global section of the bundle $S^\infty(\Sigma)$. If ∂ is a symplectic connection on Σ corresponding to a Hermitian connection on E, then the associated connection on $S^\infty(\Sigma)$ satisfies $\partial p_0 = 0$.

For a symplectic space Σ with a Hermitian structure we define the *Fock space* F as the left ideal of the algebra $S^\infty(\Sigma)$ generated by the vacuum projection. For $u = a(y) \circ p_0(y) \in F$ and $v = b(y) \circ p_0(y) \in F$ we define the inner product $(u,v) = \operatorname{tr} a(y) \circ p_0(y) \circ b^*(y)$. Completing F with respect to this inner product, we obtain a Hilbert space F^2. The elements of the form $P(z^\alpha) \circ p_0$, where P is a polynomial in the creation operators, are dense in F^2. The symbol $b(y) \in S^\infty(\Sigma)$ defines an operator in F which acts as left multiplication, that is, $bu = b(y) \circ a(y) \circ p_0(y)$, on vectors $u = a(y) \circ p_0(y)$. In particular, the symbol $p_0(y)$ defines the operator of projection on the vector $1 \circ p_0(y) \in F$, called the *vacuum vector*. It can be proved that for such a representation the symbols in S^m $(m < -2n)$ are mapped into trace class operators in F^2 and the trace in the algebra S^∞ coincides with the operator trace.

These constructions also carry over in the obvious way to a bundle Σ equipped with a Hermitian structure; the result is the *Fock bundle* over M, which we denote by the same symbol F (or F^2). The vacuum vector defines a global non-zero section p_0 of F.

The connection ∂ in $S^\infty(\Sigma)$ defined by (3.4) generates a connection on the bundle F. By analogy with (1.28), for a section $u = a \circ p_0 \in C^\infty(F)$ we set $\partial u = \partial(a \circ p_0) \circ p_0$. If ∂ is associated with a Hermitian connection, then $\partial p_0 = 0$, and we find that in a local orthonormal frame of the Hermitian bundle E

$$\partial u = du + \Gamma_w \circ a \circ p_0 - a \circ p_0 \circ \Gamma_w \circ p_0 = du + (\Gamma_w - \operatorname{tr} p_0 \circ \Gamma_w \circ p_0)u .$$

The form $\Gamma_n = \Gamma_w - \operatorname{tr} p_0 \circ \Gamma_w \circ p_0$ is called the *Wick (normal) connection form*. It is uniquely defined by the Wick normalisation condition $\Gamma_n \circ p_0 = p_0 \circ \Gamma_n = 0$. The corresponding curvature

$$\Omega_n = d\Gamma_n + \Gamma_n^2 = \Omega_w - \operatorname{tr} p_0 \circ \Omega_w \circ p_0$$

also satisfies the Wick normalisation $\Omega_n \circ p_0 = p_0 \circ \Omega_n = 0$ and is called the *Wick (normal) curvature.* If Ω_β^α is the curvature of the Hermitian connection in the complex bundle E and $\Omega_{\alpha\beta} = \delta_{\alpha\gamma}\Omega_\beta^\gamma$, then

$$\Omega_n = -\frac{1}{2\lambda}\Omega_{\alpha\beta}z^\beta \circ \bar{z}^\alpha = \Omega_w + \frac{1}{2}\Omega_\alpha^\alpha \ . \tag{3.8}$$

We now indicate a construction of a Fredholm family (see Chap. 2, Sect. 2) in Fock bundles, connected with the Bott generator. Let E be a complex Hermitian bundle and Σ a symplectic bundle obtained by the the realification of E. On the space of the bundle Σ we consider a homomorphism $b(z) = \varepsilon(z) + i(z) : \Lambda^+(E^*) \to \Lambda^-(E^*)$ (see the definition of the Bott generator in 3.3, Chap. 1). The vector z is regarded as a vector of the fibre E_x; at the same time, it can be regarded as a vector of the fibre Σ_x. In an orthonormal frame e_α of E we have $b(z) = z^\alpha i(e_\alpha) + \bar{z}^\alpha \varepsilon(e_\alpha)$.

We now regard $b(z)$ as a section of the bundle of Weyl algebras $S^\infty(\Sigma)$, that is, we assume that z^α and $\bar{z}^\alpha$ are symbols. Then we obtain the family of operators

$$\beta = \{F^2 \otimes \Lambda^+(E^*), F^2 \otimes \Lambda^-(E^*), b\} \ . \tag{3.9}$$

We claim that the kernel of $b(z)$ is one-dimensional and generated by the vacuum vector, while its cokernel is trivial. Computing the square of the operator $b(z)$ in $F^2 \otimes \Lambda(E^*)$ by means of (1.13) and the commutation relations (3.6), we obtain

$$\Delta = b \circ b = |z|^2 + \lambda\delta^{\alpha\beta}(\varepsilon(e_\beta)i(e_\alpha) - i(e_\alpha)\varepsilon(e_\beta)) \ .$$

This operator maps the space $F^2 \otimes \Lambda^k(E^*)$ into itself and coincides on it with the operator $\Delta_k = |z|^2 + \lambda(2k - n)$. Δ_k is invertible for $k \neq 0$, and its inverse admits the representation

$$\Delta_k^{-1} = \frac{1}{\lambda}\int_0^1 e^{-t|z|^2/\lambda}(1 + t)^{n-k-1}(1 - t)^{k-1}\,dt \ .$$

Using (3.7) and integration by parts, we easily verify that $\Delta_k \circ \Delta_k^{-1} = \Delta_k^{-1} \circ \Delta_k = 1$. For $k = 0$ the inverse does not exist, since Δ_0 annihilates the vacuum vector. We define a quasi-inverse operator by

$$\Delta_0^{-1} = \frac{1}{\lambda}\int_0^1 \frac{e^{-t|z|^2/\lambda} - e^{-|z|^2/\lambda}}{1 - t}(1 + t)^{n-1}\,dt \ ;$$

as can be easily verified, this operator satisfies $\Delta_0 \circ \Delta_0^{-1} = \Delta_0^{-1} \circ \Delta_0 = 1 - p_0.$

222   B.V. Fedosov

Defining Δ^{-1} as the dircet sum $\bigoplus_{k=0}^{n} \Delta_k^{-1}$ and setting

$$b^{-1} = b \circ \Delta^{-1} : F^2 \otimes \Lambda^-(E^*) \to F^2 \otimes \Lambda^+(E^*) \,,$$

we arrive at the following assertion.

Theorem 3.1. *A Fredholm family $B = \{F^2 \otimes \Lambda^+(E^*), F^2 \otimes \Lambda^-(E^*), b(z),$ $b^{-1}(z)\}$ can be split into a direct sum $B = B_0 \oplus B_1$, where $B_0 = \{1, 0, 0, 0, \}$, 1 denotes the trivial one-dimensional bundle generated by the vacuum vector in F^2, and B_1 is a trivial infinite-dimensional family in the orthogonal complement. The connection ∂ preserves this decomposition and annihilates the vacuum vector.*

1.2. The Bundle of Formal Weyl Algebras. Let Σ be a symplectic bundle of dimension $2n$ over a manifold M, and σ a symplectic form on the fibres of Σ. We define the *bundle $W = W(\Sigma)$ of formal Weyl algebras* over M as follows. The sections of W are functions $a(x, y, h)$, where $x \in M$, $y \in \Sigma_x$ and h is a formal variable, which are understood as formal power series in h; in other words,

$$a(x, y, h) = \sum h^k a_{k,l}(x, y) = \sum h^k a_{k i_1 i_2 \ldots i_l}(x) y^{i_1} y^{i_2} \ldots y^{i_l} \,, \qquad (3.10)$$

where $a_{k,l}(x, y)$ are smooth complex-valued functions on the total space of Σ, which on the fibres Σ_x are homogeneous polynomials of degree l (we adopt the convention of summation from 1 to $2n$ over the repeated indices $i_1, i_2, \ldots, i_l$).We ascribe the power 1 to the y^i and 2 to h; the terms of the series are ordered with respect to their total degree $2k + l$. The powers of h may be positive and negative, but are subject to the condition that all the total degrees $2k + l$ are bounded below and that there are only finitely many terms of any given total degree. The sections form an associative algebra with identity over $\mathbb{C}$ with respect to the usual operations of addition and scalar multiplication and a multiplication $a \circ b$ defined by means of a formula similar to (3.2), where the series are formal:

$$a(x, y, h) \circ b(x, y, h) = \sum_{k=0}^{\infty} \left(-\frac{ih}{2} \right)^k \frac{1}{k!} \left\langle \bigotimes^k \sigma^{-1}, \partial_y^k a \otimes \partial_y^k b \right\rangle$$

$$= \sum_{k=0}^{\infty} \left(-\frac{ih}{2} \right)^k \frac{1}{k!} \sigma^{i_1 j_1} \sigma^{i_2 j_2} \ldots \sigma^{i_k j_k} \frac{\partial^k a}{\partial y^{i_1} \partial y^{i_2} \ldots \partial y^{i_k}} \frac{\partial^k b}{\partial y^{j_1} \partial y^{j_2} \ldots \partial y^{j_k}} \,.$$

Fixing a point $x \in M$, we obtain the algebra W_x, which is the fibre of the bundle W at x.

$W(x)$ has much in common with the bundle $S^\infty(\Sigma)$ of Weyl symbols discussed in the previous article. The only difference between them is that the numerical parameter $\lambda > 0$ is replaced here by the formal variable h, and we consider 'functions' of h that are formal Laurent series in h whose coefficients are polynomials in y. A symbol $b(z)$ from the Bott family B can also be regarded as a section of $W(\Sigma)$, while its parametrix $b^{-1}(z) \in C^\infty(S^\infty(\Sigma))$ cannot.

To simplify the notation, we denote the algebra of the sections of $C^\infty(W)$ by W if this does not create ambiguity. The centre Z of W consists of the sections $a(x, h)$ independent of y. By $\varepsilon : W \to Z$ we denote the projection on the centre, that is, $\varepsilon(a(x, y, h)) = a(x, 0, h)$. In W there is a filtration $\ldots \supset W_{-1} \supset W_0 \supset W_1 \supset W_2 \supset \ldots$ with respect to the smallest total degree of the terms of the series. Such a filtration also exists in the centre, but only with respect to the even degrees.

In a local reference frame of Σ the coordinates y^i ($i = 1, 2, \ldots, 2n$) of u, regarded as linear forms on the fibres Σ_x, are non-commutative generators of the algebra W over the centre, in the sense that any monomial $y^{i_1} y^{i_2} \ldots y^{i_k}$ is equal to the symmetrized product $\circ$ of the generators $y^{i_1}, y^{i_2}, \ldots, y^{i_k}$, that is,

$$y^{i_1} y^{i_2} \ldots y^{i_k} = \frac{1}{k!} \sum_s y^{i_{s(1)}} \circ y^{i_{s(2)}} \circ \ldots \circ y^{i_{s(k)}} \,,$$

where s is a permutation of $1, 2, \ldots, k$. Consequently, the series (3.10) can also be written in the form

$$a = \sum a_{i_1 i_2 \ldots i_k} y^{i_1} \circ y^{i_2} \circ \ldots \circ y^{i_k} \,, \tag{3.11}$$

where the coefficients $a_{i_1 i_2 \ldots i_k} \in Z$ are symmetric in the subscripts $i_1, i_2, \ldots, i_k$, and this representation is unique. The generators y^i satisfy commutation relations analogous to (3.3), namely,

$$[y^i, y^j] = -ih\sigma^{ij}(x) \,.$$

We also consider the bundle $W \otimes \Lambda$ of differential forms with values in W. First we consider central forms. A differential p-form with values in Z is a formal series $\varphi = \sum_{k=-N}^{\infty} h^k \varphi_k$, where φ_k are scalar differential p-forms on M. The exterior multiplication and exterior differential of such forms is defined term by term.

A differential p-form a with values in W is given locally by a series of the form (3.11), where the coefficients $a_{i_1 i_2 \ldots i_k}$ are local p-forms on M with values in Z and behave like tensors under a linear change of the generators y^i. The product $a \circ b$ of two forms $a \in W \otimes \Lambda^p$ and $b \in W \otimes \Lambda^q$ is defined term by term, that is,

$$a \circ b = \sum_{k,l=0}^{\infty} a_{i_1 i_2 \ldots i_k} \wedge b_{j_1 j_2 \ldots j_l} y^{i_1} \circ y^{i_2} \circ \ldots \circ y^{i_k} \circ y^{j_1} \circ y^{j_2} \circ \ldots y^{j_l} \,,$$

and their commutator is defined by

$$[a, b] = a \circ b - (-1)^{pq} b \circ a \,.$$

A filtration W_k in W generates a filtration $W_k \otimes \Lambda$ in the algebra $W \otimes \Lambda$ of differential forms. The projection $\varepsilon : W \otimes \Lambda \to Z \otimes \Lambda$ on central forms associates with a of the form (3.11) with symmetric coefficients $a_{i_1 i_2 \ldots i_k}$ the term of the series with $k = 0$.

By a *connection on the bundle* W we understand a linear mapping $\mathcal{D} : W \otimes \Lambda^p \to W \otimes \Lambda^{p+1}$ with the following properties:

1) $\mathcal{D}(a \circ b) = (\mathcal{D}a) \circ b + (-1)^p a \circ \mathcal{D}b$, where a is a p-form;

2) for central forms, $\mathcal{D}$ coincides with exterior differentiation.

As in 1.1, with a symplectic connection ∂ on Σ we associate a connection, also denoted by ∂, which acts on the generators y^i according to the formula $\partial y^i = -\Gamma^i_j y^j$. The properties 1) and 2) guarantee the existence of a unique extension of ∂ to any sections of $W \otimes \Lambda$. In a symplectic local reference frame (for which $\sigma_{ij} = \mathrm{const}$), the connection ∂ is written in a form analogous to (3.4), namely,

$$\partial a = da + \left[\frac{i}{2h}\Gamma_{ij}y^i y^j, a\right], \qquad \Gamma_{ij} = \sigma_{ik}\Gamma^k_j \ .$$

An arbitrary connection $\mathcal{D}$ in $W \otimes \Lambda$ can be written as

$$\mathcal{D}a = d + [\Gamma, a] = \partial a + [\Delta\Gamma, a] \ ,$$

where $\Delta\Gamma$ is a globally defined 1-form with values in W and $\Gamma = (i/(2h)) \times \Gamma_{ij}y^i y^j + \Delta\Gamma$ is a local 1-form called the *connection form*. In general, Γ and $\Delta\Gamma$ are not uniquely defined, but up to central forms. Unless stated otherwise, we assume that Γ and $\Delta\Gamma$ satisfy the Weyl normalisation property $\Gamma = \varepsilon\Delta\Gamma = 0$. When it becomes necessary to emphasize this normalisation, we write Γ_w and $\Delta\Gamma_w$, respectively.

The (*Weyl*) *curvature of the connection* $\mathcal{D}$ is defined, as usual, by

$$\Omega = \Omega_w = d\Gamma_w + \Gamma_w^2 = \frac{i}{2h}\Omega_{ij}y^i y^j + \partial\Delta\Gamma_w + (\Delta\Gamma_w)^2 \ ,$$

where $\Omega_{ij} = \sigma_{ik}\Omega^k_j$ is the curvature of the symplectic connection ∂ on the bundle Σ with the upper index omitted. The curvature satisfies Bianchi's identity $\mathcal{D}\Omega = 0$ and the equality $\mathcal{D}^2 a = [\Omega, a]$, which is established by direct verification.

1.3. Abelian Connections and Quantisation. In what follows we consider connections $\mathcal{D} = \partial + [\Delta\Gamma, \cdot]$ of a more specialized form, where

$$\Delta\Gamma = \frac{i}{h}\sigma_{ij}y^i\theta^j + \zeta \ ,$$

θ^j are local scalar 1-forms on M, and ζ is a section of the bundle $W_0 \otimes \Lambda^1$ such that $\varepsilon\zeta = 0$. Such connections reduce a filtration by not more than 1. Thus,

$$\mathcal{D}a = -\delta a + \partial a + [\zeta, a] \ , \tag{3.12}$$

where $\delta a = -[(i/h)\sigma_{ij}y^i\theta^j, a] = \theta^i \wedge (\partial a/\partial y^i)$. In particular,

$$\mathcal{D}y^i = -\theta^i - \Gamma_j^i y^j + [\zeta, y^i]\,,$$

from which we see that if we change to another frame $e_i' = e_j s_i^j$ of Σ, then the θ^i are replaced by the forms $(\theta^i)'$ satisfying $\theta^i = s_j^i(\theta^j)'$. Consequently, a connection $\mathcal{D}$ of the form (3.12) defines a homomorphism $\delta : \Sigma^* \to T^*M$ of the bundle Σ^* of linear forms, given by $\delta : y^i \mapsto \theta^i$. A connection $\mathcal{D}$ of the form (3.12) is called *non-degenerate* if this homomorphism is an isomorphism. Hence, if there is a non-degenerate connection $\mathcal{D}$ on the bundle $W(\Sigma)$ over the manifold, then the tangent bundle TM is isomorphic to Σ. In particular, $\dim M = \dim \Sigma = 2n$. The bundle Σ can be identified with TM; however, for various reasons this is not convenient.

For the curvature of $\mathcal{D}$ we obtain

$$\Omega = -\frac{i}{2h}\sigma_{ij}\theta^i \wedge \theta^j + \partial\left(\frac{i}{h}\sigma_{ij}y^i\theta^j\right) + \frac{i}{2h}\Omega_{ij}y^iy^j - \delta\zeta + \partial\zeta + \zeta^2\,. \quad (3.13)$$

The quantisation construction proposed below is based on the concept of Abelian connection. A connection $\mathcal{D}$ on $W(\Sigma)$ is called *Abelian* if its curvature is a central form (see 3.1, Chap. 1). This is equivalent to the condition that $\mathcal{D}^2 a = 0$ for any section.

If $\mathcal{D}$ is non-degenerate and Abelian, then, by Bianchi's identity, $\mathcal{D}\Omega = 0$ and the dominant term $(1/(2h))\sigma_{ij}\theta^i \wedge \theta^j$ in (3.13) is of degree -2 and defines a non-degenerate closed 2-form $\delta\sigma = \frac{1}{2}\sigma_{ij}\theta^i \wedge \theta^j$ on M. Therefore, a non-degenerate Abelian connection may exist only on a symplectic manifold M. Here we prove the existence of non-degenerate Abelian connections for any symplectic manifold. We do not specifically discuss degenerate Abelian connections, although they are of considerable interest. As an auxiliary result, in the next article we construct a degenerate Abelian connection on a manifold of the form $M \times \mathbb{R}$, where M is a symplectic manifold. We mention that, in general, in the symbol bundles $S^\infty(\Sigma)$ considered in 1.1 there are no Abelian connections, which makes it necessary for us to go over to bundles of formal Weyl algebras $W(\Sigma)$.

Theorem 3.2. *If the mapping $\delta : \Sigma^* \to T^*M$ is a bundle isomorphism and the 2-form $\delta\sigma = \frac{1}{2}\sigma_{ij}\theta^i \wedge \theta^j$ is closed, then for any closed central form $\varphi \in Z_0 \otimes \Lambda^2$ there is an Abelian connection $\mathcal{D}$ on the bundle $W(\Sigma)$, with Weyl curvature*

$$\Omega = -\frac{i}{h}\delta\sigma + \varphi\,. \quad (3.14)$$

Proof. We introduce a local basis X_i of vector fields on M so that $\theta^i(X_j) = \delta_j^i$, and define two operators δ and δ^* on the bundle $W \otimes \Lambda$. The operator δ has already been introduced, being given by $\delta a = \theta^i \wedge (\partial a/\partial y^i)$. We set $\delta^* a = y^i i(X_i)a$, where $i(X_i)$ is the contraction operator of the X_i and the form a (see Chap. 1, Sect. 3.3) and multiplication by y^i is understood in the sense of the usual (commutative) product of functions. Clearly, $\delta^2 = 0$ and $(\delta^*)^2 = 0$. We also define the operator $(\delta + \delta^*)^2 = \delta\delta^* + \delta^*\delta$, which is analogous to the Laplace operator.

Any form $a \in W \otimes \Lambda$ can be written as a sum of terms of the form

$$a_{pq} = a_{i_1 i_2 \ldots i_p j_1 j_2 \ldots j_q}(x,h) y^{i_1} y^{i_2} \ldots y^{i_p} \theta^{j_1} \wedge \theta^{j_2} \wedge \ldots \wedge \theta^{j_q} \, ,$$

where the coefficients belong to the centre of the algebra W. Direct calculation yields

$$(\delta\delta^* + \delta^*\delta)a_{pq} = (p + q)a_{pq} \, .$$

We now define the operator δ^{-1} by setting $\delta^{-1}a_{pq} = (1/(p+q))\delta^* a_{pq}$ for $p + q > 0$ and $\delta^{-1}a_{00} = 0$. The operators δ and δ^{-1} have the following properties:

1) $\delta^2 = 0$, $(\delta^{-1})^2 = 0$;

2) δ reduces a filtration by 1, while δ^{-1} increases it by 1;

3) any form admits the representation

$$a = \delta\delta^{-1}a + \delta^{-1}\delta a + a_{00} \, , \tag{3.15}$$

analogous to the Hodge-de Rham decomposition (1.7). We also mention that δ is antiderivation in the algebra $W \otimes \Lambda$, that is, $\delta(a \circ b) = (\delta a) \circ b + (-1)^p a \circ \delta b$, where a is a p-form; δ^{-1} does not have this property.

Now let $\mathcal{D} = -\delta + \partial + [\zeta, \cdot]$, where $\zeta \in W_0 \otimes \Lambda^1$. We find ζ from the condition that (3.13) and (3.14) coincide; furthermore, we require ζ to satisfy the additional condition

$$\delta^{-1}\zeta = 0 \, , \tag{3.16}$$

which automatically guarantees the Weyl normalisation $\varepsilon\zeta = 0$. Then for ζ we obtain the equation

$$\delta\zeta = -\varphi + \partial\left(\frac{i}{h}\sigma_{ij}\theta^i y^j\right) + \frac{i}{2h}\Omega_{ij}y^i y^j + \partial\zeta + \zeta^2 \, . \tag{3.17}$$

Applying δ^{-1}, taking into account that $\delta^{-1}\delta\zeta = \zeta$ and using (3.15) and (3.16), we find that

$$\zeta = \zeta_0 + \delta^{-1}(\partial\zeta + \zeta^2) \, , \tag{3.18}$$

where

$$\zeta_0 = \delta^{-1}\left(-\varphi + \partial\left(\frac{i}{h}\sigma_{ij}\theta^i y^j\right) + \frac{i}{2h}\omega_{ij}y^i y^j\right) \, .$$

The last equation can be solved by means of the iterations

$$\zeta_{n+1} = \zeta_0 + \delta^{-1}(\partial\zeta_n + \zeta_n^2) \, ,$$

with initial term ζ_0. We have

$$\zeta_{n+1} - \zeta_n = \delta^{-1}\left(\partial(\zeta_n - \zeta_{n-1}) + \tfrac{1}{2}[\zeta_n + \zeta_{n-1}, \zeta_n - \zeta_{n-1}]\right) \, ,$$

from which we see that $\zeta_{n+1} - \zeta_n \in W_n \otimes \Lambda^1$, so that the iterations converge.

We claim that the form ζ defined in this way yields indeed an Abelian connection with curvature (3.14), that is, it satisfies (3.17).We denote by Ω and Ω_0 the expressions (3.13) and (3.14), respectively, and set $A = \Omega - \Omega_0$. From the construction of ζ it follows that $\delta^{-1}A = 0$. Indeed, by (3.18) and (3.15),

$$\delta^{-1}(\Omega - \Omega_0) = \zeta - \delta^{-1}\delta\zeta = \delta\delta^{-1}\zeta = 0 \ .$$

Since the form Ω_0 is closed, from Bianchi's identity we find that $\mathcal{D}A = \mathcal{D}\Omega - d\Omega_0 = 0$, which, written more explicitly, yields

$$\delta A = \partial A + [\zeta, A] \ .$$

Applying δ^{-1} and taking into account that, by (3.15), $\delta^{-1}\delta A = A$, we arrive at the equation

$$A = \delta^{-1}(\partial A + [\zeta, A]) \ ,$$

from which, in view of the same iteration arguments, it follows that $A = 0$. The theorem is proved.

We make a few remarks about the generalisations of Theorem 3.2 and the properties of the construction of an Abelian connection.

1. The construction of the connection $\mathcal{D}$ depends smoothly on a parameter; that is, if the isomorphism $\delta(t) : \Sigma^* \to T^*M$ with the closed form $\delta(t)\sigma$, the symplectic connection $\partial(t)$ on Σ and the closed 2-form $\varphi(t) \in Z_0 \otimes \Lambda^2$ depend smoothly on the parameter t, then so does $\mathcal{D}(t)$.

2. The construction of $\mathcal{D}$ has the property of 'error adjustment'. In ζ we isolate the terms of degree zero that are quadratic in y^i, that is, we rewrite ζ in the form $(i/(2h))\zeta_{ij}y^i y^j + \Delta\zeta$. Then, as follows from the construction, $\Delta\zeta \in W_1 \otimes \Lambda^1$. The quadratic terms can be included in the symplectic connection ∂. In other words, $\mathcal{D} = -\delta + \tilde{\partial} + [\Delta\zeta, \cdot]$, where $\tilde{\partial} = \partial + [(i/(2h))\xi_{ij}y^i y^j, \cdot]$ is another symplectic connection on Σ. The latter has the property that $\delta\tilde{\partial} + \tilde{\partial}\delta = 0$, and induces a torsion-free symplectic connection on M under the mapping $\Sigma \to TM$.

A closer analysis of the iterations shows that the form $\delta\sigma$ does not need to be closed, but only sufficiently close to a closed form. More precisely, Theorem 3.2 holds in the following stronger form.

Theorem 3.3. *Let* $\delta : \Sigma^* \to T^*M$ *be an isomorphism, and suppose that there is a closed form* $\kappa = \frac{1}{2}\kappa_{ij}\theta^i \wedge \theta^j$ *such that the coefficients of the form* $\delta\sigma - \kappa$ *and of its exterior differential are sufficiently small. Then for any closed form* $\varphi \in Z_0 \otimes \Lambda^2$ *the iterations of the equation (3.18) with*

$$\zeta_0 = \delta^{-1}\left(-\frac{i}{h}(\delta\sigma - \kappa) - \varphi + \partial\left(\frac{i}{h}\sigma_{ij}\theta^i y^j\right) + \frac{i}{2h}\Omega_{ij}y^i y^j\right)$$

converge and define a form $\zeta \in W_{-1} \otimes \Lambda^1$ *for which (3.12) is an Abelian connection with curvature* $-(i/h)\kappa + \varphi$.

3. The construction can be generalized to a bundle W whose own coefficients form an algebra bundle, for example, with matrix coefficients or with coefficients that are homomorphisms of vector bundles. An important case, used in what follows, is that of an algebra W with coefficients in the bundle of Weyl symbols $S^\infty(E)$, where E is some complex bundle over M. For example, we consider an algebra W with coefficients in $\mathrm{Hom}(E, E)$, where E is a complex vector bundle over M with connection ∇. We seek an Abelian connection $\mathcal{D}$ on the algebra bundle $W \otimes \mathrm{Hom}(E, E)$ in a form similar to (3.12), namely,

$$\mathcal{D}a = -\delta a + (\partial \otimes 1 + 1 \otimes \nabla)a + [\zeta, a] \,.$$

We find that ζ satisfies the equation

$$\zeta = \zeta_0 + \delta^{-1}((\partial \otimes 1 + 1 \otimes \nabla)\zeta + \zeta^2) \,,$$

which is analogous to (3.18) and where

$$\zeta_0 = \delta^{-1}\left(\varphi + \Omega_\nabla + \partial\left(\frac{i}{h}\sigma_{ij}\theta^i y^j \right) + \frac{i}{2h}\Omega_{ij}y^i y^j \right) \,.$$

We mention that in the case of a bundle W with coefficients in $\mathrm{Hom}(E, E)$ the concept of Weyl normalisation of a local connection form $\Gamma_\mathcal{D}$ needs to be refined. In what follows we assume that if $\mathcal{D}$ is rewritten locally in the form

$$\mathcal{D} = d + [\Gamma_\mathcal{D}, \cdot] \,,$$

then the Weyl normalisation means that $\Gamma_\mathcal{D}$ satisfies the condition $\varepsilon\Gamma_\mathcal{D} = \Gamma\mathcal{D}|_{y=0} = \Gamma_\nabla$, where Γ_∇ is the local form of the connection ∇ in E, so that

$$\mathcal{D} = \nabla + [\Delta\Gamma_\mathcal{D}, \cdot] \,,$$

where $\varepsilon(\Delta\Gamma_\mathcal{D}) = 0$. With such a normalisation we find that

$$\Omega_\mathcal{D} = d\Gamma_\mathcal{D} + \Gamma_\mathcal{D}^2 = \Omega_\nabla + \nabla(\Delta\Gamma_\mathcal{D}) + (\Delta\Gamma_\mathcal{D})^2 = -\frac{i}{h}\delta\sigma + \varphi \,;$$

this form is called the *Weyl curvature of the connection $\mathcal{D}$*.

4. We also define an Abelian connection $\mathcal{D}$ in a bundle W with coefficients in $\mathrm{Hom}(E^0, E^1)$. Let $\mathcal{D}_i = -\delta + (\partial \otimes 1 + 1 \otimes \nabla_i) + [\zeta_i, \cdot]$ be Abelian connections in $W \otimes \mathrm{Hom}(E^i, E^i)$, where $\delta^{-1}\zeta_i = 0$, with the same Weyl curvature $\Omega_0 = \Omega_1 = \Omega$. For a section $a \in W \otimes \mathrm{Hom}(E^0, E^1)$ we set

$$\mathcal{D}a = -\delta a + (\partial \otimes 1 + 1 \otimes \nabla)a + \zeta_1 \circ a - a \circ \zeta_0 \,,$$

where ∇ is the covariant differential in $\mathrm{Hom}(E^0, E^1)$ defined by ∇_0 and ∇_1. Then $\mathcal{D}^2 a = \Omega_1 \circ a - a \circ \Omega_0 = \Omega \circ a - a \circ \Omega = 0$, since Ω is a central form.

The remark concerning the Weyl normalisation of the curvature form $\Gamma_\mathcal{D}$ made in the preceding article can also be extended to the case of a bundle W with coefficients in $\mathrm{Hom}(E^0, E^1)$. Indeed, we rewrite $\mathcal{D}$ locally in the form

$$\mathcal{D}a = da + \Gamma_1 \circ a - a \circ \Gamma_0 \; ;$$

here the Weyl normalisation means that $\varepsilon\Gamma_1 = \Gamma_{\nabla_1}$ and $\varepsilon\Gamma_0 = \Gamma_{\nabla_0}$, where Γ_{∇_i} are the forms of the connections ∇_i in the bundles E^i, respectively. Then

$$\Omega_{\mathcal{D}} = d\Gamma_0 + \Gamma_0^2 = d\Gamma_1 + \Gamma_1^2$$

is called the *Weyl curvature of the connection* $\mathcal{D}$ in $W \otimes \operatorname{Hom}(E^0, E^1)$.

We extend the study of non-degenerate Abelian connections constructed in Theorem 3.2. An Abelian connection generates a complex

$$\mathcal{D} : 0 \to W \to W \otimes \Lambda^1 \to W \otimes \Lambda^2 \to \ldots \to W \otimes \Lambda^{2n} \to 0 \; .$$

Consequently, the question arises of the cohomologies of this complex.

Theorem 3.4. *Under the conditions in Theorem 3.2, there is a bijective mapping* $Q : W \otimes \Lambda^p \to W \otimes \Lambda^p$ *that preserves a filtration and is such that* $\mathcal{D} = -Q\delta Q^{-1}$.

Proof. We consider the equation

$$a = b + \delta^{-1}(\mathcal{D} + \delta)a$$

for the p-form a, where b is a given p-form. It is clear that the operator $\delta^{-1}(\mathcal{D} + \delta)$ increases a filtration by 1, since $\mathcal{D} + \delta$ preserves filtration. From this it follows that for any form $b \in W_k \otimes \Lambda^p$ the iteration method yields a unique solution $a \in W_k \otimes \Lambda^p$. We define an operator Q by setting $a = Qb$. Its inverse is given by

$$Q^{-1}a = a - \delta^{-1}(\mathcal{D} + \delta)a \; .$$

We prove that $Q^{-1}\mathcal{D} + \delta Q^{-1} = 0$, that is,

$$\mathcal{D}a - \delta^{-1}(\mathcal{D} + \delta)\mathcal{D}a + \delta a - \delta\delta^{-1}(\mathcal{D} + \delta)a = 0$$

for any form a. Since $\mathcal{D}^2 a = \delta^2 a = 0$, we find that $(\mathcal{D} + \delta)\mathcal{D}a = \delta\mathcal{D}a = \delta(\mathcal{D} + \delta)a$. Then the equality we want to prove can be rewritten in the form

$$(\mathcal{D} + \delta)a = \delta^{-1}\delta(\mathcal{D} + \delta)a + \delta\delta^{-1}(\mathcal{D} + \delta)a \; .$$

By (3.15), this relation holds, since the component of type $(0,0)$ for the form $(\mathcal{D} + \delta)a$ is equal to zero.

Corollary. *The complex* $\mathcal{D} : W \otimes \Lambda^p \to W \otimes \Lambda^{p+1}$ *is exact for* $p > 0$; *if* $p = 0$, *then* $\operatorname{Ker}\mathcal{D}$ *coincides with* QZ, *where* Z *is the centre of* W.

This assertion follows from the fact that the operator Q maps $\mathcal{D}$ into the complex $-\delta$, for which, by (3.15), the statement is obvious.

For a given Abelian connection $\mathcal{D}$, by *quantum observables* we understand the flat sections $a \in W$, that is, the sections for which $\mathcal{D}a = 0$. The subalgebra $W_{\mathcal{D}} = \operatorname{Ker}\mathcal{D} \subset W$ of flat sections is called the *algebra of quantum observables*.

We now consider examples. Let M be a symplectic manifold with symplectic form $\sigma = \frac{1}{2}\sigma_{ij}dx^i \wedge dx^j$, and $\Sigma = TM$. Also, let $\delta : \Sigma^* \to T^*M$ be the identity mapping and ∂ a symplectic torsion-free connection on M, and suppose that the forms θ^i coincide with dx^i, where x^i are local coordinates on M. For $\varphi = 0$, the equation (3.18) in ζ has the form

$$\zeta = \frac{i}{4h}\Omega_{ijkl}y^i y^j y^k dx^l + \delta^{-1}(\partial\zeta + \zeta^2)\,,$$

where $\Omega_{ijkl} = \sigma_{im}\Omega^m_{jkl}$ is the curvature tensor for $\mathcal{D}$ with the upper index lowered. The iterations yield

$$\zeta = \frac{i}{4h}\Omega_{ijkl}y^i y^j y^k dx^l + \cdots$$

For any $a \in Z$, the flat section $\hat{a} = Qa$ is found from the equation

$$\hat{a} = a + \delta^{-1}(\partial\hat{a} + [\zeta, a])\,,$$

from which, by iteration, we obtain the expansion

$$\hat{a} = a + (\partial_i a)y^i + \tfrac{1}{2}(\partial_i\partial_j a)y^i y^j + \tfrac{1}{6}(\partial_i\partial_j\partial_k a)y^i y^j y^k$$
$$- \tfrac{1}{24}\Omega_{ijkl}\sigma^{lm}(\partial_m a)y^i y^j y^k + \cdots\,,$$

where $\partial_i = i(X_i)\partial$ is the covariant derivative in the direction of the vector $X_i = \partial/\partial x^i$.

The inverse mapping of $Q : Z \to W_{\mathcal{D}}$ is $\varepsilon : \hat{a} \mapsto a = \varepsilon(\hat{a}) = \hat{a}|_{y=0}$. Using these mappings, we can identify the algebra of quantum observables with Z; here the multiplication $\circ$ in $W_{\mathcal{D}}$ corresponds to the multiplication $*$ in Z mentioned in the introduction to this chapter: for $a, b \in Z$ we set $a * b = \varepsilon(\hat{a} \circ \hat{b})$. From the expansion of $\hat{a} = Qa$ it follows that

$$a * b - b * a = -ih\sigma^{ij}\partial_i a \partial_j b + \cdots = -ih\{a, b\} + \cdots\,,$$

that is, the product $*$ satisfies the correspondence principle. We mention, however, that it is more convenient to work directly with the algebra $W_{\mathcal{D}}$ and $\circ$ than with $*$ in Z.

If ∂ has zero curvature, then we can write the full expansion

$$\hat{a} = \sum_{|\alpha|=0}^{\infty} \frac{1}{\alpha!}(\partial^\alpha a)y^\alpha \tag{3.19}$$

(α is a multi-index), from which for the product $*$ we obtain

$$a * b = \sum_{k=0}^{\infty} \frac{1}{k!}\left(-\frac{ih}{2}\right)^k \left\langle \bigotimes^k \sigma^{-1}, \partial_x^k a \otimes \partial_x^k b \right\rangle\,. \tag{3.20}$$

For $M = \mathbb{R}^{2n}$ or the torus $M = \mathbf{T}^{2n}$ with the standard symplectic structure and connection $\partial = d$, (3.19) coincides with the formal Taylor series for the function $a(x + y, h)$, and (3.20) with the formula (3.2) for the composition of Weyl symbols. In the general case, the flat sections do not admit such a simple interpretation. On an intuitive level, the connection $\mathcal{D}$ in the algebra W serves to replace shifts in the fibres, which are meaningless for formal Weyl algebras, by shifts in the base, and the flat sections are those for which such a replacement is admissible.

1.4. Automorphisms and Homotopies. Let $f : M \to M$ be a diffeomorphism of the manifold M and $s = s_f : \Sigma \to \Sigma$ a symplectic lifting of f to the space of the bundle Σ. This means that the fibre Σ_x is mapped into the fibre $\Sigma_{f(x)}$ and the mapping $s : \Sigma_x \to \Sigma_{f(x)}$ is an isomorphism that preserves the symplectic form σ on the fibres. The mapping s induces an automorphism A_s of the algebra of the sections of the bundle $W(\Sigma)$, given by

$$(A_s a)(x, y, h) = a(s^{-1}(x, y), h) \ . \tag{3.21}$$

In addition to the automorphisms (3.21), we consider *interior automorphisms* of W, that is, automorphisms of the form $A_U a = U \circ a \circ U^{-1}$, where $U = 1 + V$, $V \in W_1$, is an invertible element of W. The section U is not defined uniquely by A, but only up to a factor of the form $1 + z$, where $z \in Z_2$. We always assume that U is a globally given section. When it is necessary to consider local interior automorphisms, we state this explicitly.

By an *automorphism of the algebra W* we understand a composition of automorphisms of the form A_s and A_U. Those of them for which s is a symplectic lifting of the identity mapping f are called *fibrewise automorphisms*. We can extend any automorphism to the algebra $W \otimes \Lambda$ by setting $A\varphi = (f^{-1})^*\varphi$ for central forms.

Let ∂ be a fixed symplectic connection on Σ and its associated connection in W ($\partial y^i = -\Gamma^i_j y^j$), and let $A(t)$ be a smooth family of automorphisms of $W \otimes \Lambda$. Direct calculation shows that if $a(t) = A(t)a$, where $a \in W \otimes \Lambda$, then the derivative $\dot{a}(t) = (d/dt)a(t)$ can be represented in the form

$$\dot{a}(t) = [H(t), a(t)] + (i(X(t)\partial + \partial i(X(t)))a(t) \ , \tag{3.22}$$

where $X(t) = \dot{f}(t)$ is a vector field over M and $H(t) = (i/(2h))H_{ij}y^iy^j + \Delta H$; here $\Delta H \in W_1$ is a section of the bundle W, which depends on t.

The converse also holds.

Theorem 3.5. *For any vector field $X(t)$ on M and any section $H(t) = (i/(2h))H_{ij}y^iy^j + \Delta H$ with $\Delta H \in W_1$, the equation (3.22) has a unique solution $a(t)$ for a given initial condition $a(0) \in W \otimes \Lambda$. Furthermore, the mapping $a \mapsto a(t)$ is an automorphism of the algebra $W \otimes \Lambda$.*

Proof. First let $\Delta H = 0$, that is, H is quadratic in y. We define the diffeomorphism $s(t) : \Sigma \to \Sigma$ as shift along the solutions of the system of differential equations

$$\dot{x} = X(t) \ , \qquad \frac{\partial y}{\partial t} = A(t)y \ ,$$

where $\partial y/\partial t = i(X)\partial y$ is the covariant derivative of $y \in \Sigma_x$ along the curve $x(t)$ and $A(t)$ a homomorphism of Σ given by the matrix $A^i_j = -\sigma^{ik}H_{kj}$. It is easy to see that $s(t)$ is a symplectic lifting of the mapping $f(t) = x(t)$ defined by the vector field $X(t)$.

In the general case we substitute $a(t) = A_{s(t)}b(t)$ in (3.22), where $s(t)$: $\Sigma \to \Sigma$ is the mapping defined by $X(t)$ and the quadratic part of $H(t)$. Then we find that $b(t)$ satisfies the equation

$$\dot{b}(t) = [\tilde{H}(t), b(t)] \, ,$$

where $\tilde{H}(t) \in W_1$. This reduces to the integral equation

$$b(t) = b(0) + \int\limits_0^t [\tilde{H}(\tau), b(\tau)] \, d\tau \; ;$$

the latter is solved by means of iterations, which converge and define a unique solution. The result can be written in the form $b(t) = U(t)b(0)U^{-1}(t)$, where $U(t)$ is the solution of the equation

$$U(t) = 1 + \int\limits_0^t \tilde{H}(\tau) \circ U(\tau) \, d\tau \, ,$$

for which the iterations also converge, since $\tilde{H}(\tau) \in W_1$.

Therefore, the mapping $a(0) \mapsto a(t)$ is the composition of the interior automorphism $A_{U(t)}$ and the automorphism $A_{s(t)}$. The theorem is proved.

The choice of connection ∂ in (3.22) is unessential. A change $\Delta\Gamma$ in the connection form induces a change $-i(X)\Delta\Gamma$ in H.

Let $\mathcal{D}$ be a non-degenerate Abelian connection of the form (3.12) on a symplectic manifold M. For any automorphism A, the connection $\mathcal{D}_A$ defined by $\mathcal{D}_A a = A\mathcal{D}(A^{-1}a)$ is again a non-degenerate Abelian one of the form (3.12), called the *image of the connection* $\mathcal{D}$ *under the automorphism* A. Next, let $A(t)$ be a family of automorphisms and $A(0)$ the identity automorphism. This yields a family of connections $\mathcal{D}_t = \mathcal{D}_{A(t)} = \partial + [\Delta\Gamma, \cdot]$. We subject $\Delta\Gamma$ to the Weyl normalisation condition $\varepsilon(\Delta\Gamma) = 0$ and define the family of Weyl curvatures

$$\Omega(t) = \Omega_0 + \partial(\Delta\Gamma) + (\Delta\Gamma)^2 \in Z_{-2} \otimes \Lambda^2 \, .$$

All these forms are closed and non-degenerate in the sense that the terms $\Omega_0(t)$ of degree zero in the form $ih\Omega(t)$ define a symplectic form on M.

The following assertion is the main result in this subsection.

Theorem 3.6. *Let* $\Omega(t) \in Z_{-2} \otimes \Lambda^2$ *be a family of non-degenerate closed forms, where* $\Omega(0)$ *is the curvature of a non-degenerate Abelian connection* $\mathcal{D}$. *A family of automorphisms* $A(t)$, $A(0) = 1$, *such that the form* $\Omega(t)$ *is the* *curvature of the connection $\mathcal{D}_t = \mathcal{D}_{A(t)}$ exists if and only if there is a family of central forms $\psi(t) \in Z_{-2} \otimes \Lambda^1$ such that $\Omega(t) = d\psi(t)$.*

Proof. Let $A(t)$ be the family of automorphisms defined by (3.22) and $\mathcal{D}_t = \mathcal{D}_{A(t)}$ the family of images of the connection $\mathcal{D}$ under the automorphisms $A(t)$. Replacing ∂ in (3.22) by $\mathcal{D}_t$, we arrive at an equation of a similar type

$$\dot{a}(t) = [H(t), a(t)] + (i(X(t))\mathcal{D}_t + \mathcal{D}_t i(X(t)))a(t) , \qquad (3.23)$$

where $H(t)$ is a section of W of the form

$$H(t) = \frac{i}{h} H_i y^i + \frac{i}{2h} H_{ij} y^i y^j + \Delta H ,$$
$$\Delta H(t) \in W_1 , \qquad H(t) + i(X(t))\Delta\Gamma(t) \in W_0 .$$

Such a section is defined uniquely by the family of automorphisms if we impose the Weyl normalisation condition $\varepsilon H(t) = 0$.

We consider the form $A(t)\mathcal{D}a = \mathcal{D}_t a(t)$. We have $(d/dt)\mathcal{D}_t a(t) = [\Delta\dot{\Gamma}(t), a(t)] + \mathcal{D}_t \dot{a}(t)$. Replacing $\dot{a}$ by (3.23) and using the fact that $\mathcal{D}_t^2 = 0$, we obtain

$$\frac{d}{dt}\mathcal{D}_t a(t) = [\Delta\dot{\Gamma}(t), a(t)] + [\mathcal{D}_t H(t), a(t)] + [H(t), \mathcal{D}_t a(t)] + \mathcal{D}_t i(X)\mathcal{D}_t a(t) .$$

On the other hand, rewriting (3.23) directly for the forms $A(t)\mathcal{D}a$, we find that

$$\frac{d}{dt} A(t)\mathcal{D}a = [H(t), \mathcal{D}_t a] + \mathcal{D}_t i(X)\mathcal{D}_t a(t) .$$

Comparing these two expressions, we conclude that

$$[\mathcal{D}_t H(t) + \Delta\dot{\Gamma}(t), a(t)] = 0$$

for any section $a(t)$. Consequently, the 1-form

$$\psi(t) = \mathcal{D}_t H(t) + \Delta\dot{\Gamma}(t) \in W_{-2} \otimes \Lambda^1$$

is central. From this, the formula for the variation of the curvature of a connection $\dot{\Omega}(t) = \mathcal{D}_t \Delta\dot{\Gamma}(t)$ (see 3.1, Chap. 1) and the equality $\mathcal{D}_t^2 H = 0$ it follows that $\dot{\Omega}(t) = d\psi(t)$. Hence, one side of the theorem is proved.

To prove the other side, we first establish an auxiliary result, which is also used in what follows.

Lemma 3.1. *Let $\mathcal{D}$ be a non-degenerate Abelian connection with Weyl curvature Ω, and $\Omega(t) \in Z_{-2} \otimes \Lambda^2$ a family of non-degenerate closed 2-forms such that $\Omega(0) = \Omega$. Then there is a family of non-degenerate Abelian connections $\mathcal{D}_t$, $\mathcal{D}_0 = \mathcal{D}$, with Weyl curvature $\Omega(t)$.*

To prove this, it suffices to construct isomorphisms $\delta(t) : \Sigma^* \to T^*M$, $\delta(0) = \delta$, such that $\Omega + \frac{i}{h}\delta(t)\sigma \in Z_0 \otimes \Lambda^2$ (see Remark 1 on Theorem 3.2). We define $\delta(t)$ as the composition $B(t)\delta$, where $B(t) : T^*M \to T^*M$ is the solution of the differential equation $\dot{B} = C(t)B$, $B(0) = 1$. $C(t)$ is given by the matrix $C^i_j(t) = \frac{1}{2}\omega^{ik}(t)\dot{\omega}_{kj}(t)$, where ω_{ij} is the matrix of the coefficients of $\Omega_0(t)$ (the term of degree zero in the form $ih\Omega(t)$) and ω^{ij} are the elements of the inverse matrix. It is easy to verify that $\delta(t)$ is the required isomorphism.

Let $\mathcal{D}_t = \partial + [\Delta\Gamma(t), \cdot]$ be the family of connections constructed in Lemma 3.1. We claim that if $\dot{\Omega} = d\psi$, then this family has the form $\mathcal{D}_{A(t)}$. We find $H(t)$ from the equation

$$\mathcal{D}_t H(t) = \psi(t) - \Delta\dot{\Gamma}(t) .$$

By the corollary to Theorem 3.4, this equation is solvable if and only if the form on the right-hand side is a flat one. This condition is satisfied, since

$$\mathcal{D}_t(\psi(t) - \Delta\dot{\Gamma}(t)) = d\psi(t) - \mathcal{D}_t\Delta\dot{\Gamma}(t) = d\psi(t) - \dot{\Omega}(t) = 0 .$$

As $H(t)$ we can take

$$\mathcal{D}_t^{-1}(\psi(t) - \Delta\dot{\Gamma}(t)) = -Q_t\delta^{-1}Q_t^{-1}(\psi(t) - \Delta\dot{\Gamma}(t)) .$$

The vector field $X(t)$ is determined uniquely from the condition $i(X)\Delta\Gamma + H(t) \in W_0$. For this choice of $H(t)$ and $X(t)$, the equation (3.23) reduces to (3.22) and, at the same time, determines the family of automorphisms $A(t)$.

It remains to show that the connections $\mathcal{D}_t$ and $\mathcal{D}_{A(t)}$ coincide, that is, $\mathcal{D}_t A(t)a = A(t)\mathcal{D}_0 a$ for any section $a \in W \otimes \Lambda$. Since $\mathcal{D}_t H(t) + \Delta\dot{\Gamma}(t)$ is a central form, it follows that the equations (3.23) for $\mathcal{D}_t A(t)a$ and $A(t)\mathcal{D}_0 a$ coincide. The required equality now follows from the uniqueness of the solution. The theorem is proved.

We apply the above theorem to the study of automorphisms that preserve a non-degenerate Abelian connection $\mathcal{D}$, that is, $\mathcal{D}_{A(t)} = \mathcal{D}$. The equation

$$\dot{a} = [H(t), a] + (i(X(t))\mathcal{D} + \mathcal{D}i(X(t)))a$$

defines an automorphism $A(t)$ that preserves the connection $\mathcal{D}$ if and only if $\mathcal{D}H$ is a central form. In particular, if

$$H = \frac{i}{h}H_0 + \frac{i}{h}H_i y^i + \cdots \in W_{\mathcal{D}} ,$$

then $\mathcal{D}H = 0$, and $X(t)$ is a Hamiltonian vector field on M with Hamiltonian H_0. The second term in the above equation vanishes on the sections $a \in W_{\mathcal{D}}$, and we obtain *Heisenberg's equation*

$$\dot{a} = [H, a] . \tag{3.24}$$

Consequently, Heisenberg's equation is uniquely solvable in the algebra $W_{\mathcal{D}}$ for given initial conditions.

In the general case, the condition $\mathcal{D}H = \varphi \in Z \otimes \Lambda^1$ means the following. The form φ is closed, since $d\varphi = \mathcal{D}^2 H = 0$, therefore, locally it has the form dH_0, where $H_0 \in Z$. Then $H - H_0$ belongs locally to the algebra $W_{\mathcal{D}}$. The component H_0 does not influence Heisenberg's equation. We arrive at the following assertion.

Corollary. *Any automorphism $A(t)$ of the algebra W which preserves the connection $A(t)$ is defined localy by a Hamiltonian $H(t) \in W_{\mathcal{D}}$.*

In conclusion, we give an interesting interpretation to the solutions of Heisenberg's equation in $W_{\mathcal{D}}$. Let $H(t) = (i/h)H_0 + (i/h)H_i y^i + (i/(2h)) \times H_{ij} y^i y^j + \Delta H \in W_{\mathcal{D}}$ and $\Delta H \in W_1$. We lift Σ and $W(\Sigma)$ to the manifold $M \times \mathbb{R}$. On the bundle W over $M \times \mathbb{R}$ we introduce the connection $\tilde{\mathcal{D}} = \mathcal{D} + dt \wedge (\partial/\partial t - [H(t), \cdot])$. It is easy to see that this connection is Abelian. The algebra of quantum observables $W_{\tilde{\mathcal{D}}}$ over $M \times \mathbb{R}$ consists of the sections $a(x, y, t, h)$ such that $\mathcal{D}a = 0$ and $\dot{a} - [H, a] = 0$. Hence, Heisenberg's equation of motion can be rewritten as $\tilde{\mathcal{D}}a = 0$ in the bundle W over $M \times \mathbb{R}$.

1.5. The Trace in the Algebra of Quantum Observables. Let M, Σ, W, $\mathcal{D}$ and $W_{\mathcal{D}}$ be the same as above, and let $W_{\mathcal{D}}^{\mathrm{comp}}$ be the ideal in the algebra $W_{\mathcal{D}}$ of the flat sections with compact support. We denote by $L(h)$ the set of formal Laurent series in h with constant complex coefficients and finitely many terms with negative powers of h. We construct a linear functional $\mathrm{tr} : W_{\mathcal{D}}^{\mathrm{comp}} \to L(h)$ which satisfies the following two conditions:

1) tr is a *functional of local type*, that is, it has the form $\int\limits_{S} G(a, \mathcal{D})$, where the density G depends polynomially on the coefficients of a, those of $\mathcal{D}$ and their derivatives;

2) if A is an automorphism of the algebra W and $\mathcal{D}_A$ the image of the connection $\mathcal{D}$ under this automorphism, then

$$\int\limits_{M} G(a, \mathcal{D}) = \int\limits_{M} G(Aa, \mathcal{D}_A) \,.$$

In other words, the trace is invariant under automorphisms. In particular, if A does not change $\mathcal{D}$, then a and Aa have the same trace in the algebra $W_{\mathcal{D}}$. Since any automorphism preserving the connection $\mathcal{D}$ is infinitesimally defined by a Hamiltonian H with $\mathcal{D}H \in Z \otimes \Lambda^1$, we are led to the condition $\mathrm{tr}[H, a] = 0$. From this it follows that $\mathrm{tr}[a, b] = 0$ for any $b \in W_{\mathcal{D}}$ and $a \in W_{\mathcal{D}}^{\mathrm{comp}}$.

In view of its linearity and local character, it suffices to define the trace for the sections $a \in W_{\mathcal{D}}$ with small support. Indeed, if $\rho_i(x)$ is a partition of unity on M, then, setting $\hat{\rho}_i = Q\rho_i(x)$ (see the definition of Q in 1.3), we obtain a partition of unity in the algebra $W_{\mathcal{D}}$, and the trace of the section a is equal to the sum of the traces of the $\hat{\rho}_i \circ a$ with small supports.

Thus, we consider the sections $a \in W_{\mathcal{D}}^{\mathrm{comp}}$ with support in a sufficiently small coordinate neighbourhood U of the manifold M with local coordinates x^i $(i = 1, 2, \ldots, 2n)$. We choose a reference frame in the bundle Σ over U in which the symplectic form σ has constant coefficients σ_{ij}. A non-degenerate

Abelian connection $\mathcal{D}$ in W can be rewritten with respect to this frame in the form

$$\mathcal{D} = -\delta + d + [\zeta, \cdot]\,,$$

where $\delta y^i = \theta^i = \theta^i_j(x)dx^j$ and $\zeta \in W_0 \otimes \Lambda^1$. Let $\Omega = -(i/h)\delta\sigma + \varphi$ be the Weyl curvature of $\mathcal{D}$, $\delta\sigma = \frac{1}{2}\sigma_{ij}\theta^i \wedge \theta^j$ a closed 2-form, and $\varphi \in Z_0 \otimes \Lambda^2$ also a closed form. Without loss of generality, we assume that the local coordinates are chosen so that $\theta^i_j(0) = \delta^i_j$. We introduce the standard Abelian connection $\mathcal{D}_0 = -\delta_0 + d$, where $\delta_0 y^i = dx^i$, in the bundle W over U with curvature

$$\Omega_0 = -\frac{i}{h}\delta_0\sigma = -\frac{i}{2h}\sigma_{ij}dx^i \wedge dx^j\,.$$

We set

$$\delta(t) = (1-t)\delta + t\delta_0\,, \qquad \Omega(t) = (1-t)\Omega + t\Omega_0\,.$$

In a sufficiently small neigbourhood U, the mapping $\delta(t)$ is non-degenerate and the coefficients of the form $\delta(t)\sigma - ih\Omega(t)$ and of its differential are sufficiently small; therefore, by Theorem 3.3, there is an Abelian connection $\mathcal{D}_t$ with curvature $\Omega(t)$ which links the connections $\mathcal{D}$ and $\mathcal{D}_0$. In addition, $\Omega(t) = \Omega - \Omega_0 = d\psi$, since in the neighbourhood U any closed form is exact. By Theorem 3.6, there is a family of automorphisms $A(t)$ of the bundle W over U such that $\mathcal{D}_t$ is the image of $\mathcal{D}_0$ under $A(t)$.

In view of the property 2) of the trace, we deduce that the traces of $a \in W_\mathcal{D}^{\text{comp}}$ and $A(1)a \in W_{\mathcal{D}_0}^{\text{comp}}(U)$ must coincide. Hence, the problem of constructing the trace in the algebra $W_\mathcal{D}$ reduces to the local problem of constructing the trace in the algebra $W_{\mathcal{D}_0}(U)$, which is invariant with respect to the automorphisms $A(t)$ that preserve the connection $\mathcal{D}_0$. By the corollary to Theorem 3.6, the invariance condition is equivalent to $\text{tr}[a, b] = 0$ for any $a \in W_{\mathcal{D}_0}^{\text{comp}}(U)$ and $b \in W_{\delta_0}(U)$. The solution of this problem is unique up to normalisation, and is given by

$$\text{tr}\, a = \frac{1}{(2\pi h)^n} \int_U \varepsilon(a) \frac{(\delta_0\sigma)^n}{n!}\,,$$

where the formal series $\varepsilon(a) \in Z$ is integrated term by term. Indeed, for the connection $\mathcal{D}_0$ the algebra $W_{\mathcal{D}_0}$ with multiplication $\circ$ is isomorphic to the algebra Z with multiplication $*$ (see 1.3). We have $x^i * a - a * x^i = -ih(\partial a/\partial x^j)\sigma^{ij}$. Since the trace of the commutator is equal to zero, this means that the trace of elements of the form $(\partial a/\partial x^i) \in Z$ is zero. But then $\text{tr}\, a$ must have the form $c(h) \int_U a(x, h)\, dx$. The factor $c(h)$ is chosen so that the expression of the trace coincides with that in the algebra of Weyl symbols $S^\infty(\mathbb{R}^{2n})$ where the parameter λ is replaced by the formal parameter h.

The trace defined in this way has indeed the property that $\text{tr}[a, b] = 0$, since $\varepsilon(a \circ b) = \varepsilon(a) * \varepsilon(b)$, which coincides with the usual commutative product $\varepsilon(a)\varepsilon(b)$ up to terms that are total derivatives.

The construction of the trace and its properties also remain valid for the algebra $W_\mathcal{D}$ with coefficients in the algebra of matrices, or in the bundles $\mathrm{Hom}(E, E)$. A natural difference here is the fact that we need to integrate $\mathrm{tr}_0\, \varepsilon(a)$ instead of $\varepsilon(a)$, where tr_0 is the trace in the algebra of the coefficients.

§2. The Index Theorem in the Algebra of Quantum Observables

2.1. The Index in the Algebra $W_\mathcal{D}$ and Its Properties. Below we consider bundles of formal Weyl algebras $W(\Sigma) \otimes \mathrm{Hom}(E^i, E^j)$ $(i, j = 0, 1)$ over a symplectic manifold M with coefficients in the bundle $\mathrm{Hom}(E^i, E^j)$, where E^i and E^j are vector bundles over M. For brevity, we denote them by $W(E^i, E^j)$. Let $\delta : \Sigma^* \to T^*M$ be an isomorphism, and suppose that $\delta\sigma = \frac{1}{2}\sigma_{ij}\theta^i \wedge \theta^j$ is a closed 2-form. Then (see Remark 4 on Theorem 3.2) there is a non-degenerate Abelian connection in $W(E^i, E^j)$

$$\mathcal{D} = -\delta + \partial \otimes 1 + 1 \otimes \nabla + [\zeta, \cdot]\,,$$

where ∇ is the connection in $\mathrm{Hom}(E^i, E^j)$, $\zeta \in W_0 \otimes \Lambda^1$ and $\varepsilon\zeta = 0$, with Weyl curvature $\Omega = -(1/h)\delta\sigma$. The corresponding algebra of flat sections $W_\mathcal{D}(E^i, E^j)$ is called the *algebra of quantum observables with coefficients in* $\mathrm{Hom}(E^i, E^j)$.

A section $A \in W_\mathcal{D}(E^0, E^1)$ is called *elliptic* if there is an $R \in W_\mathcal{D}(E^1, E^0)$ for which $1 - R \circ A \in W_\mathcal{D}^{\mathrm{comp}}(E^0, E^0)$ and $1 - A \circ R \in W_\mathcal{D}^{\mathrm{comp}}(E^1, E^1)$. Therefore, an elliptic section defines a collection $\Xi = \{1, 1, A, R\}$ analogous to a family of Fredholm operators (see Sec. 2, Chap. 2). By analogy with (1.8), we define

$$\mathrm{ind}\, A = \mathrm{tr}(1 - R \circ A) \quad \mathrm{tr}(1 \quad A \circ R)\,,$$

where the trace is taken in $W_\mathcal{D}^{\mathrm{comp}}$. We also use the notation $\mathrm{ind}\, \Xi$.

We can also consider more general collections $\Xi = \{P^0, P^1, A, R\}$, where $P^i \in W_\mathcal{D}(E^i, E^i)$ are idempotent, that is, $(P^i)^2 = P^i$, and $A \in W_\mathcal{D}(E^0, E^1)$ and $R \in W_\mathcal{D}(E^1, E^0)$ satisfy conditions similar to (2.10) and (2.11). For simplicity, we confine ourself to the case $P^i = 1$. The results also remain valid in the general case.

Let $\varepsilon A = h^{-N}a(x) + \cdots$, where the dots denote terms of higher order in h. The function $a(x)$ with values in $\mathrm{Hom}(E^0, E^1)$ is called the *leading term of the section A*. It is easy to see that A is elliptic if and only if $a(x)$ is an isomorphism outside a compactum. Consequently, the leading term is analogous to the principal symbol of an elliptic operator.

Just as for elliptic operators, we can prove that the index has the logarithmic and stability properties (see 1.2, Chap. 1). A specific property of the index is its stability under deformations of the Abelian connection $\mathcal{D}$ which preserve the class of cohomologies of its curvature and its non-degeneracy. This follows from the corresponding property of the trace. Hence, the index

238 B.V. Fedosov

depends only on the cohomology class of the Weyl curvature, that is, the class of cohomologies of the symplectic form $\delta\sigma$ on M, and on the class of virtual triples $\xi = \{E^0, E^1, a(x)\}$ in $K^{\mathrm{comp}}(M)$.

The multiplicative property also holds, but in a formulation closer to that in Sec. 2, Chap. 2. We consider two symplectic manifolds with corresponding quantum structures M_i, Σ_i, E_i^j, $\mathcal{D}_i$ and $W_{\mathcal{D}_i}(E_i^j, E_i^k)$ ($i = 1, 2$, $j, k = 0, 1$). Let $\mathcal{E}^0 = (E_1^0 \otimes E_2^0) \oplus (E_1^1 \otimes E_2^1)$ and $\mathcal{E}^1 = (E_1^1 \otimes E_2^0) \oplus (E_1^0 \otimes E_2^1)$. Then over the manifold $M = M_1 \times M_2$ we can define bundles of formal Weyl algebras $W(\mathcal{E}^i, \mathcal{E}^j)$, which can be given by means of the tensor products $W(E_1^i, E_1^j) \otimes W(E_2^k, E_2^l)$ and their direct sums. Abelian connections $\mathcal{D}$ on the bundles $W(\mathcal{E}^i, \mathcal{E}^j)$ are defined as connections in tensor products, that is, $\mathcal{D} = \mathcal{D}_1 \otimes 1 + 1 \otimes \mathcal{D}_2$. For elliptic collections $\Xi_i = \{1, 1, A_i, R_i\}$ in the algebras $W_{\mathcal{D}_i}$ over the manifolds M_i we define the tensor product by means of formulas analogous to (2.6), namely,

$$\Xi = \Xi_1 \otimes \Xi_2 = \{1, 1, \mathcal{A}, \mathcal{R}\},$$

where

$$\begin{aligned}
\mathcal{A} &= \begin{pmatrix} A_1 \otimes (1 - R_2 \circ A_2) & -1 \otimes R_2 \\ 1 \otimes A_2 & R_1 \otimes 1 \end{pmatrix}, \\
\mathcal{R} &= \begin{pmatrix} R_1 \otimes 1 & 1 \otimes R_2 \\ -1 \otimes A_2 & A_1 \otimes (1 - A_2 \circ R_2) \end{pmatrix}.
\end{aligned} \tag{3.25}$$

Then, taking into account the obvious equality $\operatorname{tr} T_1 \otimes T_2 = \operatorname{tr} T_1 \operatorname{tr} T_2$, where $T_i \in W_{\mathcal{D}_i}^{\mathrm{comp}}$, we deduce that $\operatorname{ind} \Xi = \operatorname{ind} \Xi_1 \operatorname{ind} \Xi_2$.

We now formulate the index theorem in the algebra of quantum observables, which is the analogue of the Atiyah-Singer theorem. Let $d(A)$ be the class in $K^{\mathrm{comp}}(M)$ of the virtual bundle $\xi = \{E^0, E^1, a(x)\}$ with compact support defined by the leading term $a(x)$. As usual, we denote by ω the form $-(1/(2\pi i))\Omega = \delta\sigma/(2\pi h)$.

Theorem 3.7. *The index of A is*

$$\operatorname{ind} A = \int_M \operatorname{ch} d(A) e^\omega A(\Sigma), \tag{3.26}$$

where $A(\Sigma)$ is the A-class of the bundle Σ.

Here e^ω is a power series in $1/(2\pi h)$, which terminates because M is finite-dimensional. Therefore, $\operatorname{ind} A$ is a polynomial in $1/h$ of degree not higher than $n = \frac{1}{2} \dim M$. The formula (3.26) also holds in the general case, when $\Omega = -(i/h)\delta\sigma + \varphi$, $\varphi \in Z_0 \otimes \Lambda^2$; here the index is a formal Laurent series that begins with the power h^{-n}.

The formula (3.26) can be rewritten in another form, which sometimes is more convenient in applications. We recall that a symplectic bundle Σ admits a Hermitian structure, that is, it is the realification of an n-dimensional complex Hermitian bundle E. Then the complexification of Σ is isomorphic to $E \oplus E^*$ and, since the A-class is multiplicative, we find that $A(\Sigma) = A(E)A(E^*)$.

Since the function $(t/2)\,\mathrm{sh}(t/2)$ is even, it follows that $A(E^*) = A(E)$ (see the properties of characteristic classes in 3.1, Chap. 1). Consequently, $A(\Sigma)$ is given by the form

$$\det \frac{\omega(E)}{e^{\omega(E)/2} - e^{-\omega(E)/2}} = \det e^{-\omega(E)/2}\mathcal{T}(E) = e^{-c_1(E)/2}\mathcal{T}(E)\,,$$

where $\omega(E)$ is the curvature of the Hermitian connection in E divided by $-2\pi i$, and $c_1(E)$ is the first Chern class of E. We can rewrite (3.26) in the form

$$\mathrm{ind}\, A = \int_M \mathrm{ch}\, d(A) e^{\omega - c_1(E)/2}\mathcal{T}(E)\,. \tag{3.27}$$

We claim that if M is the space of the cotangent bundle T^*X of an n-dimensional manifold X with symplectic form $d\xi_i \wedge dx^i$ (we assume summation from 1 to n), then (3.27) reduces to the Atiyah-Singer formula. Indeed, Σ is isomorphic to $TM = T(T^*X) \approx TX \oplus TX$. The form $\delta\sigma = d\xi_i \wedge dx^i$ is exact, so that its cohomology class in $H(M)$ is equal to zero. E is isomorphic to the real bundle TX, which implies that $c_1(E) = 0$. Consequently, the form $\omega - \frac{1}{2}c_1(E)$ is cohomologic to 0, and we recover the Atiyah-Singer formula. In this case the index does not depend on h and is an integer.

2.2. Outline of the Proof of the Theorem. The main steps in the proof of the theorem are the same as in that of the Atiyah-Singer theorem (Atiyah and Singer (1968a)). However, in this case there are specific details connected with the fact that the elements of the algebra $W_\mathcal{D}$ are not operators; therefore, all the arguments based on the existence of a finite-dimensional kernel and cokernel are inapplicable. On the other hand, some questions arising in a purely algebraic approach are easier than in an operator method. Thus, for example, the usual difficulties stemming from the non-availability of a spinor structure (see Berline and Vergne (1985) and Connes and Moscovici (1982)) do not arise here.

The proof is based on the construction of a *Thom homomorphism* in the algebra of quantum observables, analogous to that in K-theory (see 3.3, Chap. 1).

Step 1. Over a symplectic manifold M we consider a quantum structure consisting of a symplectic bundle Σ, an isomorphism $\delta : \Sigma^* \to T^*M$ such that the form $\delta\sigma = \frac{1}{2}\sigma_{ij}\theta^i \wedge \theta^j$ is closed, and the bundles of coefficient algebras $\mathrm{Hom}(E^i, E^j)$, and the bundles of formal Weyl algebras $W = W(M)$ with coefficients in the former. In addition, in W we consider the non-degenerate Abelian connection $\mathcal{D} = -\delta + \partial + [\zeta, \cdot]$ with Weyl curvature $\Omega = \Omega_W = -(i/h)\delta\sigma$.

We also consider a complex Hermitian bundle E over M with Hermitian connection ∇. Then (see 1.1) the Fock bundle $F(E)$ and the vacuum vector p_0 are defined. The algebra bundle $S^\infty(E)$ with coefficients in $\mathrm{Hom}(\Lambda(E^*), \Lambda(E^*))$ acts in $F = F(E) \otimes \Lambda(E^*)$. With the connection ∇ on

E we associate connections on F and $S^\infty(E)$, which we denote by the same symbol ∇. The curvature $\Omega_\nabla \in S^\infty(E) \otimes \Lambda^2$ satisfies the Wick normalisation $\Omega_\nabla \circ p_0 = p_0 \circ \Omega_\nabla = 0$, where p_0 is the vacuum projection in $S^\infty(E)$. The local connection form Γ_∇ on F satisfies the same condition.

We now consider the bundle of Weyl algebras $W(M) \otimes S^\infty(E)$ with coefficients in $S^\infty(E)$. Its sections a are 'functions' $a(x, y, z, h, \lambda)$ (where $x \in M$, $y \in \Sigma_x$, $z \in E_x$, and $\lambda > 0$ is a parameter of the algebra $S^\infty(E)$)), which are formal power series in y and h. In $W(M) \otimes S^\infty(E)$ we construct a non-degenerate Abelian connection $\mathcal{D}_1$ with the same curvature $\Omega_{\mathcal{D}_1} = \Omega_{\mathcal{D}} = -(1/h)\delta\sigma$, by setting

$$\mathcal{D}_1(a \otimes b) = \mathcal{D}a \otimes b + a \otimes \nabla b + [\zeta_1, a \otimes b] \,,$$

where ζ_1 is a 1-form on M with values in $W_0 \otimes S^\infty(E)$ and such that $\delta^{-1}\zeta_1 = 0$. Just as in the proof of Theorem 3.2 (see also Remark 3 on this theorem), ζ_1 is found by means of iterations from the equation

$$\zeta_1 = \delta^{-1}(1 \otimes \Omega_\nabla) + \delta^{-1}\{((\mathcal{D} + \delta) \otimes 1 + 1 \otimes \nabla)\zeta_1 + \zeta_1^2\} \,.$$

From (3.8) we see that the initial iteration $\delta^{-1}(1 \otimes \Omega_\nabla)$ is a combination of Weyl symbols of the form $z^\alpha \circ \bar{z}^\beta$, where the first factor is a creation operator and the second one an annihilation operator. From this it follows that the form ζ_1 also consists of terms containing products of the form $z^{\alpha_1} \circ \bar{z}^{\beta_1} \circ \cdots \circ z^{\alpha_k} \circ \bar{z}^{\beta_k}$, where the first factor is a creation operator and the last one an annihilation operator. From this we also see that the parameter λ occurs in the terms of ζ_1 of a given degree in y and h as a Laurent polynomial. If $a \in W$ and $p_0 \in S^\infty(E)$ is the vacuum projection, then from what has been said above it follows that $\zeta_1 \circ (a \otimes p_0) = (a \otimes p_0) \circ \zeta_1 = 0$. Consequently,

$$\mathcal{D}_1(a \otimes p_0) = \mathcal{D}a \otimes p_0 + a \otimes \nabla p_0 = 0 \tag{3.28}$$

for any flat section $a \in W_{\mathcal{D}}$, that is, $a \otimes p_0 \in (W \otimes S^\infty)_{\mathcal{D}_1}$ is a flat section with respect to the connection $\mathcal{D}_1$. It is also clear that $\mathrm{tr}(a \otimes p_0)$ in the algebra $(W \otimes S^\infty)_{\mathcal{D}_1}$ coincides with the trace $\mathrm{tr}\, a$ in the algebra $W_{\mathcal{D}}$ for $a \in W_{\mathcal{D}}^{\mathrm{comp}}$, since $\mathrm{tr}\, p_0 = 1$ in the coefficient algebra $S^\infty(E_x)$.

Now let $\Xi = \{1, 1, A, R\}$ be an *elliptic collection* in $W_{\mathcal{D}}$, and $B = \{F(E) \otimes \Lambda^+(E^*), F(E) \otimes \Lambda^-(E^*), b(z), b^{-1}(z)\}$ the Fredholm family corresponding to the Bott generator (see 1.1). By Theorem 3.1, the latter can be written as a direct sum $B_0 \oplus B_1$, where $B_0 = \{p_0, 0, 0, 0\}$ and $B_1 = \{1 - p_0, 1, b, b^{-1}\}$ is a trivial infinite-dimensional family. Then the tensor product $\Xi \otimes B$ in the bundle $W \otimes S^\infty$, defined by (3.25), can also be written as the direct sum of $\Xi \otimes B_0$ and $\Xi \otimes B_1$. Generally speaking, the sections $\mathcal{A}$ and $\mathcal{R}$ in $\Xi \otimes B$ are not flat, that is, they do not belong to the algebra $(W \otimes S^\infty)_{\mathcal{D}_1}$. Therefore, we define the section $\hat{\mathcal{A}}$ by means of the quantisation procedure (Theorem 3.4), that is, by means of the iterations

$$\hat{\mathcal{A}} = \mathcal{A} + \delta^{-1}((\mathcal{D}_1 + \delta)\mathcal{A}) \,.$$

As a result, we obtain an elliptic collection $\Xi \hat{\otimes} B \in (W \otimes S^\infty)_{\mathcal{D}_1}$. Here the term $\Xi \otimes B_0$ does not change since, by (3.28), it is already flat with respect to $\mathcal{D}_1$, while the term $\Xi \otimes B_1$ changes but remains trivial. From this we find that

$$\operatorname{ind} \Xi \otimes B = \operatorname{ind} \Xi \otimes B_0 = \operatorname{ind} \Xi , \qquad (3.29)$$

where the left-hand side is computed in the algebra $(W \otimes S^\infty)_{\mathcal{D}_1}$ and the right-hand side in the algebra $W_{\mathcal{D}}$. We mention that, in view of (3.29), the left-hand side does not depend on the parameter λ.

Step 2. First we replace the exact parametrix $b^{-1}(z)$ constructed in 1.1 by an approximate parametrix $r(z)$ where the parameter λ occurs as a Laurent polynomial. To this end, we multiply $b^{-1}(z)$ by a truncating symbol $\rho(z)$ equal to zero in the neighbourhood of $z = 0$ and independent of λ. Then, as $\lambda \to 0$, $\rho \circ b^{-1}$ admits an asymptotic expansion in powers of λ. We choose the approximate parametrix $r(z, \lambda)$ to be a sufficiently large finite segment of this expansion. Since the index is stable, such a change does not influence $\operatorname{ind} \Xi \otimes B$, but now $\Xi \otimes B$ is a Laurent polynomial in λ.

In every fibre E_x of E we consider the bundle $W(E_x)$ of formal Weyl algebras with respect to the symplectic form $\sigma(\eta_1, \eta_2) = \operatorname{Im}\langle \eta_1, \eta_2 \rangle$, where $\langle \, , \, \rangle$ is a Hermitian inner product on the fibres. The sections of $W(E_x)$ are formal series $a(z, \eta, \lambda)$ of the form (3.10) (here we write z, η and λ instead of x, y and h). We also consider the standard Abelian connection on the fibres of $W(E_x)$

$$\mathcal{D}_f = d + \left[\frac{i}{\lambda} \operatorname{Im}\langle \eta, dz \rangle, \cdot \right] ,$$

where d is the differential with respect to z and $\bar{z}$. With the Weyl symbol $a(z, \lambda) \in S^\infty(E_x)$, which is a Laurent polynomial in λ, we associate the flat section $a(z + \eta, \lambda) \in W_{\mathcal{D}_f}(E_x)$, regarded as a formal Taylor series in η and $\bar{\eta}$. This mapping is called a *formalisation*. The connection $\mathcal{D}_f$ remains invariant under unitary changes $z' = U(x)z$, $\eta' = U(x)\eta$; therefore, the formalisation is also defined for the algebra bundles $S^\infty(E)$. Thus, we have found a mapping of $W(M) \otimes S^\infty(E)$ into the bundle of formal Weyl algebras with coefficients in the algebra bundle $W_{\mathcal{D}_f}(E_x)$, which we denote by $W(M) \otimes W_{\mathcal{D}_f}(E_x)$. By means of the formalisation (that is, the substitution $z \mapsto z + \eta$), from the Abelian connection $\mathcal{D}_1$ on $W(M) \otimes S^\infty(E)$ we obtain an Abelian connection $\mathcal{D}_2$ on the bundle $W(M) \otimes W_{\mathcal{D}_f}(E_x)$. Consequently, the formalisation yields a mapping $(W(M) \otimes S^\infty(E))_{\mathcal{D}_1} \to (W(M) \otimes W_{\mathcal{D}_f}(E_x))_{\mathcal{D}_2}$. The algebra $(W(M) \otimes W_{\mathcal{D}_f}(E_x))_{\mathcal{D}_2}$ coincides with that of the flat sections of $W(M)$ with coefficients in $W(E_x)$, which in what follows we denote by $W(M) \otimes W(E_x)$ with respect to the Abelian connection $\mathcal{D}_3 = \mathcal{D}_2 \otimes 1 + 1 \otimes \mathcal{D}_f$. Hence, we finally obtain a mapping $(W(M) \otimes S^\infty(E))_{\mathcal{D}_1} \to (W(M) \otimes W(E_x))_{\mathcal{D}_3}$ that associates with every flat section $a(x, y, z, \lambda, h)$, which is a Laurent polynomial in λ, the flat section $a(x, y, z + \eta, \lambda, h) \in (W(M) \otimes W(E_x))_{\mathcal{D}_3}$. Thus, by means of the formalisation we obtain an elliptic collection $\Xi \hat{\otimes} B$ in the algebra $(W(M) \otimes W(E_x))_{\mathcal{D}_3}$, since the traces in $S^\infty(E_x)$ and $W_{\mathcal{D}_f}(E_x)$ are compatible (see 1.5).

Since the index of $\Xi \oplus B$ does not depend on λ, it is natural to set $\lambda = h$. Under this identification, $W(M) \otimes W(E_x)$ corresponds to the bundle of formal Weyl algebras $W(E)$ over the total space of E, and the algebra of flat sections $(W(M) \otimes W(E_x))_{\mathcal{D}_3}$ to the algebra $E_{\tilde{\mathcal{D}}}(E)$ of flat sections of $W(E)$ with respect to the connection $\tilde{\mathcal{D}} = \mathcal{D}_3|_{\lambda \doteq h}$. In addition, the coefficients of a given power of h turn out to be infinite power series in z and $\bar{z}$, but they converge for $|z|$ sufficiently small. Hence, we need to consider not the algebra $W_{\tilde{\mathcal{D}}}(E)$, which is meaningless, but $W_{\tilde{\mathcal{D}}}(U)$, where U is a neighbourhood of the zero section of E. Since the support of $\Xi \otimes B$ is contained in a sufficiently small neigbourhood of $z = 0$, as a result we obtain an elliptic collection $\Xi \otimes B$ in $W_{\tilde{\mathcal{D}}}(U)$.

In this way, the combination of steps 1 and 2 associates with every elliptic collection Ξ in $W_{\mathcal{D}}(M)$ an elliptic collection $\Xi \otimes B$ in $W_{\tilde{\mathcal{D}}}(U)$ with the same index. This mapping is called the *Thom homomorphism* in the algebra of quantum observables.

We compute the Weyl curvature of the connection $\tilde{\mathcal{D}}$. We rewrite $\mathcal{D}_1$ in the form

$$\mathcal{D}_1 = \mathcal{D} \otimes 1 + [\Gamma(z), \cdot] \,,$$

where $\Gamma(z) = 1 \otimes \Gamma_\nabla + \zeta_1$ and Γ_∇ is the connection form on the bundle $S^\infty(E)$ satisfying the Wick normalisation, that is,

$$\Gamma_\nabla = -\frac{1}{2\lambda} \Gamma_{\alpha\beta} z^\beta \circ \bar{z}^\alpha = -\frac{1}{2\lambda} \Gamma_{\alpha\beta} \bar{z}^\alpha z^\beta + \frac{1}{2} \Gamma_\alpha^\alpha \,;$$

here $\Gamma_{\alpha\beta} = \delta_{\alpha\gamma} \Gamma_\beta^\gamma$ is the form of the Hermitian connection ∇ in an orthonormal reference frame on E. From the construction of $\mathcal{D}_1$ it follows that

$$(\mathcal{D} \otimes 1)\Gamma(z) + \Gamma(z) \circ \Gamma(z) = 0 \,.$$

The formalisation yields the connection $\mathcal{D}_2$, which we rewrite in the form

$$\mathcal{D}_2 = \mathcal{D} \otimes 1 + [\Gamma(z + \eta), \cdot] \,,$$

and, as before,

$$(\mathcal{D} \otimes 1)\Gamma(z + \eta) + \Gamma(z + \eta) \circ \Gamma(z + \eta) = 0 \,. \tag{3.30}$$

Next, we rewrite $\mathcal{D}_3$ as

$$\mathcal{D}_3 = \mathcal{D} \otimes 1 + 1 \otimes \mathcal{D}_f + [\Gamma(z + \eta), \cdot] \,,$$

where

$$(1 \otimes \mathcal{D}_f)\Gamma(x + \eta) = 0 \,. \tag{3.31}$$

The connection $\tilde{\mathcal{D}}$ has the same form as $\mathcal{D}_2$, with λ replaced by h. Let

$$\varepsilon\Gamma(z + \eta) = \Gamma(z + \eta)|_{y=0,\,\eta=0,\,\lambda=h} = -\frac{1}{2h} \Gamma_{\alpha\beta} \bar{z}^\alpha z^\beta + \frac{1}{2} \Gamma_\alpha^\alpha \,.$$

Then $\tilde{\mathcal{D}}$ can also be rewritten as

$$\tilde{\mathcal{D}} = \mathcal{D} \otimes 1 + 1 \otimes \mathcal{D}_f + [\Gamma(x + \eta) - \varepsilon\Gamma, \cdot] \,,$$

where we have already set $\lambda = h$. Here $\Gamma(z + \eta) - \varepsilon\Gamma$ satisfies the Weyl normalisation; hence, by (3.30) and (3.31), we find that the Weyl curvature $\Omega_{\tilde{\mathcal{D}}}$ is given by

$$\Omega_{\tilde{\mathcal{D}}} = \Omega_{\mathcal{D}} \otimes 1 + 1 \otimes \Omega_{\mathcal{D}_f} - (\mathcal{D} \otimes 1)\varepsilon\Gamma - \mathcal{D}_f\varepsilon\Gamma$$

$$= -\frac{i}{h}\left\{ \delta\sigma - \frac{1}{2i}d\bar{z}^\alpha \wedge dz^\alpha + \frac{1}{2i}\Gamma_{\alpha\beta} \wedge (d\bar{z}^\alpha z^\beta + \bar{z}^\alpha dz^\beta) - \frac{1}{2i}d\Gamma_{\alpha\beta}\bar{z}^\alpha z^\beta \right\} - \frac{1}{2}\Omega_\alpha^\alpha$$

$$= -\frac{i}{h}\left\{ \delta\sigma - \frac{1}{2i}\overline{(dz^\alpha + \Gamma_\beta^\alpha z^\beta)} \wedge (dz^\alpha + \Gamma_\beta^\alpha z^\beta) - \frac{1}{2i}\Omega_{\alpha\beta}\bar{z}^\alpha z^\beta \right\} - \frac{1}{2}\Omega_\alpha^\alpha \,,$$

where Ω_β^α, $\Omega_{\alpha\beta} = \delta_{\alpha\gamma}\Omega_\beta^\gamma$, is the curvature of the connection ∇ on E. The term in braces yields a symplectic form on the space of the bundle E. We mention that, beside the leading term of degree -2, the curvature $\Omega_{\tilde{\mathcal{D}}}$ also contains a term of degree zero, equal to $-\frac{1}{2}\operatorname{tr}\Omega_\nabla$. Thus, the Thom homomorphism maps the homogeneous Weyl curvature $\Omega_{\mathcal{D}} = -\dfrac{i}{h}\delta\sigma$ on to the non-homogeneous form $\Omega_{\tilde{\mathcal{D}}}$. This is why from the very beginning we considered non-homogeneous curvatures.

Denoting by σ_E a symplectic form on U, for $\omega_{\tilde{\mathcal{D}}} = -(1/(2\pi i))\Omega_{\tilde{\mathcal{D}}}$ we find the expression

$$\omega_{\tilde{\mathcal{D}}} = \frac{1}{2\pi h}\sigma_E - \frac{1}{2}c_1(E) \,, \tag{3.32}$$

where $c_1(E) = \operatorname{tr}\omega_\nabla$ is the first Chern form of the bundle E.

Step 3. Just as in the proof of the Atiyah-Singer formula, we use the Thom homomorphism to reduce the problem to the case where M is a domain U in $\mathbb{R}^{2k}$. Let X be a domain with compact closure in M, which contains the support of the elliptic collection Ξ. Also, let $i_1 : X \to \mathbb{R}^k$ be a smooth embedding with normal bundle N. Then we also have the embedding $T^*X \to T^*\mathbb{R}^k = \mathbb{R}^{2k}$. Incorporating X in T^*X as the zero section, we obtain

$$i_2 : X \to T^*X \to \mathbb{R}^{2k} \,.$$

The normal bundle under the embedding i_2 is $N_2 = N \oplus N \oplus T^*X$, on which we can introduce a complex structure in a natural way. Hence, the Thom homomorphism is defined in the algebra of quantum observables, and we complete the desired reduction, where U is a neighbourhood of X embedded in $\mathbb{R}^{2k}$.

Step 4. Let $\Xi = \{1, 1, A, R\}$ be an elliptic collection in the algebra of quantum observables $W_{\tilde{\mathcal{D}}}(U)$ in the domain $U \subset \mathbb{R}^{2k}$. Without loss of generality, we may assume that the coefficient algebra is the matrix algebra. It is easy to show that the index is expressed as the functional

$$\operatorname{ind}\Xi = \int_U F(a, r, \Omega_{\tilde{\mathcal{D}}}) \,,$$

where the density F depends polynomially on the leading terms a and r of A, $R \in W_{\tilde{\mathcal{D}}}(U)$ and their derivatives, as well as on the coefficients of the form $\Omega_{\tilde{\mathcal{D}}}$ and their derivatives. In addition, the density vanishes outside a compactum $U_0 \subset U$.

By means of a deformation that depends analytically on a parameter, we can contract the curvature form $\tilde{\Omega}$ in terms of non-degenerate closed 2-forms on U, to a form Ω_0 with constant coefficients. Consequently, in what follows we restrict our attention to forms $\tilde{\Omega}$ close to Ω_0, that is, $\tilde{\Omega} = \Omega_0 + \Omega_1$, where Ω_1 is a closed form with small variable coefficients. The general case follows from this by the principle of analytic continuation.

Let g be a linear transformation of $\mathbb{R}^{2n}$ which differs slightly from the identity, so that $gU_0 \subset U$. The forms Ω_0 and $g^*\Omega_0$ are exact in U (since their coefficients are constant); therefore, the cohomology classes of $\Omega_0 + \Omega_1$ and $g^*\Omega_0 + \Omega_1$ in U coincide. Hence, in view of the stability of the index (see 2.1), from this we find that

$$\operatorname{ind} \Xi = \int F(a, r, g^*\Omega_0 + \Omega_1) \, .$$

The integrand depends analytically on g, so the equality holds for any $g \in \operatorname{GL}(2n, \mathbb{C})$. Averaging over the unitary group (see Fedosov (1974), Atiyah, Bott and Patody (1973) and Hörmander (1979)), we arrive at (3.27) in the case where $M = U \subset \mathbb{R}^{2k}$. Applying the inverse of the Thom isomorphism in cohomologies (that is, integrating along the fibres of the normal bundle N_2) and using (3.32), we arrive at (3.27) in the general case. This completes the proof.

2.3. The Asymptotic Operator Representation. Let $W_{\mathcal{D}}$ be the algebra of quantum observables with numerical coefficients on a symplectic manifold M, and $W_{\mathcal{D}}^s \subset W_{\mathcal{D}}$ the subalgebra of the flat sections $a \in W_{\mathcal{D}}$ that stabilize outside a compactum in M, that is, $a(x, y, h) \equiv a(h)$ outside a compactum. Instead of the stabilisation condition we can consider a more general one, which guarantees the 'correct' behaviour at infinity, but we confine ourselves to the simpler version. Let Λ be a given bounded subset of positive numbers for which 0 is a limit point. Next, suppose that with every quantum observable $a \in W_{\mathcal{D}}^s$, for any natural number N and any $\lambda \in \Lambda$ we associate a bounded operator $\operatorname{Op}_N(a)$ in a Hilbert space, so that the mapping $a \mapsto \operatorname{Op}_N(a)$ is linear and satisfies the following conditions:

1) $\|\operatorname{Op}_N(a)\| \leq C$ and $\|\operatorname{Op}_N(a) - \operatorname{Op}_{N+1}(a)\| \leq C\lambda^{N+1}$;
2) $\|\operatorname{Op}_N(a)\operatorname{Op}_N(b) - \operatorname{Op}_N(a \circ b)\| \leq C\lambda^{N+1}$;
3) if $a \in W_{\mathcal{D}}^{\mathrm{comp}}$, then $\operatorname{Op}_N(a)$ is a trace class operator and

$$\|\operatorname{Op}_N(a)\|_{\mathrm{tr}} \leq C\lambda^{-n} \, , \qquad \left| \operatorname{tr} \operatorname{Op}_N(a) - \operatorname{tr} a|_N \right| \leq C\lambda^{-n+N+1} \, .$$

Here $\| \; \|_{\mathrm{tr}}$ is the trace norm, $\operatorname{tr} \operatorname{Op}_N(a)$ the operator trace, $\operatorname{tr} a|_N$ a truncation of the formal series $\operatorname{tr} a$ with h replaced by a number $\lambda \in \Lambda$, and

$n = \frac{1}{2} \dim M$. The constants C in the estimates depend on N, a and b. The map $a \mapsto \mathrm{Op}_N(a)$ is called an *asymptotic operator representation* (AOR).

We give a trivial example of AOR. Let $M = \mathbb{R}^{2n}$ with the standard symplectic structure and connection $\mathcal{D} = -\delta + d$. The quantum observables have the form $a(x + y, h)$ (see 1.3). We take Λ to be the interval $(0, 1)$. With a quantum observable $a \in W_{\mathcal{D}}^s$ we associate a PDO in $L^2(\mathbb{R}^n)$ with Weyl symbol $a|_N$ (see Hörmander (1979)). Then all the conditions in the definition are satisfied.

The index theorem enables us to derive necessary conditions for the existence of an AOR. For simplicity, we consider an algebra $W_{\mathcal{D}}^s$ on M where the curvature of the connection $\mathcal{D}$ consists only of the highest term, that is, $\Omega = -(i/h)\delta\sigma$, with $\delta\sigma$ a symplectic form on M. Then for any $\xi \in K^{\mathrm{comp}}(M)$ the expression (3.26) is a polynomial in $1/h$ of degree not higher than $n = \frac{1}{2} \dim M$. In particular, the substitution $h = \lambda$ in $\mathrm{ind}\, A$ is meaningful.

Theorem 3.8. *If there is an AOR of the algebra $W_{\mathcal{D}}^s$, then for any $\xi \in K^{\mathrm{comp}}(M)$ the right-hand side of the formula (3.26) on the set Λ takes integer values modulo $O(\lambda^\infty)$ as $\lambda \to 0$.*

To prove this, we remark that if there is an AOR of $W_{\mathcal{D}}^s$ with coefficients in $\mathbb{C}$, then there is also an AOR of the algebra $W_{\mathcal{D}}^s$ with matrix coefficients. Any element $\xi \in K^{\mathrm{comp}}(M)$ can be realized in the form $\{1, 1, a(x)\}$ with a matrix-valued function $a(x)$. Associating with it, by means of quantisation, an elliptic element $A \in W_{\mathcal{D}}^s$ with leading term $a(x)$ and then using the AOR, we obtain the Fredholm operators $A_N = \mathrm{Op}_N(a)$ with index

$$\mathrm{ind}\, A_N = \mathrm{tr}(1 - R_N A_N) - \mathrm{tr}(1 - A_N R_N) \in \mathbb{Z}\,.$$

For N sufficiently large and λ small, $\mathrm{ind}\, A_N$ does not depend on N, since the difference $A_N - A_{N+1}$ has a small norm. On the other hand, by the property 3) of an AOR, $|\mathrm{tr}(1 - R_N A_N) - \mathrm{tr}(1 - R \circ A)|_N - \mathrm{tr}(1 - A_N R_N) + \mathrm{tr}(1 - A \circ R)|_N \leq C\lambda^{-n+N+1}$. From this we see that, as $\lambda \to 0$ in Λ, the fractional part of $\mathrm{ind}\, A_N$ does not exceed $C\lambda^{-n+N+1}$ for any N, that is, it is of order $O(\lambda^\infty)$.

With a few changes, this proof also carries over to more general elliptic collections $\Xi = \{P^0, P^1, A, R\}$ in the algebra of quantum observables $W_{\mathcal{D}^s}$.

For every additive generator of $K^{\mathrm{comp}}(M)$ Theorem 3.6 yields an integer value condition, which is necessary for the exisence of an AOR. Thus, we obtain a collection of necessary conditions of the form $P(1/\lambda) \in \mathbb{Z} \bmod O(\lambda^\infty)$ as $\lambda \to 0$ in Λ, where P is a polynomial of degree not higher than n. We mention that these conditions do not depend on the construction of the AOR.

We describe how from Theorem 3.8 we can obtain simple integer value conditions with a linear polynomial P. Let E be an n-dimensional complex bundle over M whose realification coincides with TM.

Theorem 3.9. *A necessary condition for the existence of an AOR is that the form*

$$\omega_n = \frac{\delta\sigma}{2\pi h} - \frac{1}{2}c_1(E) \tag{3.33}$$

should be asymptotically integer-valued as $\lambda \to 0$ in the set Λ, that is, for any orientated two-dimensional compact submanifold $S \subset M$ the integral of this form over S should be an integer modulo $O(\lambda^\infty)$.

Proof. Let V be a neighbourhood of S with compact closure in M, and $i : V \to \mathbb{R}^k$ an embedding with normal bundle N. Then, as in Sect. 2.2, we have an embedding $i_2 : v \to TV \to \mathbb{R}^{2k}$ with normal bundle $N_2 = N \oplus N \oplus TM$, which has complex structure. In addition, we have embeddings $i_3 : S \to V \to \mathbb{R}^k$ with normal bundle N_3 and $i_4 : S \to TS \to \mathbb{R}^{2k}$ with normal bundle $N_4 = N_3 \oplus N_3 \oplus TS$, which also has complex structure, since such a structure exists on TS. Consequently, by means of the Thom isomorphism, i_2 and i_4 define isomorphisms $i_{2!} : K^{\mathrm{comp}}(V) \to K^{\mathrm{comp}}(\mathbb{R}^{2k})$ and $i_{4!} : K(S) \to K^{\mathrm{comp}}(\mathbb{R}^{2k})$. Let $\xi = (i_{2!})^{-1} i_{4!} 1 \in K^{\mathrm{comp}}(V)$, and let $\Xi = \{P^0, P^1, A, R\}$ be an elliptic collection in the algebra $W_{\mathcal{D}^s}$ with support in V, whose leading terms $\{p^0, p^1, a, r\}$ yield the element ξ. Then, by (3.27),

$$\mathrm{ind}\, \Xi = \int\limits_V \mathrm{ch}\, \xi \mathcal{T}(E) e^{\omega_n} \ .$$

By (1.17), $\mathrm{ch}\, \xi \mathcal{T}(E) = (i_{2*})^{-1} \mathrm{ch}\, i_{4!} 1$, where $(i_{2*})^{-1}$ is the Thom isomorphism in cohomologies (fibre integration along the fibres of N_2). Since $i_{2!} \xi = i_{4!} 1$, it follows that

$$\mathrm{ind}\, \Xi = \int\limits_V (i_{2*})^{-1} \mathrm{ch}\, i_4 1 \cdot e^{\omega_n} = \int\limits_S (i_{4*})^{-1} \mathrm{ch}\, i_{4!} 1 \ ,$$

which, again by (1.17), reduces to

$$\mathrm{ind}\, \Xi = \int\limits_S \mathcal{T}(S) e^{\omega_n} = \int\limits_S \omega_n + \frac{1}{2} \int\limits_S c_1(TS) \ ,$$

where $\mathcal{T}(S) = 1 + \frac{1}{2} c_1(TS)$ is the Todd class of the one-dimensional complex tangent bundle of the Riemann surface S. The last term is equal to half of the Euler characteristic of S, so it is an integer. By Theorem 3.8, from this it follows that $\int\limits_S \omega_n$ has an asymptotic integer value.

The condition that the form ω_n should be integer-valued on Λ is also sufficient for the existence of an AOR (the proof of this assertion will be published soon in the journal 'Functional Analysis and Its Applications'). From this we easily deduce the well-known integer value theorem (Palais (1965)), which is the converse of Theorem 3.9. It asserts that if the form ω_n is integer-valued, then the number (3.26) is an integer for any $\xi \in K^{\mathrm{comp}}(M)$.

We also mention that the integer value condition for the form ω_n has occured repeatedly in other quantisation conditions, for example, in the asymptotic quantisation of Karasev and Maslov (1984), as well as in the geometric quantisation of Costant and Kirillov (1972).

2.4. Examples. A fairly large number of examples of symplectic manifolds and quantum structures on them is supplied by Kirillov's method of orbits (Kirillov (1972)) in representation theory. The two examples considered below are orbits of a unitary group.

1. Let $M = M_n = M_n(\lambda_1, \lambda_2, \ldots, \lambda_n)$ be a manifold of Hermitian matrices of order n with given simple eigenvalues $\lambda_1, \lambda_2, \ldots, \lambda_n$ (a generic orbit). Let $E_1, E_2, \ldots, E_n$ be the corresponding eigenspaces and $P_1, P_2, \ldots, P_n$ the orthogonal projections on these eigenspaces. The latter define complex linear bundles over M. The sections of E_i are vector functions $u(x)$ with values in $\mathbb{C}^n$ such that $P_i(x)u(x) = u(x)$. We introduce a connection in E_i by setting $\partial u = P_i du$. The curvature of this connection is equal to $P_i\, dP_i\, dP_i$, and we set

$$c_1(E_i) = -\frac{1}{2\pi i}\,\mathrm{tr}\, P_i\, dP_i\, dP_i \; .$$

These forms define the first Chern class of the bundles E_i.

We consider a symplectic structure on M defined by the *Kirillov form*

$$\sigma = \sum_{i=1}^{n} \lambda_i c_1(E_i) = \sum_{i=1}^{n-1} (\lambda_i - \lambda_{n-1}) c_1(E_i) \; ,$$

which, as is easily seen, is non-degenerate on M. Let $\mathcal{D}$ be an Abelian connection with curvature $\Omega_w = -(i/h)\sigma$ on the bundle of formal Weyl algebras, and let $W_{\mathcal{D}}$ be the algebra of quantum observables.

The manifold M_n is given in the space $\mathbb{R}^{n^2}$ of all Hermitian matrices by the equations

$$f_k(A) = \mathrm{tr}\, A^k = \lambda_1^k + \lambda_2^k + \cdots + \lambda_n^k \; , \qquad k = 1, 2, \ldots, n \; .$$

The differentials of these functions are linearly independent, since the eigenvalues $\lambda_1, \lambda_2, \ldots, \lambda_n$ are distinct. From this it follows that the normal bundle under the embedding $M_n \to \mathbb{R}^{n^l}$ is trivial and that the A-class of the tangent bundle TM is equal to 1. The formula (3.26) for the index of an elliptic collection $\varXi$ in the algebra $W_{\mathcal{D}}$ takes the form

$$\mathrm{ind}\, \varXi = \int\limits_{M} \mathrm{ch}\, \xi e^{\sigma/(2\pi h)} \; , \tag{3.34}$$

where $\xi \in K(M)$ is the distinguishing element of $\varXi$.

We choose $\xi = \xi_1^{n-1}\xi_2^{n-2}\ldots\xi_{n-2}^2 \in K(M)$, where ξ_i is the virtual bundle $\{1, E_i^*\}$. Since the Chern character is multiplicative, we find that the integrand in (3.34) is equal to

$$c_1^{n-1}(E_1) \wedge c_1^{n-2}(E_2) \wedge \ldots \wedge c_1^2(E_{n-2}) \wedge c_1(E_{n-1})\frac{\lambda_{n-1} - \lambda_n}{2\pi h} \; .$$

To compute the integral, we remark that the manifold $M = M_n(\lambda_1, \lambda_2, \ldots, \lambda_n)$ is a bundle space with base $\mathbf{CP}^{n-1}$ and fibre $M_{n-1}(\lambda_2, \lambda_3, \ldots, \lambda_n)$. The projection $M_n \to \mathbf{CP}^{n-1}$ associates with the matrix $A \in M_n$ the one-dimensional subspace E_1 in $\mathbf{C}^n$. Similarly, M_{n-1} is a bundle with base $\mathbf{CP}^{n-2}$ and fibre M_{n-2}, and so on. Using the formula

$$\int_{\mathbf{CP}^{n-1}} c_1^{n-1}(E_1) = 1 \,, \tag{3.35}$$

we obtain

$$\operatorname{ind} \Xi = \frac{\lambda_{n-1} - \lambda_n}{2\pi h} \,.$$

Theorem 3.8 and symmetry considerations lead to the integer value conditions $\frac{\lambda_i - \lambda_j}{2\pi\lambda} \in \mathbb{Z}$. Under these conditions, we can use the method of orbits to construct unitary representations of the group $\mathrm{SU}(n)$. From Kirillov's character formula it follows that the dimension of such a representation is equal to

$$\operatorname{tr} 1 = \int_M e^{\sigma/(2\pi\lambda)} \,,$$

which coincides with the index of the elliptic collection $\Xi = \{1, 0, 0, 0\}$ computed by means of the formula (3.34) with $h = \lambda$.

2. Let $M = \mathbf{CP}^n$. This is a degenerate orbit of the group $\mathrm{SU}(n+1)$ with $\lambda_1 = 1$, $\lambda_2 = \lambda_3 = \ldots \lambda_{n+1} = 0$. The symplectic structure is given by the form $\sigma = c_1(E_1)$. Let $W_{\mathcal{D}}$ be the algebra of quantum observables with $\Omega_w = -(i/h)\sigma$.

The ring $K(\mathbf{CP}^n)$ is generated by one generator $\xi = \{1, E_1^*\}$ and the relation $\xi^{n+1} = 0$ (Atiyah (1967)). The complex tangent bundle E to the manifold $\mathbf{CP}^n$ is stably isomorphic to $(n+1)E_1$. From this we deduce that $c_1(E) = (n+1)c_1(E_1) = (n+1)\sigma$ and $\mathcal{T}(E) = (\sigma/(1 - e^{-\sigma}))^{n+1}$.

Applying (3.27) to compute the index of the elliptic collection Ξ_k with distinguishing element $\xi^k \in K(M)$, we find that

$$\operatorname{ind} \Xi_k = \int_M e^{(1/(2\pi h) - (n+1)/2)\sigma}(1 - e^{-\sigma})^k \left(\frac{\sigma}{1 - e^{-\sigma}}\right)^{n+1} \,.$$

By (3.35), the integral is equal to the residue of $e^{(1/(2\pi h) - (n+1)/2)z}(1 - e^{-z})^{k-n-1}$ at $z = 0$. For $k = n - 1$ we obtain the integer value condition

$$\frac{1}{2\pi\lambda} - \frac{n-1}{2} = m \in \mathbb{Z} \,. \tag{3.36}$$

The fact that $\operatorname{ind} \Xi_k$ is an integer for the remaining values of k follows from (3.36). For the index of the elliptic collection $\Xi_0 = \{1, 0, 0, 0\}$ under the condition (3.36), from (3.27) we find that

$$\operatorname{tr} 1 = \int_M e^{(m-1)\sigma}\mathcal{T}(E) \,. \tag{3.37}$$

The form $(m-1)\sigma$ defines the first Chern class of the tensor power E_1^{m-1}. Consequently, (3.37) coincides externally with the formula for the multiplicity of the spectrum of operators with periodic bicharacteristics given in Boutet de Monvel and Guillemin (1989). The computation of the integral (3.37) yields

$$\operatorname{tr} 1 = \frac{m(m+1)\dots(m+n-1)}{n!} \,,$$

which is the multiplicity of the eigenvalues of an $(n+1)$-dimensional oscillator.

References[*]

Alvarez-Gaumé, L. (1984): Supersymmetry and the Atiyah-Singer index theorem. Physica A *124*, 29–45. Zbl. 599.58042

Atiyah, M.F. (1967): *K*-Theory. Benjamin, New York. Zbl. 159,533

Atiyah, M.F. (1974): Elliptic Operators and Compact Groups. Lect. Notes Math. *401*. Zbl. 297.58009

Atiyah, M.F. (1976): Elliptic operators, discrete groups and von Neumann algebras. Astérisque *32-33*, 43–72. Zbl. 323.58015

Atiyah, M.F., Bott, R.A. (1967): A Lefschetz fixed point formula for elliptic complexes. I. Ann. Math., II. Ser. *87*, 347–407. Zbl. 161,432

Atiyah, M.F., Bott, R.A. (1984): The moment map and equivariant cohomology. Topology *23*, 1–28. Zbl. 521.58025

Atiyah, M.F., Bott, R.A., Patody, V.K. (1973): On the heat equation and the index theorem. Invent. Math. *19*, 279–330. Zbl. 257.58008. Errata (1975) *29*, 277-280. Zbl. 301.58018

Atiyah, M.F., Schmid, W. (1977): A geometric construction of the discrete series for semi-simple Lie groups. Invent. Math. *42*, 1–62. Zbl. 373.22001

Atiyah, M.F., Segal, G.B. (1968): The index of elliptic operators. II. Ann. Math., II. Ser. *87*, 531–545. Zbl. 164,242

Atiyah, M.F., Singer, I.M. (1968a): The index of elliptic operators. I. Ann. Math., II. Ser. *87*, 484–530. Zbl. 164,240

Atiyah, M.F., Singer, I.M. (1968b): The index of elliptic operators. III. Ann. Math., II. Ser. *87*, 546–604. Zbl. 164,243

Atiyah, M.F., Singer, I.M. (1971a): The index of elliptic operators. IV. Ann. Math., II. Ser. *93*, 119–138. Zbl. 212,286

Atiyah, M.F., Singer, I.M. (1971b): The index of elliptic operators. V. Ann. Math., II. Ser. *93*, 139–149. Zbl. 212,286

Bayen, F., Flato, M., Fronsdal, C., Lichnerowicz, A., Sternheimer, D. (1978): Deformation theory and quantization. Ann. Phys. *111*, 61–110. Zbl. 377.53024

Berline, N., Vergne, M. (1985): The equivariant index and Kirillov's character formula. Am. J. Math. *109*, 1159–1190. Zbl. 604.58046

Bismut, J.-M. (1984): The Atiyah-Singer theorems: a probabilistic approach. I: The index theorem. II: The Lefschetz fixed point formulas. J. Funct. Anal. *57*, 56–99; 329–348. I: Zbl. 538.58033. II: Zbl. 556.58027

---

[*] For the convenience of the reader, references to reviews in Zentralblatt für Mathematik (Zbl.), compiled by means of the MATH database, have, as far as possible, been included in this bibliography.

Booss, B. (1977): Topologie und Analysis. Eine Einführung in die Atiyah-Singer Indexformel. Springer, Berlin Heidelberg New York. Zbl. 364.58017

Bott, R. (1967a): Lectures on $K(X)$. Harvard Univ. preprint (W.A. Benjamin, 1969. Zbl. 194,239)

Bott, R. (1967b): A residue formula for holomorphic vector fields. J. Differ. Geom. 1, 311–330. Zbl. 179,288

Boutet de Monvel, L. (1971): Boundary problems for pseudodifferential operators. Acta Math. 126, 11–51. Zbl. 206,394

Boutet de Monvel, L. (1979): On the index of Toeplitz operators of several complex variables. Invent. Math. 50, 249–272. Zbl. 398.47018

Boutet de Monvel, L., Guillemin, V. (1989): The spectral theory of Toeplitz operators. Ann. Math. Stud. 99. Zbl. 469.47021

Brenner, A.V., Shubin, M.A. (1981): The Atiyah-Bott-Lefschetz theorem for manifolds with boundary. Funkts. Anal. Prilozh. 15, no. 4, 67–68. English transl.: Funct. Anal. Appl. 15, 286–287 (1982). Zbl. 483.58015

Breuer, M. (1968,1969): Fredholm theories in von Neumann algebras. I, II. Math. Ann. 178, 243–254; 180, 313–325. I: Zbl.162,187, II: Zbl. 175,441

Chern, S.S. (1956): Complex Manifolds. University Press, Chicago. Zbl. 74,303

Coburn, L.A., Moyer, R.D., Singer, I.M. (1973): The C^*-algebras of almost periodic pseudo-differential operators. Acta Math. 130, 279–307. Zbl. 263.47042

Colin de Verdier, Y. (1979): Sur le spectre des opérateurs à bicaractéristiques périodiques. Comment. Math. Helv. 54, 508–522. Zbl. 459.58014

Connes, A., Moscovici, H. (1982): The L^2-index theorem for homogeneous spaces of Lie groups. Ann. Math., II Ser. 115, 291–330. Zbl. 515.58031

Dupont, J.L. (1978): Curvature and Characteristic Classes. Lect. Notes Math. 640. Zbl. 373.57009

Emch, G. (1972): Algebraic methods in Statistical Mechanics and Quantum Field Theory. Wiley-Interscience, New York. Zbl. 235.46085

Fedosov, B.V. (1974): Analytic index formulae. Tr. Mosk. Mat. O.-va 30, 159–240. English transl.: Trans. Mosc. Math. Soc. 30, 159–240 (1976). Zbl. 349.58006

Fedosov, B.V. (1978): A periodicity theorem in the symbol algebra. Mat. Sb., Nov. Ser. 105, 431–462. English transl.: Math. USSR, Sb. 34, 382–410 (1978). Zbl. 391.47033

Fedosov, B.V. (1986): Quantisation and the index. Dokl. Akad. Nauk SSSR 291, 82–86. English transl.: Sov. Phys., Dokl. 31, 877–878 (1986). Zbl. 635.58019

Fedosov, B.V. (1989): The index theorem in the algebra of quantum observables. Dokl. Akad. Nauk SSSR 305, 835–838. English transl.: Sov. Phys., Dokl. 34, 319–321 (1989)

Fedosov, B.V., Shubin, M.A. (1978): The index of stochastic elliptic operators. I, II. Mat. Sb., Nov. Ser. 106, 108–140, 455–483. English transl.: Math. USSR, Sb. 34, 671–699 (1978); 35, 131–156 (1979). I: Zbl. 409.47030, II: Zbl. 392.58010

Forster, O. (1977): Riemannsche Flächen. Springer-Verlag, Berlin Heidelberg New York. Zbl. 381.30021

Gel'fand, I.M. (1960): On elliptic equations. Usp. Mat. Nauk 15, No. 3, 121–132. English transl.: Russ. Math. Surv. 15, No. 3, 113–123 (1960). Zbl. 95,78

Getzler, E. (1983): Pseudodifferential operators on supermanifold and the Atiyah-Singer index theorem. Commun. Math. Phys. 92, 163–178. Zbl. 543.58026

Gohberg, I.C., Krein, M.G. (1965): Introduction to the Theory of Linear Non-selfadjoint Operators in a Hilbert Space. Nauka, Moscow. English transl.: Am. Math. Soc., Providence 1969. Zbl. 138,78

Hörmander, L. (1971): Fourier integral operators. I. Acta Math. 127, 79–183. Zbl. 212,466

Hörmander, L. (1979): The Weyl calculus of pseudodifferential operators. Commun. Pure Appl. Math. 32, 358–443. Zbl. 388.47032

Hörmander, L. (1984–1985): The Analysis of Linear Partial Differential Operators. 1–4. Springer-Verlag, Berlin Heidelberg New York. I,II: Zbl. 521.35001/2, III: Zbl. 601.35001, IV: Zbl. 612.35001

Husemöller, D. (1966): Fibre Bundles. McGraw-Hill, New York. Zbl. 144,448

Karasev, M.V., Maslov, V.P. (1984): Asymptotic and geometric quantisation. Usp. Mat. Nauk *39*, No. 6, 115–173. English transl.: Russ. Math. Surv. *39*, No. 6,133–205 (1984). Zbl. 588.58031

Kirillov, A.A. (1972): Elements of Representation Theory. Nauka, Moscow. English transl.: Springer-Verlag, Berlin Heidelberg New York 1976. Zbl. 264.22011

Kirillov, A.A. (1985): Geometric Quantisation. (Itogi Nauki Tekh., Sovr. Probl. Mat., Fundam. Napr. *4*.) Moscow, 141–178. English transl. in: Encycl. Math. Sci. *4*, Springer-Verlag, Berlin Heidelberg New York 1988, 137–172. Zbl. 591.58014

Lecomte, P., De Wilde, M. (1983): Existence of star products and of formal deformations of the Poisson Lie algebra of arbitrary symplectic manifolds. Lett. Math. Phys. *7*, 487–496. Zbl. 526.58023

Lichnerowicz, A. (1955): Théorie Globale des Connexions et des Groups d'Holonomie. Edizioni Cremonese, Rome. Zbl. 116,391

Palais, R.S. (1965): Seminar on the Atiyah-Singer Index Theorem. Princeton Univ. Press, Princeton. Zbl. 137,170

Rempel, S., Schulze, B.-W. (1982): Index Theory of Elliptic Boundary Problems. Akademie-Verlag, Berlin. Zbl. 504.35002

Schwarz, A.S. (Shvarts) (1981): Elliptic Operators in quantum field theory. (Itogi Nauki Tekh., Ser. Sovrem. Probl. Mat. *17*.) Moscow, 113–173. English transl.: J. Sov. Math. *21*, 551–601 (1983). Zbl. 482.35080

Shubin, M.A. (1978): Pseudodifferential Operators and Spectral Theory. Nauka, Moscow. Zbl. 451.47064

Shubin, M.A. (1979): Spectral theory and the index of elliptic operators with almost periodic coefficients. Usp. Mat. Nauk *34*, No. 2, 95–135. English transl.: Russ. Math. Surv. *34*, No. 2, 109–157 (1979). Zbl. 431.47027

Springer, G. (1957): Introduction to Riemann Surfaces. Addison-Wesley, Reading, Mass. Zbl. 78,66

Treves, F. (1980): Introduction to Pseudodifferential and Fourier Integral Operators. Plenum, New York. Zbl. 453.47027

Author Index

Subject Index

Encyclopaedia of Mathematical Sciences
Editor-in-Chief: R. V. Gamkrelidze

Dynamical Systems

Volume 1: **D.V. Anosov, V.I. Arnol'd** (Eds.)
Dynamical Systems I
Ordinary Differential Equations and Smooth
Dynamical Systems
2nd printing 1994. IX, 233 pp. 25 figs. ISBN 3-540-17000-6

Volume 2: **Ya.G. Sinai** (Ed.)
Dynamical Systems II
Ergodic Theory with Applications to Dynamical
Systems and Statistical Mechanics
1989. IX, 281 pp. 25 figs. ISBN 3-540-17001-4

Volume 3: **V.I. Arnold, V.V. Kozlov, A.I. Neishtadt**
Dynamical Systems III
Mathematical Aspects of Classical and Celestial
Mechanics
2nd ed. 1993. XIV, 291 pp. 81 figs. ISBN 3-540-57241-4

Volume 4: **V.I. Arnol'd, S.P. Novikov** (Eds.)
Dynamical Systems IV
Symplectic Geometry and its Applications
1990. VII, 283 pp. 62 figs. ISBN 3-540-17003-0

Volume 5: **V.I. Arnol'd** (Ed.)
Dynamical Systems V
Bifurcation Theory and Catastrophe Theory
1994. IX, 271 pp. 130 figs. ISBN 3-540-18173-3

Volume 6: **V.I. Arnol'd** (Ed.)
Dynamical Systems VI
Singularity Theory I
1993. V, 245 pp. 55 figs. ISBN 3-540-50583-0

Volume 16: **V.I. Arnol'd, S.P. Novikov** (Eds.)
Dynamical Systems VII
Nonholonomic Dynamical Systems.
Integrable Hamiltonian Systems.
1994. VII, 341 pp. 9 figs. ISBN 3-540-18176-8

Vol. 39: **V.I. Arnol'd** (Ed.)
Dynamical Systems VIII
Singularity Theory II. Applications
1993. V, 235 pp. 134 figs. ISBN 3-540-53376-1

Vol. 66: **D.V. Anosov** (Ed.)
Dynamical Systems IX
Dynamical Systems with Hyperbolic Behavior
1995. VII, 235 pp. 39 figs. ISBN 3-540-57043-8

Partial Differential Equations

Volume 30: **Yu.V. Egorov, M.A. Shubin** (Eds.)
Partial Differential Equations I
Foundations of the Classical Theory
1991. V, 259 pp. 4 figs. ISBN 3-540-52002-3

Volume 31: **Yu.V. Egorov, M.A. Shubin** (Eds.)
Partial Differential Equations II
Elements of the Modern Theory. Equations
with Constant Coefficients
1995. VII, 263 pp. 5 figs. ISBN 3-540-52001-5

Volume 32: **Yu.V. Egorov, M.A. Shubin** (Eds.)
Partial Differential Equations III
The Cauchy Problem. Qualitative Theory
of Partial Differential Equations
1991. VII, 197 pp. ISBN 3-540-52003-1

Volume 33: **Yu.V. Egorov, M.A. Shubin** (Eds.)
Partial Differential Equations IV
Microlocal Analysis and Hyperbolic Equations
1993. VII, 241 pp. 6 figs. ISBN 3-540-53363-X

Volume 63: **Yu.V. Egorov, M.A. Shubin** (Eds.)
Partial Differential Equations VI
Elliptic and Parabolic Operators
1994. VII, 325 pp. 5 figs. ISBN 3-540-54678-2

Volume 64: **M.A. Shubin** (Ed.)
Partial Differential Equations VII
Spectral Theory of Differential Operators
1994. V, 272 pp. ISBN 3-540-54677-4

Volume 65: **M.A. Shubin** (Ed.)
Partial Differential Equations VIII
Overdetermined Systems. Dissipative Singular
Schrödinger Operator. Index Theory
1996. VII, 258 pp. 1 fig. ISBN 3-540-57036-5

Preisänderungen vorbehalten

Tm.BA95.06.13

Encyclopaedia of Mathematical Sciences
Editor-in-Chief: R. V. Gamkrelidze

Analysis

Volume 13: **R.V. Gamkrelidze** (Ed.)
Analysis I
Integral Representations and Asymptotic Methods
1989. VII, 238 pp. 3 figs. ISBN 3-540-17008-1

Volume 14: **R.V. Gamkrelidze** (Ed.)
Analysis II
Convex Analysis and Approximation Theory
1990. VII, 255 pp. 21 figs. ISBN 3-540-18179-2

Volume 26: **S.M. Nikol'skiĭ** (Ed.)
Analysis III
Spaces of Differentiable Functions
1991. VII, 221 pp. 22 figs. ISBN 3-540-51866-5

Volume 27: **V.G. Maz'ya, S.M. Nikol'skiĭ** (Eds.)
Analysis IV
Linear and Boundary Integral Equations
1991. VII, 233 pp. 4 figs. ISBN 3-540-51997-1

Volume 19: **N.K. Nikol'skij** (Ed.)
Functional Analysis I
Linear Functional Analysis
1992. V, 283 pp. ISBN 3-540-50584-9

Volume 20: **A.L. Onishchik** (Ed.)
Lie Groups and Lie Algebras I
Foundations of Lie Theory.
Lie Transformation Groups
1993. VII, 235 pp. 4 tabs. ISBN 3-540-18697-2

Volume 41: **A.L. Onishchik, E.B. Vinberg** (Eds.)
Lie Groups and Lie Algebras III
Structure of Lie Groups and Lie Algebras
1994. V, 248 pp. 1 fig., 7 tabs. ISBN 3-540-54683-9

Volume 22: **A.A. Kirillov** (Ed.)
Representation Theory and Non-commutative Harmonic Analysis I
Fundamental Concepts. Representations
of Virasoro and Affine Algebras
1994. VII, 234 pp. 11 figs. ISBN 3-540-18698-0

Volume 15: **V. P. Khavin, N. K. Nikol'skij** (Eds.)
Commutative Harmonic Analysis I
General Survey, Classical Aspects
1991. IX, 268 pp. 1 fig. ISBN 3-540-18180-6

Volume 72: **V. P. Havin, N. K. Nikol'skij** (Eds.)
Commutative Harmonic Analysis III
Generalized Functions. Applications
1995. VII, 266 pp. 34 fig. ISBN 3-540-57034-9

Volume 42: **V. P. Khavin, N. K. Nikol'skij** (Eds.)
Commutative Harmonic Analysis IV
Harmonic Analysis in $\mathbb{R}^n$
1992. IX, 228 pp. 1 fig. ISBN 3-540-53379-6

Several Complex Variables

Volume 7: **A.G. Vitushkin** (Ed.)
Several Complex Variables I
Introduction to Complex Analysis
1990. VII, 248 pp. ISBN 3-540-17004-9

Volume 8: **G.M. Khenkin, A.G. Vitushkin** (Eds.)
Several Complex Variables II
Function Theory in Classical Domains. Complex
Potential Theory
1994. VII, 260 pp. 19 figs. ISBN 3-540-18175-X

Volume 9: **G.M. Khenkin** (Ed.)
Several Complex Variables III
Geometric Function Theory
1989. VII, 261 pp. ISBN 3-540-17005-7

Volume 10: **S.G. Gindikin, G.M. Khenkin** (Eds.)
Several Complex Variables IV
Algebraic Aspects of Complex Analysis
1990. VII, 251 pp. ISBN 3-540-18174-1

Volume 54: **G.M. Khenkin** (Ed.)
Several Complex Variables V
Complex Analysis in Partial Differential Equations
and Mathematical Physics
1993. VII, 286 pp. ISBN 3-540-54451-8

Volume 69: **W. Barth, R. Narasimhan** (Eds.)
Several Complex Variables VI
Complex Manifolds
1990. IX, 310 pp. 4 figs. ISBN 3-540-52788-5

Volume 74: **H. Grauert, T. Peternell, R. Remmert** (Eds.)
Several Complex Variables VII
Sheaf-Theoretical Methods in Complex Analysis
1994. VIII, 369 pp. ISBN 3-540-56259-1

Preisänderungen vorbehalten

Tm.BA95.03.17

Springer-Verlag and the Environment

We at Springer-Verlag firmly believe that an international science publisher has a special obligation to the environment, and our corporate policies consistently reflect this conviction.

We also expect our business partners – paper mills, printers, packaging manufacturers, etc. – to commit themselves to using environmentally friendly materials and production processes.

The paper in this book is made from low- or no-chlorine pulp and is acid free, in conformance with international standards for paper permanency.